Nonlinear Physics for Beginners

Cover photograph: Multiple morphological transitions in an electrodeposition pattern, in a linear cell of copper sulfate solution with two parallel copper electrodes, discovered by Professor Lui Lam and his students of the Nonlinear Physics Group at San Jose State University, California.

Nonlinear Physics for Beginners

Fractals, Chaos, Solitons, Pattern Formation,
Cellular Automata and Complex Systems

Lui Lam

San Jose State University

World Scientific
Singapore • New Jersey • London • Hong Kong

Published by
World Scientific Publishing Co. Pte. Ltd.
P O Box 128, Farrer Road, Singapore 912805
USA office: Suite 1B, 1060 Main Street, River Edge, NJ 07661
UK office: 57 Shelton Street, Covent Garden, London WC2H 9HE

Library of Congress Cataloging-in-Publication Data
Lam, Lui.
    Nonlinear physics for beginners : fractals, chaos, solitons,
  pattern formation, cellular automata, complex systems / Lui Lam.
       p.   cm.
    Includes bibliographical references and index.
    ISBN 9810201400. -- ISBN 9810201419 (pbk.)
    1. Nonlinear theories.   2. Mathematical physics.   I. Title.
QC20.7.N6L36   1998
530.1'5--dc20
                                                                                              96-35744
                                                                                               CIP

British Library Cataloguing-in-Publication Data
A catalogue record for this book is available from the British Library.

The editor and publishers would like to thank the authors and the publishers of various journals and books for their assistance and permission to reproduce the selected reprints found in this volume:

    1. American Association for the Advancement of Science
    2. American Association of Physics Teachers
    3. American Institute of Physics
    4. European Physical Society
    5. IOP Publishing Ltd.
    6. Macmillan Journals Ltd.
    7. Macmillan Magazines Ltd.
    8. Scientific America

While every effort has been made to contact the publishers of reprinted papers prior to publication, we have not been successful in some cases. Where we could not contact the publishers, we have acknowledged the source of the material. Proper credit will be accorded to these publishers in future editions of this work after permission is granted.

Copyright © 1998 by World Scientific Publishing Co. Pte. Ltd.

All rights reserved. This book, or parts thereof, may not be reproduced in any form or by any means, electronic or mechanical, including photocopying, recording or any information storage and retrieval system now known or to be invented, without written permission from the Publisher.

For photocopying of material in this volume, please pay a copying fee through the Copyright Clearance Center, Inc., 222 Rosewood Drive, Danvers, MA 01923, USA. In this case permission to photocopy is not required from the publisher.

Printed in Singapore by Uto-Print

12.3 Oblique Roll Instability in an Electroconvective
Anisotropic Fluid 172
      R. Ribotta, A. Joets and Lin Lei (L. Lam)
      [Phys. Rev. Lett. **56**, 1595–1597 (1986);
      ibid **56**, 2335(E) (1986)]

12.4 Critical Behavior in the Transitions to Convective
Flows in Nematic Liquid Crystals 176
      A. Joets, R. Ribotta and L. Lam [unpublished (1988)]

12.5 Chemical Waves 187
      J. Ross, S. C. Müller and C. Vidal
      [Science **240**, 460–465 (1988)]

13 Cellular Automata and Complex Systems   193

13.1 More is Different 193
      P. W. Anderson [Science **177**, 393–396 (1972)]

13.2 Cellular Automata as Models of Complexity 197
      S. Wolfram [Nature **311**, 419–424 (1984)]

13.3 Catastrophes and Self-Organized Criticality 203
      P. Bak [Comput. Phys. **5**, 430–433 (1991)]

13.4 Active-Walker Models: Growth and Form in
Nonequilibrium Systems 207
      L. Lam and R. D. Pochy
      [Comput. Phys. **7**, 534–541 (1993)]

13.5 Active Walks and Path Dependent Phenomena in
Social Systems 215
      L. Lam, C.-Q. Shu and S. Bödefeld [unpublished (1997)]

PART III   PROJECTS   223

14 Computational   229

14.1 Fractals 229
    14.1.1 The Chaos Game and Sierpinski Gasket 229
          P. U. Pendse
    14.1.2 Iteration Maps and the Sierpinski Fractals 231
          R. D. Freimuth
    14.1.3 Calculating the Box Dimension 237
          V. M. Castillo
    14.1.4 Diffusion-Limited Aggregates in Radial Geometry 240
          R. D. Pochy
    14.1.5 The Dielectric Breakdown Model with Noise
         Reduction 245
          R. D. Pochy

14.2	Chaos		257
	14.2.1	The Tent Map	257
		R. D. Pochy, Y. S. Yung and W. A. Baldwin	
	14.2.2	The Waterwheel	260
		R. D. Freimuth	
14.3	Pattern Formation	271
	14.3.1	Biased Random Walks	271
		M. A. Guzman and R. D. Pochy	
	14.3.2	Surface Tension and the Evolution of Deformed Water Drops	275
		V. M. Castillo	
	14.3.3	Ising-like Model of Ferrofluid Patterns	279
		V. M. Castillo	
14.4	Cellular Automata	283
	14.4.1	One-Dimensional Totalistic Cellular Automata	283
		R. D. Pochy	
	14.4.2	Two-Dimensional Cellular Automata: Formation of Clusters	286
		R. D. Pochy	

15 Theoretical   291

15.1	Curve Length and the Scaling Parameter	291
	T. H. Watkins	
15.2	Analysis of the Back-Propagating Neural Network for the XOR Problem	294
	V. M. Castillo	

16 Experimental   301

16.1	Instabilities of Finite Water Columns	301
	M. C. Fallis, M. M. Masuda, R. C. LeRoy and N. Neisan	
16.2	Viscous Fingering in Optical Cement Displaced by Water	308
	J. M. Hillendahl	
16.3	The Fractal Nature of Shock-Wave Induced Fractures	315
	R. G. Klingler	

Epilogue: The Real World   319

Appendices   321

A1	Computer Program for Active Walk	321
A2	Publications from Nonlinear Physics Group of SJSU	328

Acknowledgments   333

Index   335

Contents

Prologue:	**The Ground Has Shifted**	**1**

PART I OVERVIEW — 3

- **1 Introduction** — 5
  - 1.1 A Quiet Revolution — 5
  - 1.2 Nonlinearity — 6
- **2 Fractals** — 11
- **3 Chaos** — 17
- **4 Solitons** — 23
- **5 Pattern Formation** — 27
- **6 Cellular Automata** — 35
- **7 Complex Systems** — 37
- **8 Remarks and Further Reading** — 43

PART II REPRINTS — 47

- **9 Fractals** — 51
  - 9.1 Fractal Growth Processes — 51
    L. M. Sander [Nature **322**, 789–793 (1986)]
  - 9.2 Fractal Geometry in Crumpled Paper Balls — 56
    M. A. F. Gomes [Am. J. Phys. **55**, 649–650 (1987)]
  - 9.3 Fractal of Large Scale Structures in the Universe — 58
    L. Z. Fang [Mod. Phys. Lett. **A1**, 601–605 (1986)]
  - 9.4 The Devil's Staircase — 63
    P. Bak [Phys. Today **39**(12), 38–45 (1986)]
  - 9.5 Multifractal Phenomena in Physics and Chemistry — 71
    H. E. Stanley and *P. Meakin*
    [Nature **335**, 405–409 (1988)]

9.6 Simple Multifractals with Sierpinski Gasket Supports 76
L. Lam, R. D. Freimuth and *J. L. Drake*
[unpublished (1992)]

10 Chaos 92

10.1 Chaos 92
J. P. Crutchfield, J. D. Farmer, N. H. Packard and *R. S. Shaw* [Sci. Am. **254**(12), 46–58 (1986)]

10.2 Chaos in a Dripping Faucet 104
H. N. Nunez Yepez, A. L. Salas Brito, C. A. Vargas and *L. A. Vicente* [Eur. J. Phys. **10**, 99–105 (1989)]

10.3 Chaos, Strange Attractors, and Fractal Basin Boundaries in Nonlinear Dynamics 111
C. Grebogi, E. Ott and *J. A. Yorke*
[Science **238**, 632–638 (1987)]

10.4 Nonlinear Forecasting as a Way of Distinguishing Chaos from Measurement Error in Time Series 118
G. Sugihara and *R. M. May* [Nature **344**, 734–741 (1990)]

10.5 Controlling Chaos 125
E. Ott and *M. Spano* [Phys. Today **48**(5), 34–40 (1995)]

10.6 Quantum Chaos 132
M. C. Gutzwiller [Sci. Am. **266**(1), 78–84 (1992)]

10.7 How Random is a Coin Toss? 139
J. Ford [Phys. Today **36**(4), 40–47 (1983)]

11 Solitons 147

11.1 Solitons 147
R. K. Bullough [Phys. Bulletin, 78–82 (Feb. 1978)]

11.2 Soliton Propagation in Liquid Crystals 152
Lin Lei (L. Lam), Shu Changqing, Shen Juelian, P. M. Lam and *Huang Yun* [Phys. Rev. Lett. **49**, 1335–1338 (1982); *ibid.* **52**, 2190(E) (1984)]

11.3 Possible Relevance of Soliton Solutions to Superconductivity 157
T. D. Lee [Nature **330**, 460–461 (1987)]

12 Pattern Formation 159

12.1 Dendrites, Viscous Fingers, and the Theory of Pattern Formation 159
J. S. Langer [Science **243**, 1150–1156 (1989)]

12.2 Tip Splitting Without Interfacial Tension and Dendritic Growth Patterns Arising from Molecular Anisotropy 166
J. Nittmann and *H. E. Stanley*
[Nature **321**, 663–668 (1986)]

Prologue
The Ground Has Shifted

Two things stand out in the physics department of San Jose State University. First, our teaching load is 12 units plus five office hours per week. Second, undergraduate research and publications with students as coauthors are very much encouraged. To get some fun out of these challenging demands and to maintain my own vitality as a research physicist, in the Fall of 1988 onwards, I created two new courses in nonlinear physics to teach. At that time, there were no suitable textbooks; reviews and research papers were used as teaching materials or recommended reading. There was much excitement in the classroom, for both the students and the instructor, mostly due to the freshness and novelty of the material we learned together.

Today, ten years later, with nonlinear science enjoying much more publicity and so many chaos and fractals books on the market, there is still no single textbook that contains all the topics in nonlinear physics that I would like covered. Then, seemingly quite suddenly, but actually long in the making, the physics profession finds itself in an employment crisis. In response, a broadening of research and curriculum in physics was urged. What else, if not nonlinear science, is more suited to answer this call for action? In fact, nonlinear science is so broad that it covers *all* the disciplines, in both natural and social sciences. Judging by the titles of papers published in physics journals alone — where the words DNA, traffic, river and evolution frequently appear — there is strong indication that the frontier of physics has shifted.

And shifted indeed. A quick survey discovered the existence of eight special journals and two magazines serving nonlinear science. Most began their publications in the last few years. The journals are *Physica D, Nonlinearity, Nonlinear Science, Chaos, Chaos Solitons and Fractals, International Journal of Bifurcation and Chaos, Fractals* and *Complex Systems*. The magazines are *Nonlinear Science Today* and *Complexity*. In the last 10 to 15 years, a number of centers around the world, which caters to the study of nonlinear and complex systems, has popped up one after the other. More recently, a graduate program for the study of complex systems was established in Ann Arbor. A Topical Group in statistical physics and nonlinear physics has just been established within the American Physical Society.

We hope that this book will help the reader to understand this important shift in paradigm, and to share the excitement of new developments in nonlinear science. The basic principles expounded in the book are very general and applicable beyond physics. The book can be used for self-study, as a textbook for a one-semester course, or as a supplement to other courses in linear or nonlinear systems. No mathematical knowledge beyond calculus and no computer literacy are required. (On the other hand, with the help of the listed programs in Part III, it may be a good time for the student to learn some computer skills.) Beginners referred to in the book title are those who have a background in introductory college physics.

With a few exceptions, the chapters in Part I are quite independent of each other. It is recommended that Chapter 2 be read before Chapters 3 and 5, or, the reader may start from any chapter and go back to the earlier sections when the need arises. Consequently, the book can also be used for a one- or two-unit course if a subset of the topics is covered. Supplementary materials can be found from the further-reading list in Chapter 8.

PART I

OVERVIEW

1 Introduction

1.1 A Quiet Revolution

Over the last two decades or so, something very important happened in the development of science. It was a revolution, albeit a quiet one. Like a revolution these developments touched the soul of many people, changed their outlooks of the world, and in some cases, even their lives. Unlike in revolutions, these changes did not always happen abruptly; but there are some important years, if not dates, that one can quote. It is quiet, because no one called a press conference to announce it and so there were no headlines in the newspapers.

Here we are talking about what is now called *nonlinear science*. Nonlinear science is not a new branch of science in the usual sense. It does not add a new subject of study, such as chemical physics being added to chemistry or physics. Rather, nonlinear science encompasses *all* the existing disciplines in science — in both natural and social sciences.

To put things in perspective, consider quantum mechanics and relativity, the two discoveries in physics which were developed at the beginning of this century. They are justifiably recognized as revolutions. These two revolutions present unexpected concepts and insights by going beyond the classical domains (Fig. 1.1). In fact, new results are obtained in quantum mechanics when one goes to the microscopic level ($< 10^{-8}$ cm) and, in the case of relativity, when the speed of the object is close to that of light ($\sim 10^{10}$ cm/s).

Nonlinear science, like quantum mechanics and relativity, delivers a whole set of fundamentally new ideas and surprising results. Yet, unlike quantum mechanics and relativity, nonlinear science covers systems of *every* scale and objects moving with *any* speed, i.e., the whole area displayed in Fig. 1.1. Then, by the same standard, nonlinear science is more than qualified to be called a revolution. The fact that nonlinear science delivers within the conventional system sizes and speed limits should not be counted as negative toward its novelty. On the contrary, in view of

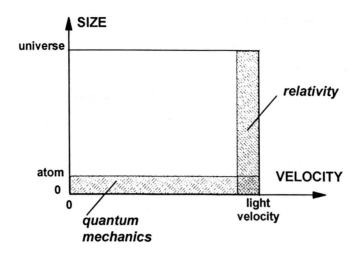

Fig. 1.1. Sketch of the limited domains of applicability of quantum mechanics and relativity theory, the two well-recognized revolutions in physics. In contrast, nonlinear science applies to the whole domain displayed here (not to scale).

its wide applicability, nonlinear science is made more important and powerful as a true revolution.

As a consequence of the multidisplinary nature of nonlinear science, people working in very different disciplines such as in economics and earthquakes now have some common vocabularies and can communicate with each other. Moreover, the applicability of nonlinear science in a broad sprectrum of scales implies that one can study the same phenomenon in very different systems with corresponding experimental tools. For example, one can study fractals on the kitchen table by photographing potato chips with an ordinary camera, while someone else with a sophisticated and expensive electron microscope will do it in a clean room with semiconductor chips. And, amazingly, they could be both working in the forefront of research in nonlinear science. In short, nonlinear science can really bring people together! Nonlinear science is a game everyone can play!

For pedagogical purposes, one may divide the content of nonlinear science into six categories, viz., fractals, chaos, solitons, pattern formation, cellular automata, and complex systems. The common theme underlying this diversity of subjects is the nonlinearity of the systems under study.

1.2 Nonlinearity

A system is nonlinear if the output from the system is not proportional to the input (Fig. 1.2). For example, a dielectric crystal becomes nonlinear if the output light intensity is no longer proportional to the incident light intensity. The examination system used by a professor is nonlinear if the grade points earned by a student do not increase linearly as a function of the number of hours put in by the student, which is usually the case.

1.2 Nonlinearity

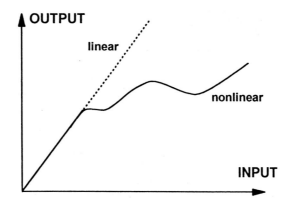

Fig. 1.2. Definition of a nonlinear system. The broken line represents a linear system with the output proportional to the input. The solid line represents a nonlinear system.

It is not that difficult to see that nonlinearity is more common than linearity. Consider the differential equation

$$dx/dt = ax^\alpha \tag{1.1}$$

where a and α are constants. Equation (1) states that the rate of increase of a certain quantity $x(t)$ is proportional to the present value raised to some power α, a rather common occurrence in real systems. Among all the possible choices of α, the solution of Eq. (1.1) is linear in t (given by $x = at + b$) only when $\alpha = 0$. For other values of α, $x(t)$ is a nonlinar function of t. [For $\alpha = 1$, the solution becomes $x = b\exp(at)$. For $\alpha \neq 1$, $x = [(1-\alpha)(at+b)]^{1/(1-\alpha)}$, which becomes a power law when $b = 0$. Here b is a constant.]

In fact, almost all known systems in natural or social sciences are nonlinear when the input is large enough. A well-known example is a spring. When the displacement of the spring becomes large, Hook's law breaks down and the spring becomes a nonlinear oscillator. A second example is a simple pendulum. Only when the displacement angle of the pendulum is small does the pendulum behave linearly. There are important qualitative differences between the behavior of a system in its linear and nonlinear regimes. For example, the period of the pendulum oscillation does not depend on the amplitude (the maximum displacement angle) in the linear regime, but it does so in the nonlinear regime.

Mathematically, the signature of a nonlinear system is the breakdown of the superposition principle which states that the sum of two solutions of the equation(s) describing the system is again a solution. The physical consequence is that in a nonlinear system, the behavior of the whole is more than the sum of its parts. (Life is an example that easily comes to mind.)

There are two ways that the superposition principle may break down. First, the equation itself is nonlinear. For example, the equation of motion for the point mass in a simple pendulum is given by

$$d^2\theta/dt^2 + (g/L)\sin\theta = 0 \tag{1.2}$$

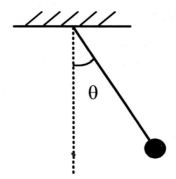

Fig. 1.3. Sketch of a simple pendulum.

where θ is the angle between the vertical and the pendulum, g the acceleration due to gravity, and L the length of the pendulum (Fig. 1.3). It is easy to show that if $\theta_1(t)$ and $\theta_2(t)$ are each a solution of Eq. (1.2), then the sum $\theta_1(t) + \theta_2(t)$ is *not* a solution, a consequence of the simple fact that $\sin\theta_1 + \sin\theta_2 \neq \sin(\theta_1 + \theta_2)$. Consequently, Eq. (1.2) is a nonlinear equation due to the presence of the nonlinear term $\sin\theta$. [In contrast, in the linear regime where θ is small, one could replace $\sin\theta$ by θ, a linear term, in Eq. (1.2) and the superposition principle becomes valid.]

Second, the equation itself may be linear but the boundary is unknown or moving. For example, in the viscous fingering problem of pattern formation in a Hele-Shaw cell (which is nothing but a volume of liquid enclosed between two narrowly spaced parallel plates), one tries to determine the shape and movement of a single, unknown interface separating two immiscible liquids when one of them is pushed into the other. The pressure field P in each type of liquid is simply given by the Laplace equation, $\nabla^2 P = 0$, which is a linear equation. However, the superposition of two solutions of this problem (corresponding to different external conditions set at the far ends of the cell) contain two "interfaces," and obviously does not represent a solution of the original problem.

The nonlinearity of a system makes the system highly nontrivial and its analysis difficult. For example:

(i) For a nonlinear system, a small disturbance such as a slight change of the initial conditions, can result in a big difference in the behavior of the system at a later time. This could make the behavior of a nonlinear system very complex [as in the case of chaos (see Chapter 3)].

(ii) If the equations describing the nonlinear system are known, the breakdown of the superposition principle renders the Fourier transform technique — which makes the analysis of a linear problem so "easy" — inapplicable. And there is no similar systematic method in solving nonlinear equations (for example, the celebrated inverse scattering method in soliton theory is applicable only to a subset of integrable systems, and there is still no way to know, a priori, which integrable system is susceptible to this method).

(iii) In many cases, from the simple diffusion-limited aggregation model of fractal pattern growth (Section 9.1) to many examples in complex systems such as the economic system (Section 10.4), the equations are not even known or simply may not exist.

All these complications make the use of computers an invaluable tool in the study of nonlinear systems since computers do not differentiate linear equations from nonlinear equations, can be used for direct simulations and can display complex results for easy visualization. The important role played by computers is partly responsible for the fact that the rise of nonlinear science is a quite recent phenomenon, correlated with the widespread accessibility of personal computers. The other reason for the late coming of nonlinear science is that it takes time for the "easy" problems of linear systems to be exhausted first, especially because many linear problems such as the propagation of electromagnetic waves in telecommunications, are technologically very important in our daily lives.

2 Fractals

Many spatial structures in nature result from the self-assembly of a large number of identical components. To be efficient, the self-assembly process takes advantage of and occurs via some simple prescriptions, which we call the principles of organization. The two simplest principles are the principle of regularity and the principle of randomness. With the former, the components arrange themselves in a periodic or quasiperiodic regular fashion, resulting in crystals, alloys, a formation of soldiers in a parade, etc. Examples of structures (or nonstructures) resulting from the latter are those in gases and the distribution of animal hairs.

Between these two extremes there is the principle of self-similarity, leading to self-similar structures called fractals. In a self-similar fractal, part of the system, when blown up in scale (with the *same* magnification in different directions), resembles the whole. A fractal usually has a fractional dimension. These concepts can be illustrated by the example of a Sierpinski gasket (SG). To construct the SG, in the first step ($n = 0$) let us start with an equilateral triangle with each side equal to one. In the next step ($n = 1$) cut out the middle inverted triangle; in step $n = 2$, do the same for each of the three triangles left over from the previous step (Fig. 2.1). Repeat this cutting procedure until $n = \infty$. (Of course, this can be done only in your mind but not in practice.) The set of "triangles" left at step $n = \infty$ is the Sierpinski gasket. It is easy to see that every small part of the SG has the same shape as the whole; the SG is thus a self-similar fractal. [The SG can be generated by a computer in at least four different ways (Sections 9.6, 13.2, 14.1.1 and 14.1.2).]

The dimension of an object D is given by

$$N_\varepsilon \sim \varepsilon^{-D} \qquad (2.1)$$

where N_ε is the minimal number of identical small objects (of linear size ε each) needed to cover the original object. Here the tilde denotes "proportional to when $\varepsilon \to 0$." The dimension D so defined by Eq. (2.1) is called the box dimension

(Section 14.1.3). Note that Eq. (2.1) is equivalent to

$$D = \lim_{\varepsilon \to 0} (-\log N_\varepsilon / \log \varepsilon) \qquad (2.2)$$

[That Eq. (2.1) indeed gives $D = 2$ for a square can be checked easily by using small squares to cover up the original square. The procedure is similar to what is shown in Fig. 2.1 below.]

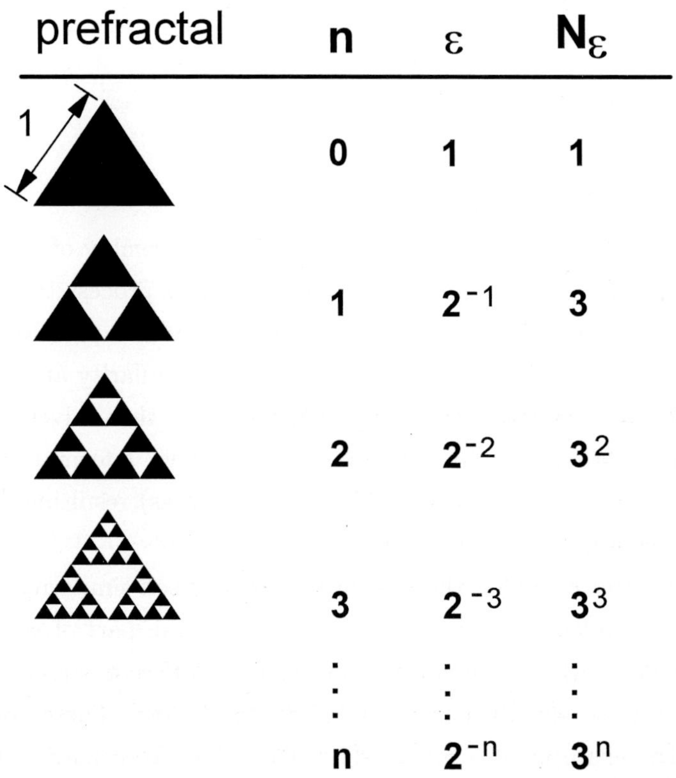

prefractal	n	ε	N_ε
(triangle, size 1)	0	1	1
(n=1 gasket)	1	2^{-1}	3
(n=2 gasket)	2	2^{-2}	3^2
(n=3 gasket)	3	2^{-3}	3^3
⋮	⋮	⋮	⋮
	n	2^{-n}	3^n

Fig. 2.1. The construction of the Sierpinski gasket, and the procedure in determining its (fractal) dimension. Equilateral triangles of linear size ε are used to cover the gasket. The figure at $n = 0$ is called the initiator; the one at $n = 1$ is called the generator; each figure at step n is called a prefractal.

To determine the dimension of the SG, let us try to cover it with small equilateral triangles. Note that the SG cannot be drawn out explicitly (it exists only at $n = \infty$). However, one can still proceed as shown in Fig. 2.1. For each ε, the number of small triangles N_ε, shown on each row, definitely cover the figure on that row and hence the SG itself, since the SG is a subset of this figure. One therefore has $N_\varepsilon = 3^n$ for $\varepsilon = 1/2^n$. Using Eq. (2.2) and the fact that $\log N_\varepsilon / \log \varepsilon = (n \log 3)/(-n \log 2)$, one obtains $D = \log 3 / \log 2 \approx 1.58$, which is *not* an integer.

For fractals generated from growth processes, one can define the fractal dimension by

$$M \sim R^D \qquad (2.3)$$

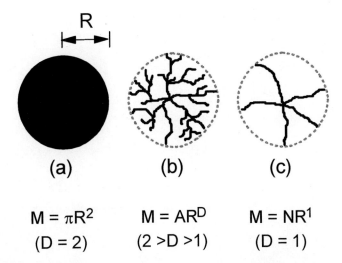

$$M = \pi R^2 \qquad M = AR^D \qquad M = NR^1$$
$$(D = 2) \qquad (2 > D > 1) \qquad (D = 1)$$

Fig. 2.2. The definition of dimension D and the origin of a fractional dimension in a growth process. M is the mass which is proportional to the black area. R is the radius, the linear size of the object. A in (b) is a constant; N is the number of lines in (c). In (a) the growth object is a solid circle; in (b) a fractal "tree" with branching; and in (c), a simple tree without branching. We thus see that branching is an essential ingredient in the formation of a fractal tree.

where M is the mass of the object when its linear size is R. Here the tilde means "proportional to when $R \to \infty$." The reason that Eq. (2.3) can give rise to a nonintegral value of D is illustrated in Fig. 2.2. [The dimensions obtained from Eqs. (2.1) and (2.3) are usually equal to each other.] Two well-known fractal growth models are the diffusion-limited aggregation model (Sections 9.1 and 14.1.4) and the dielectric breakdown model (Sections 12.2 and 14.1.5). In these models, self-similarity is valid in the statistical sense.

The ubiquitous existence of fractals in natural and mathematical systems became widely known to scientists in the early 1980s after the book *The Fractal Geometry of Nature* by Benoit Mandelbrot was published. Examples of fractals include crumpled paper balls (Section 9.2), aggregates and colloids, trees, rocks, mountains, clouds, galaxies (Section 9.3), polymers, materials with lock-in properties (Section 9.4), fractures (Section 16.3), and the stock market.

When it is required to blow up a part of the object with *different* magnifications in different directions for it to resemble the whole, the object is said to be a self-affine fractal. Interfaces and rough surfaces are such examples (Fig. 2.3). Many fractals are also multifractals, which can be roughly considered as a collection of fractals (Sections 9.5 and 9.6).

A hint of the secret of fractals lies in their power-law behavior, such as the one in Eq. (2.1) or Eq. (2.3). The power law

$$y = Ax^a \qquad (2.4)$$

is equivalent to

$$y(\lambda x) = \lambda^a y(x), \quad \text{for all } \lambda > 0 \qquad (2.5)$$

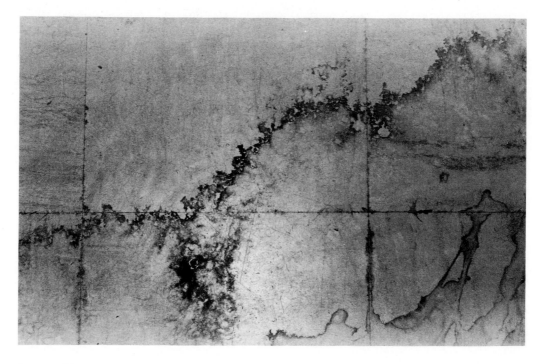

Fig. 2.3. The upper irregular curve is a self-affine fractal. The image resulted from the dried liquid splashed on the floor of the author's laboratory after the liquid underwent a free fall from a leaking pipe at the ceiling of the room.

The fact that Eq. (2.4) implies Eq. (2.5) is established by direct substitution. Since λ is arbitrary, one may choose $\lambda = 1/x$ and Eq. (2.5) reduces to $y(x) = y(1)x^a$, which is Eq. (2.4) with $A = y(1)$. The two equations are thus equivalent to each other. (The positiveness of λ ensures that λ^a is always a real number.)

Mathematically, any function $y(x)$ that satisfies Eq. (2.5) is called a homogeneous function. A homogeneous function is scale invariant, i.e., if we change the scale of measuring x so that $x \to x'$ ($\equiv \lambda x$), the new function $y'(x')$ [$\equiv y(x)$] still has the same *form* as the old one $y(x)$. This fact is guaranteed since $y(x) = \lambda^{-a} y(x')$ by Eq. (2.5), and hence $y'(x') \sim y(x')$.

Scale invariance means that if a part of a system is magnified to the size of the original system, this magnified part and the original system will look similar to each other. In other words, there is no intrinsic scale in the original system. A scale-invariant system must be self-similar and vice versa.

Thus we see that self-similarity, spatial power laws and scale invariance are three equivalent ways of expressing the fact that the system lacks a characteristic length scale. Similarly, the absence of a characteristic time scale in the system leads to temporal power laws (e.g., the $1/f$ noise, another ubiquitous phenomenon in nature). It must be noted that power laws are nonlinear equations except when the exponent is unity. To explain the widespread existence of fractals and scale-free behaviors in nonequilibrium systems, the hypothesis of self-organized criticality was proposed by Per Bak, Chao Tang and Kurt Wiesenfeld in 1987, which is supposed to be applicable to sandpiles and many other natural and social systems (Section 13.3).

Lastly, let us note that the fractal concept has found applications in the social sciences, as evidenced by Hans-Jürgan Warnecken's book *The Fractal Company: A Revolution in Corporate Culture* (Springer, 1993).

3 Chaos

In the realm of science, chaos is a technical word representing the phenomenon that the behavior of some nonlinear systems depend sensitively on the initial conditions (Section 10.1). This usage of the word obviously differs from that adopted in our daily lives, in which chaos is synonymous to "a state of utter confusion" (Fig. 3.1). (The word work is another example of this kind of free borrowing by the scientists. When you breathe heavily after carrying a heavy object up ten stories in the school building and back to the same spot, and your teacher says that you have done zero work, then you know you and your teacher are not speaking the same language.) Chaos as envisioned by the artists is sampled in Fig. 3.2.

Chaos has been investigated by Henri Poincaré at about the turn of the century and subsequently by a number of mathematicians. Recent frenzy about chaos occurred in the late 1970s, after Mitchell Feigenbaum discovered the universality properties of some simple maps, which was preceded by the important but obscure work of Edward Lorentz related to weather predictions. Not every nonlinear system is chaotic, but chaos does occur in many mathematical (Section 14.2.1) and real systems such as a dripping faucet (Section 10.2), a waterwheel (Section 14.2.2), thermal convection of liquids, electronic circuits, chemical reactions, heart beats, etc.

The signature of chaos in a dissipative system is the existence of strange attractor(s) in the phase space, which is a fractal. In contrast, the ordinary attractors existing in nonchaotic systems have simple structures and integral dimensions (Fig. 3.3). The basins of attraction could also be fractals (Section 10.3). These linkages between chaos and fractals are not fully understood.

Two findings of chaotic systems are particularly significant:

(i) In the chaotic regime, the behavior of a deterministic system appears random. This single finding forces every experimentalist to reexamine their data to determine whether some random behavior attributed to noise is due to deterministic chaos instead.

APS News   March 1996

INSIDE THE BELTWAY

Political Chaos And Uncertainty Prod Scientists Into Action
by Michael S. Lubell, APS Director of Public Affairs

San Francisco Chronicle
MONDAY, JUNE 10, 1996

Creative Control Lacking

Technology confab verges on chaotic

BY LAURA EVENSON
Chronicle Staff Writer

What sounded like a high-tech, artsy love-fest called "The Imagination Conference" starring Brian Eno, Laurie Anderson and Spike Lee may more aptly have been called "Out of Control," after Kevin Kelly's book about technology and the future.

The event Saturday evening at Bill Graham Civic Auditorium drew a visibly hip crowd of 3,000 who paid a steep $45 each to hear and see record producer Eno, multimedia shaman Anderson and innovative director Lee present a cohesive evening of performance and discussion about the interaction of creativity and technology. What they got instead were musings so random as to look like an exercise in chaos theory.

San Jose Mercury News • **Business** • Wednesday, September 27, 1995

'Let chaos reign,' Grove urges execs

SPARTAN DAILY ■ San José State University ■ Friday, August 28, 1992

Class chaos: Troubles with adding

Fig. 3.1. Samples of the use of the word chaos in newspaper headlines.

3 Chaos

Fig. 3.2. CHAOS II. An oil painting by George Cladis, 1989; photographed by Jennifer Kotter. Original size: 84" × 60". Medium: acrylic, rope and steel plates on canvas.

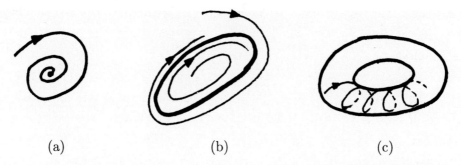

(a)   (b)   (c)

Fig. 3.3. Ordinary attractors in nonchaotic dissipative systems. (a) A point; dimension $D = 0$. (b) Limit cycle; $D = 1$. (c) Torus; $D = 2$. Other tori could have higher integral dimensions. (The limit cycle is like a besieged fortress: Outsiders want to get in; insiders want to get out. Incidentally, *The Besieged Fortress* is the title of a 1940s novel by the Chinese writer, Qian Zhongshu, in which marriage is metaphored as a fortress besieged.)

(ii) Nonlinear systems with only a few degrees of freedom can be chaotic and appear to be very complex. This finding gives hope that the complex behavior observed in many real systems may have a simple origin and may indeed be discernable.

The *apparent* unpredictability of a chaotic, deterministic, *real* system (such as the weather) arises from the system's sensitive dependence on initial conditions *and* the fact that the system's initial conditions can be measured or determined only approximately *in practice*, due to the finite resolution of any measuring instrument. Even if the physical initial conditions are known exactly, the rounding error introduced by a computer in numerical calculations amounts to producing inaccurate initial conditions for the subsequent steps in that calculation. These difficulties preclude the long-term predictability of any chaotic, real system. On the other hand, for a system that is deterministic and chaotic in nature, there is order behind its seemingly complex behavior and short-term predictability is possible. The problems are how to determine whether there is a chaotic origin behind a complicated behavior and how to do a short-term prediction.

For systems such as the weather or the stock market, due to the insurmountable complexity, the complete equations describing the system, if they exist, may never be known. Or, when the equations can be written down, there may not be a computer powerful enough to solve them. Besides, for practical reasons a successful short-term prediction for these systems is usually good enough (for example, one only needs to know the trend of the stocks slightly ahead of time — without knowing the mechanisms of the market — to make a killing in the market). Short-term prediction of the behavior of complex systems (Section 10.4) has become one of the two most exciting practical applications of chaos.

The other important application is controlling chaos (Section 10.5). This application is based on the fact there are many unstable periodic orbits embedded within a strange attractor, and one of these, if desirable, can be made stable and reached by the chaotic system with a small perturbation applied to the system, without knowing the system's dynamics in advance. The technique has been applied successfully in the control of mechanical systems, electronics, lasers, chemical systems and heart tissues.

For classical conservative systems — in particular, Hamiltonian systems — chaotic behavior can still manifest as irregular orbits in the phase space while attractors of any kind do not exist. In the quantum mechanical regime of such Hamiltonian systems, orbits in phase space are no longer well defined due to the uncertainty principle. Then, will there be chaos in the quantum regime? If yes, how will it show up? Quantum chaos is now at the very forefront of chaos research; the whole question of the correspondence between the classical and quantum regimes of a dynamical system is being studied again from new perspectives (Section 10.6).

Two other recent developments should be mentioned. One is the application of chaos to the practical problems in materials science (see the article by Alan Markworth *et al.*, in *MRS Bulletin*, July 1995). The other is the merging of chaos theory with social sciences [see *Chaos Theory in Psychology and the Life Sciences*, edited by Robin Robertson and Allan Combs (Lawrence Erlbaum Asssociated, Mahwah, NJ, 1995); and *Chaos and Order: Complex Dynamics in Literature and Science*, edited by N. K. Hayles (University of Chicago Press, Chicago, 1991)].

Finally, one should remember that there are still unsolved fundamental questions raised by chaos, such as the relationship between the deterministic and probabilistic descriptions of Newtonian dynamics (Section 10.7).

4 Solitons

Solitons are spatially localized waves traveling with constant speeds and shapes. They are special solutions of some partial differential equations (Section 11.1).

In some nonlinear media, such as a layer of shallow water or an optical fiber, under suitable conditions, the widening of a wave packet due to dispersion could be balanced exactly by the narrowing effects due to the nonlinearity of the medium. In these cases, it is possible to have solitons. For example, the equation describing wave propagation in shallow water is given by the Korteweg-deVries equation

$$\partial\theta/\partial t - \alpha\theta(\partial\theta/\partial x) + \partial^3\theta/\partial x^3 = 0 \qquad (4.1)$$

where α is a constant. The second term on the left hand side of Eq. (4.1) is the nonlinear term and the third one, the dispersion term. A soliton solution is given by

$$\theta(x,t) = -(12/\alpha)a^2 \operatorname{sech}^2[a(x - 4a^2 t - x_0)] \qquad (4.2)$$

where a and x_0 are arbitrary constants. Equation (4.2) represents a solitary wave (Fig. 4.1), i.e., a traveling wave whose transition from the asymptotic state at $\tau = -\infty$ to the other asymptotic state at $\tau = \infty$ is localized in τ. Here $\tau \equiv x - ct$, with $c = \text{const}$ ($= 4a^2$ in this case). Note that the amplitude (a^2), the wave width ($1/a$) and the velocity ($4a^2$) are related to each other — a property shared by many solitons. In this case, the "tall and thin" soliton travels faster.

Another example of a solitonic equation is the nonlinear diffusion equation

$$\partial\theta/\partial t = \partial^2\theta/\partial x^2 - \theta(\theta - a)(\theta - 1) \qquad (4.3)$$

where $0 < a < 1$. One of the possible soliton solutions [Fig. 4.2(a)] is given by

$$\theta = \{1 + \exp[(x - ct)/\sqrt{2}]\}^{-1} \qquad (4.4)$$

with

$$c = (1 - 2a)/\sqrt{2} \qquad (4.5)$$

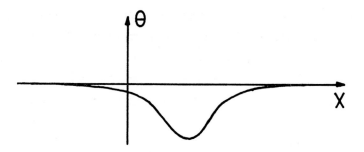

The existence of this solution may be understood as follows. Let $\theta(x,t) = X(\tau)$ and define $V(\theta)$ through $\theta(\theta - a)(\theta - 1) \equiv -\partial V(\theta)/\partial \theta$. Equation (4.3) becomes

$$\partial^2 X/\partial \tau^2 = -c\partial X/\partial \tau - \partial V/\partial X \qquad (4.6)$$

which represents the motion of a particle of unit mass moving in a potential V with damping coefficient c, with X and τ being the "displacement" of the particle and "time," respectively. When c is suitably chosen, the particle may roll down from the high hilltop (at $X = 1$) with zero velocity ($\partial X/\partial \tau = 0$), pass through the valley (at $X = a$), and then stop exactly at the lower hilltop (at $X = 0$) [Fig. 4.2(b)].

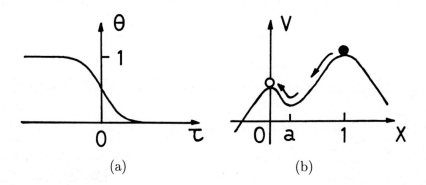

Fig. 4.2. (a) A soliton solution of Eq. (4.3). (b) Physical picture of the soliton solution.

Thus we see that solitons could appear either in the bell shape (Fig. 4.1) or in the form of a kink [Fig. 4.2(a) is actually an antikink]. In mathematics literature, these localized waves are called solitons only if they further possess the elastic collision property that two such waves will emerge from a head-on collision with their velocities and shapes unchanged. It turns out that such an elastic collision property is related to the integrability of the system (such as the Korteweg-deVries equation) and appears only rarely in nonintegrable systems (such as the nonlinear diffusion equations). Since most real systems are nonintegrable, in physics literature, the elastic collision property is dropped as a requirement in the definition of solitons which, under perturbations, may even have their velocities and shapes slightly distorted during propagation.

Even though solitons were first observed by John Scott Russell in 1834, it was only after 1965 — the year that the word soliton was coined by Norman Zabusky and Martin Kruskal — that the significance of solitons was widely appreciated and the study of solitons as a discipline took shape. Since then solitons, as both nonlinear waves and nonlinear excitations in materials, have been intensely studied in various systems including liquid crystals (Section 11.2 and Fig. 4.3), conducting polymers, high-temperature superconductors (Section 11.3), optical fibers, and even Jupiter (Fig. 4.4). In particular, optical solitons in glass fibers are becoming very important because of their demonstrated applicability in multigigabits optical transmissions over very long distances, say, four thousand times around the world.

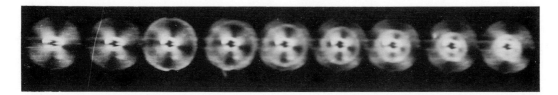

Fig. 4.3. Propagation of a two-dimensional soliton, in the form of a white ring, generated in a liquid crystal cell of radius 5 cm and thickness 20 μm. The pressure at the rim of the radial cell was maintained at atmospheric pressure. The pressure at the center was first increased above and then decreased below atmospheric pressure. The long liquid crystal molecules in the cell were vertical initially, and were out of the vertical plane within the white-ring region. The white ring propagated inward by shrinking in size, as shown from left to right in the series of photographs displayed here.

Fig. 4.4. The Great Red Spot on Jupiter's atmosphere is a soliton, coexisting with a turbulent environment.

5 Pattern Formation

One can hardly fail to notice the striking similarity between the ramified patterns formed by rivers, trees, leaf veins and lightning. These branching patterns are different from compact patterns observed in snowflakes, clouds and algae colonies. How does nature generate these patterns? Is there a simple principle or universal mechanism behind these pattern-forming phenomena? These are the profound questions that interest lay people and experts alike. Athough final answers to these questions are still lacking, tremendous progress has been made in the last 15 years.

Patterns existing in nature and laboratories may be classified into two types: (A) those involving an interface, and (B) those that do not. Type A patterns can be further separated into two classes, viz., (A1) filamentary, or (A2) compact. Of course, when filamentary patterns are much magnified, they appear as compact. The distinction between A1 and A2 depends on the scale used in the observations.

Some A1 patterns, apart from those mentioned in the beginning of this chapter, are shown in Figs. 5.1 and 5.2. To generate A1 patterns, models for aggregation and diffusive growth have been much studied and are quite successful in mimicking many real systems, which are often fractals (Sections 9.1 and 14.1.5). Examples from the biased random walk model (Section 14.3.1), in conjunction with experimental electrodeposit patterns, are shown in Fig. 5.3. Recently, a unified way of generating filamentary patterns (and others) is provided by the active walk model (AWM), as proposed by Lui Lam and his coworkers (Section 13.4). The model is based on the observations that (i) a filament may be represented by the track of a walker; (ii) to grow a track, one has only to specify how the walker chooses its next step; and (iii) the walker may distort the landscape (or environment) as it walks, and its next step is influenced by the changed landscape. It is in the sense of point (iii) that the walker is active. A representative result from the AWM is shown in Fig. 5.4. (A computer program of the AWM is given in Appendix A1.)

Physical examples of A2 patterns include those in electrodeposits (Fig. 5.5) and solidifications, viscous fingers in Hele-Shaw cells (Sections 12.1 and 16.2), snowflakes

28 Part I Overview 5 Pattern Formation

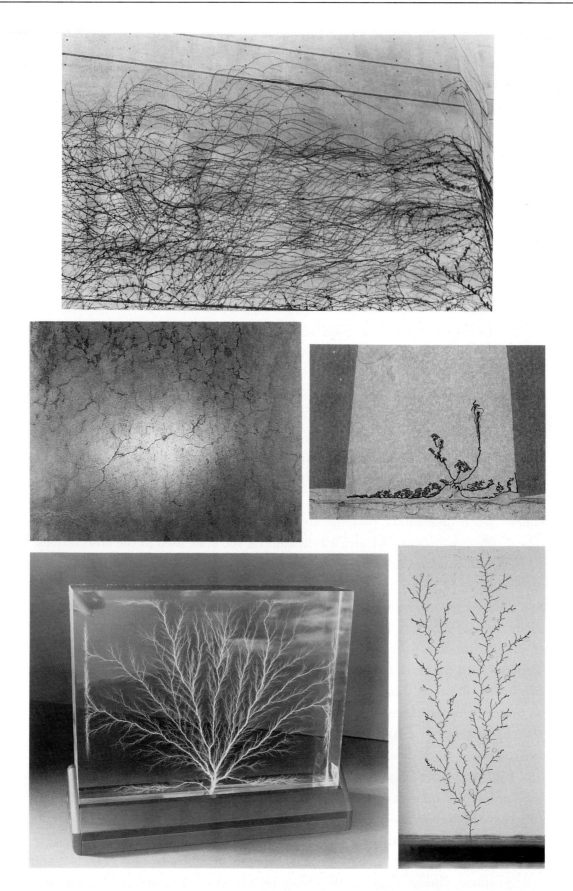

Fig. 5.1

5 Pattern Formation

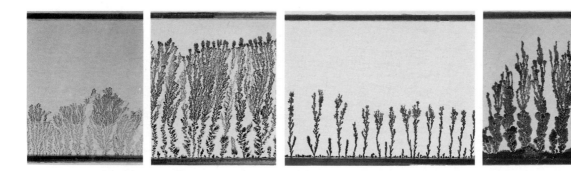

Fig. 5.2. Sensitive dependence of electrodeposit morphology on cell thickness. The voltage (20 V) and concentration of $CuSO_4$ solution (0.05 M) are the same in the four cells shown here. The cell thickness, from left to right, is 0.1 mm, 0.4 mm, 0.6 mm and 0.7 mm, respectively.

(Section 12.2), and water columns (Section 16.1). Some of these compact patterns can be generated from aggregation models, such as those in Sections 12.2 and 14.3.2, and the BPAW model in Section 13.4. (See also Section 14.3.3.) Others can be studied by theoretical or numerical solutions of the underlying equation of the interface. In particular, a unified description of the formation of viscous fingers and solidification interfaces seems to have been achieved (Section 12.1).

Type B patterns include the thermal (Rayleigh-Bénard) convection patterns in fluids and the very similar electroconvection patterns in liquid crystals (Sections 12.3 and 12.4), and those in chemical waves (Section 12.5). For type B, the patterns are the self-organized structures of certain quantities (such as the fluid velocity or the chemical densities), resulting from the linear instabilities of a homogeneous state. Secondary and higher instabilities also appear, giving rise to a series of patterns as the control parameters are varied. The similarity among type B patterns from various systems is explained by the unified theory of amplitude equations. The spirit of this approach is really simple. In this theory, the various quantities $X_j(\mathbf{r},t)$

Fig. 5.1. Physical examples of filamentary A1 patterns. TOP: Ivy on the wall of Walquist Library, San Jose State University. MIDDLE LEFT: Cracks on the basement floor of Science Building, SJSU. MIDDLE RIGHT: Chemical reaction tracks induced by dielectric breakdown in a thin layer of mineral oil. The oil was placed between two glass plates, with their inner surfaces coated by conductive indium tin oxide. The tracks appeared on the inner surfaces after a uniform electric field above a threshold was applied perpendicular to the cell. BOTTOM LEFT: An acrylic *beamtree* created by placing the acrylic in the path of an electron accelerator, then discharging the electric charge through a conductive spike placed in the acrylic. *Beamtrees* like the one shown here are made at the Stanford Linear Accelerator Center, California, as gifts to their retiring staff. BOTTOM RIGHT: An electrodeposit "tree," formed by the aggregation of positive Zn ions attracted towards the Zn cathode, shown near the bottom. The Zn anode at the top, not shown here, was parallel to the cathode. $ZnSO_4$ solution of concentration 0.01 M was used in this cell (thickness 0.1 mm). A voltage of 5 V is applied between the two electrodes.

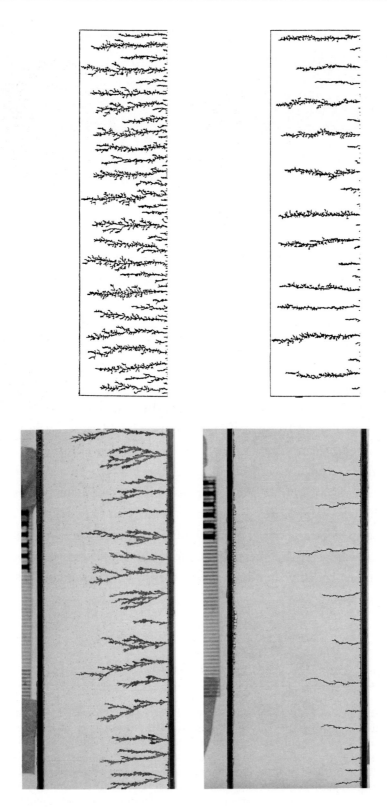

Fig. 5.3. RIGHT: Computer generated patterns from the biased random walk model. LEFT: Experimental electrodeposit patterns. Note the similarity between the patterns shown on the same row.

5 Pattern Formation

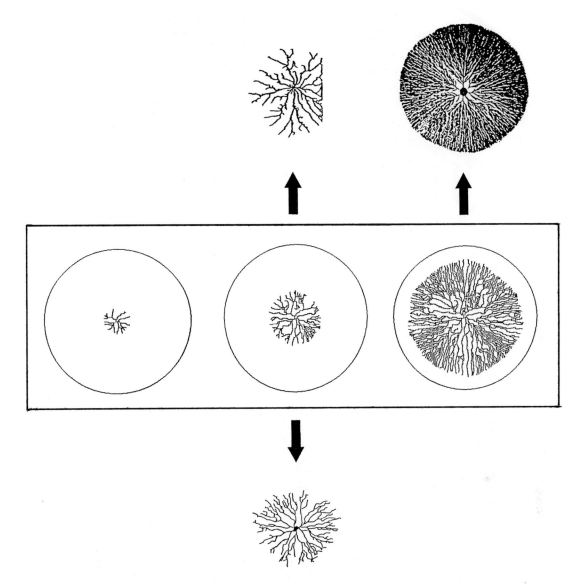

Fig. 5.4. MIDDLE: Patterns enclosed in the rectangle, from left to right, represent the time development of a computer generated pattern from the active walk model. The arrows outside the rectangle point toward the experimental patterns, with which the corresponding computer patterns should be compared. TOP LEFT: Chemical reaction pattern induced by dielectric breakdown in a thin layer of oil. The experimental setup is similar to that described in Fig. 5.1. TOP RIGHT: A dense radial morphology (DRM) from electrodeposit in a $ZnSO_4$ cell. BOTTOM: A retinal neuron. Note the similarities between the corresponding patterns. It is interesting to see whether, given enough time, the chemical reaction pattern and the neuron pattern will each grow into the DRM.

Fig. 5.5. Physical examples of compact A2 patterns. LEFT: Electrodeposit pattern formed in an open cell of CuSO$_4$ solution. RIGHT: Viscous finger pattern formed in a radial cell. Air pumped into the center of the cell displaced nematic liquid crystals which initially filled up the whole cell. (The circular grooves in the cell plate did not seem to have much effect.) Note the existence of tip splitting in both cases.

($j = 1, 2, \ldots$) involved in the equations of motion are forced to share the same amplitude function $A(\mathbf{r}, t)$. For example, it is assumed that

$$X_j(\mathbf{r}, t) = A(\mathbf{r}, t) \exp[i(kx - \omega t)] + \text{complex conjugate} \qquad (5.1)$$

for *all* j. Clearly, this cannot be exactly valid. By forcing these expressions for different j to be consistent with each other, to the first few orders in a perturbation calculation, one obtains a consistency requirement (called the solvability condition) in the form of an equation governing $A(\mathbf{r}, t)$ — the amplitude equation — which may look like this:

$$\partial A/\partial t = \alpha A + \beta(\partial^2 A/\partial x^2) - \gamma |A|^2 A \qquad (5.2)$$

where the coefficients may be complex numbers. Equation (5.2) is called the complex Ginzburg-Landau equation. The exact form of the equation depends largely on the symmetry of the problem, and the details of the system affect only the coefficients of the equation. Consequently, through the amplitude equation, a unified description of a whole class of systems is possible.

In conclusion, we see that some unified pattern-formation principle has emerged in *each* of the three types of patterns, but a universal principle valid for *all* patterns is still absent.

We note that two types of abnormal pattern growth — transformational and irreproducible growths — are often observed. Examples of the former in electrodeposit experiments are shown in Fig. 5.6. The possibility that both types of abnormal growth can be caused by noise and hence are instrinsic in nature, is demonstrated in computer models (Section 13.4). Experiments seem to support this conjecture.

5 Pattern Formation

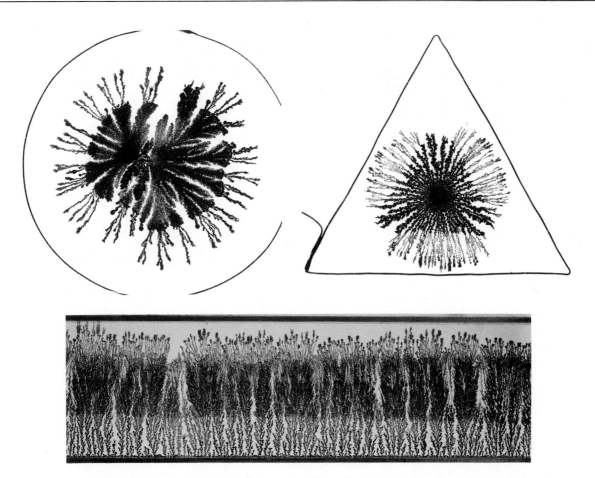

Fig. 5.6. Examples of transformational growth in electrodeposit experiments, in which morphological changes appeared while control parameters (voltage and cell thickness) were kept constant. In upper right and bottom, note the existence of multiple morphology transitions, the origin of which is not yet completely understood.

6 Cellular Automata

Cellular automata are discrete dynamical systems whose evolution is dictated by local rules. In practice, they are usually realized on a lattice of cells, with a finite number of discrete states associated with each cell, and with local rules specifying how the state of each cell should be updated in discrete time steps (Section 14.4.1). Because of the discreteness of all the quantities involved, cellular automata calculations obtained from computers are exact. Note that a cellular automaton is just a computer algorithm and is not a real machine.

Cellular automata were introduced by John von Neumann in the late 1940s, soon after the creation of electronic computers. He used cellular automata to prove that, in principle, self-reproducing machines were possible. Then, in the early 1970s, the "game of life" — a very simple two-dimensional cellular automaton capable of creating lifelike "creatures" on the lattice — invented by John Conway, became very popular with the public. But scientifically, nothing serious happened.

In 1986 Uriel Frisch, Brosl Hasslacher and Yves Pomeau demonstrated that it is possible to simulate the Navier-Stokes equations of fluid flows by using a cellular automaton of gas particles on a hexagonal lattice, with extremely simple translation and collision rules governing the movement of the particles. That a microscopic system of interacting particles with oversimplified dynamics could result in the physically correct macroscopic Navier-Stokes equations (the generally accepted equations describing fluid flow) is due to the fact that the Navier-Stokes equations are the consequence of appropriate conservation laws and are quite insensitive to microscopic details, as long as the appropriate symmetries are obeyed by the microscopic dynamics. The spirit of this lattice gas approach in "simulating" nature runs opposite to that of the molecular dynamics or Monte Carlo approach, where the more realistic the molecular interactions are, the better. By using very simple rules in the cellular automata, the lattice gas approach provides gains in computational speed, is error free, and can easily take care of very irregular boundaries. It is one of the rare occasions that nature can be "cheated", if only on the computer.

Presently, lattice gas automata provide the most efficient method in calculating fluid flow through real porous media and has found application in the oil industry. Armed with the same "cheating" principle, lattice gas automata with different updating rules have been devised to simulate various partial differential equations.

Moreover, cellular automata are being investigated as complex systems per se (Section 13.2), and as simple devices in simulating real processes in biological, physical and chemical systems (Section 14.4.2). Finally, it is interesting to note that the "game of life" has been credited with leading Christopher Langton to the very concept of artificial life that is now established as a vigorous discipline of its own.

7 Complex Systems

The fact that there is one and only one doctoral degree, the Doctor of Philosophy (Ph.D.), but not the Doctor of Physics or Doctor of Economics, attests to the fact that not too long ago, science was considered and studied as a whole. There was no division of social and natural sciences, not to mention no fragmentation of the natural science into physics, chemistry, biology, etc. As suspected by some, this compartmentalization of science is due more to administrative convenience than to the nature of science itself.

The hope of being able to return to the appealing state of a unified science was rekindled in the 1970s (as exemplified by the works of Hermann Haken and Ilya Prigogine) and early 1980s (such as the establishment of the Santa Fe Institute). The more recent development was influenced by the success of chaos theory. At that time chaos was better understood and time series obtained in almost every discipline — from both social and natural sciences — were subjected to the same analyses as inspired by chaos theory. The importance of this development was that chaos theory seemed to offer scientists a handle or an excuse, if one was needed, to tackle problems from any field of their liking. A psychological barrier was broken; no complex system was too complex to be touched.

A secondary but crucial influence that helps to propel and sustain complex systems as a viable research discipline is the prevalence of personal computers and the availability of powerful computational tools such as parallel processing computers. While theoretical study of complex systems is usually quite difficult and sometimes appears impossible, one can always resort to some form of computer simulation, or computer experiment, as some like to call it.

What is a complex system? A complex system could be one which consists of a large number of simple elements or "intelligent" agents, interacting with each other and the environment. The elements/agents may or may not evolve in time and the behavior of the system cannot be learned by reduction methods (Section 13.1). But such a definition is not without problems. For example, a system may appear complex only because we do not understand it yet, by reduction methods or not. Once understood, it becomes a simple system. Moreover, whether a system is complex or not may depend on the aspect of it that one wants to study. If we want to know the inner structure and formation mechanism of a piece of rock, the rock could be a complex system. But if we want only to know how the rock will move when given a kick, then Newtonian dynamics will do and the rock is simple.

A precise definition of a complex system is thus difficult to come by, as is frequently the case in the early stages of a new research field. However, this difficulty does not seem to hinder the study of complex systems much. In practice, it is safe to say that almost all the subjects covered in the various departments of a university — except for those in the conventional curriculums of the physics, chemistry and engineering departments — are in the realm of complex systems. The topics studied span a wide spectrum, including human languages, the origin of life, DNA and information, evolutionary biology and spin glasses, economics, psychology, ecology, ant swarms, earthquakes, immunology, self-organization of nonequilibrium systems, cellular automata (Section 13.2), neural networks (Section 15.2), etc.

While some general concepts such as complex adaptive systems and symmetry breaking have been found to be useful in their descriptions, no unifying theory governing all complex systems exists yet. However, two simple ideas capable of explaining the behavior of many complex systems have emerged. One is self-organized criticality (Section 13.3) and the other is the principle of active walks (Section 13.4). The former asserts that large dynamical systems tend to drive themselves to a critical state with no characteristic spatial and temporal scales. The latter describes how elements in a complex system communicate with their environment and with each other, through the interaction with the landscape they share. The principle of active walks has been applied successfully to very different problems such as the formation of surface reaction patterns (Fig. 5.4), ion transport in glasses, the cooperation of ants in food collection, and increasing returns in economics (Section 13.5).

In the study of complex systems such as biological evolution and human history, the relative importance of chance in determining the outcome is constantly under debate. This problem is sometimes described as the interplay of nurture and nature, or, of chance and necessity. The difficulty in solving this problem is due to (i) the scarcity of data from the field, (ii) the impossibility of recreating the events, and (iii) the absence of realistic mathematical models for these rather complex systems. Scientists are then frequently forced to come up with the best educated guess, which may put the role of chance as extremely important, non-consequential, or somewhere in between these two extremes. However, through our study of the active walk models, we have discovered that there could be a completely new answer to this question: it depends. This means that the relative importance of chance depends on the state of the system under study (specifically, to what region in the parameter space the system belongs; see below).

This new phenomenon is best demonstrated by the Boundary Probabilistic Active Walk (BPAW) model (see Section 13.4). In the BPAW model, one starts with a single particle. The landscape around this particle is changed by a landscaping rule; one of the perimeter sites of the aggregate is chosen with a probabilistic rule; then a new particle is added to this chosen site. This is followed by the change of the landscape (with the same landscaping rule) around this new particle. In a special version of this model, there are two parameters η and ρ (the exact meaning of these parameters is immaterial to our discussion here). When η and ρ are varied, the model is capable of producing five types of morphologies: blob, jellyfish, diamond, lollipop, and needle (Fig. 7.1). As shown in the morphogram (or the state diagram) in Fig. 7.2, there is a region in the middle bounded by the two solid lines — the sensitive zone — within which, for the same set of model parameters, more than one type of morphology is generated from different computer runs. The mechanism behind this is the active role played by noise (i.e. the source of contingency) within the sensitive zone.

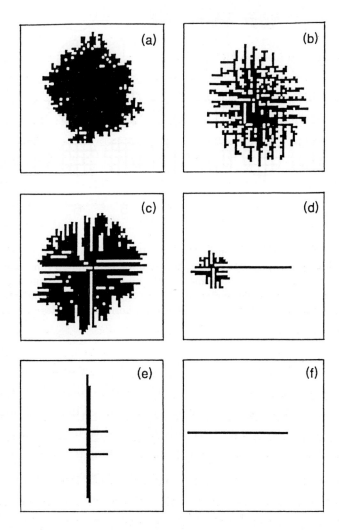

Fig. 7.1. The five types of morphologies obtained from the BPAW model. (a) Blob; (b) jellyfish; (c) diamond; (d) lollipop; (e) and (f) needle.

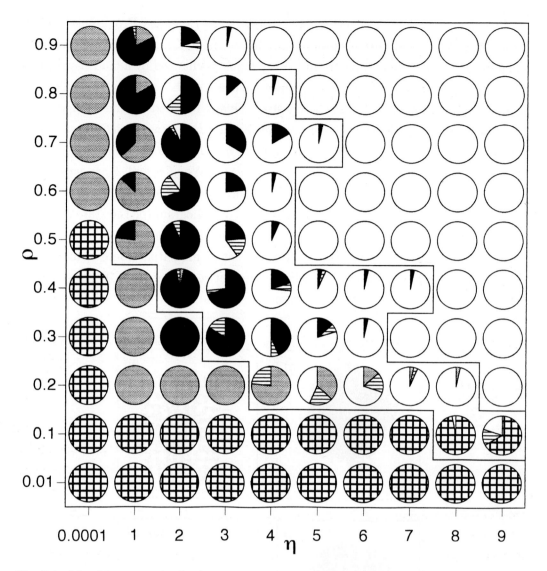

Fig. 7.2. Morphogram in the (η, ρ) plane from the BPAW model. The pie chart at each point represents the percentage of each type of morphology obtained from 30 runs of the algorithm. The parameters used in these 30 runs are the same, but the random number sequence (noise) used in the probabilistic rule in selecting the perimeter sites of the aggregate varies from run to run. The grid corresponds to blob, grey to jellyfish, black to diamond, horizontal shades to lollipop, and white to needle.

In other words, if experiments are performed and repeated with the same controlling parameters inside the sensitive zone, qualitatively different experimental results can appear — the result is intrinsically irreproducible. This could be the reason why some cold fusion experiments are currently observed to be irreproducible. (Scientifically, there is nothing wrong with an experiment being irreproducible when noise is involved. Whether the experimental result is technologically useful — which usually requires reproducibility — is a separate issue altogether.)

The existence of sensitive zones in biological, chemical and physical experiments are known to exist. An example from the empirical data on the survival and reproduction of the plant common teasel is presented in Fig. 7.3.

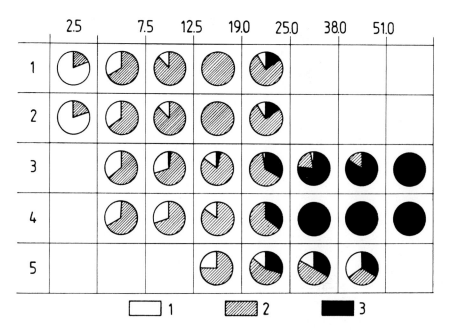

Fig. 7.3. The empirically determined state diagram of common teasel. Each pie chart represents the probabilities that the three states are found. The three possible states are (1) dying, (2) remaining vegetative, and (3) flowering. The two parameters are age (1 to 5 years, given in the first column) and the size of the rosette (in centimeters, given in the first line) in the preceding year.

Histories, be it the evolution history in biology or the human history in our real world, result from the combined effect of chance and some basic rules. This all important coexistence of chance and necessity is a feature captured by the probabilistic versions of the active walk models, of which the BPAW is a special case. Now, if our real world is believed to be such an active walk system, and the algorithm of life is allowed to be replayed — like the algorithm in the BPAW model being rerun — we may or may not recover a similar (but never identical) history of life, according to whether our world happens to sit outside or inside a sensitive zone. This observation differs from that of Stephen Gould, who argued in *Wonderful Life* (Norton, 1989) that history would always be very different if life's tape was ever replayed.

Some remarks are in order. First, the history of life (or the world) is identified with the *morphology* of the pattern output by an algorithm run, not the *pattern* itself in all its details. Second, the identification of a life's replay as the rerun of an algorithm (with noise as an ingredient) adopted here, seems to be more appropriate than identifying it as a replay of a (video)tape, the metaphor used by Gould. In this regard, life is more like a game, whereby there are some basic laws the players have to follow while the actual outcome of each game differs in details and in the final scores.

Needless to say, we are still far from a quantitative theory of history. But active walk could be helpful as a model in reaching this goal, as well as in modeling other path dependent, branching phenomena.

8 Remarks and Further Reading

Nonlinear science involves the interplay of order and disorder, as well as the simple and the complex. Technically, what makes the fascinating outcomes possible is nonlinearity. It is the nonlinearity that makes the behavior of the systems nontrivial and interesting, and the world as complex as it is.

The basic theories and principles expounded in this book are applicable to *all* branches of science, although most of the examples covered are taken from physics.

Nonlinear science is still a science in the making. Exciting results keep appearing in leading research journals such as *Physical Review Letters*, and in special journals such as *Physica D*, *Chaos*, and *Complexity*. Conference proceedings are also good sources for the latest developments.

Due to the multidisciplinary nature of nonlinear science, it is not easy for newcomers or even practitioners to be knowledgeable in all the subjects covered in the applications. In this regard, the general magazine *Scientific American* is a valuable source of information. For further reading beyond this book, a list is provided here.

Further Reading

The elementary books listed below are at the undergraduate level; the advanced books are at graduate level; but overlaps do exist. All the popular books, except the last one, are available in paperbacks. They provide perspectives linking up different topics and enjoyable reading. However, watch out for occasional inaccuracies and hypes.

Elementary

- J.-F. Gouyet, *Physics and Fractal Structures* (Springer, New York, 1996). Clear exposition of concepts, contains a lot of real examples.

- G. L. Baker and J. P. Gollub, *Chaotic Dynamics: An Introduction* (Cambridge University, Cambridge, 1996). Uses the damped driven pendulum as the primary model, contains a list of computer programs in True BASIC.

- M. Remoissenet, *Waves Called Solitons: Concepts and Experiments* (Springer, New York, 1994). Includes some simple experiments performable by the readers; covers solitons in transmission lines, hydrodynamics, mechanical systems, Josephson junctions, optical fibers, and lattice dynamics.

- D. Kaplan and L. Glass, *Understanding Nonlinear Dynamics* (Springer, New York, 1995). Four chapters on chaos, including a long one on time series analysis; one chapter on cellular automata; another on fractals; calculus is the only prerequisite; most examples from life sciences.

Advanced

- L. Lam (ed.), *Introduction to Nonlinear Physics* (Springer, New York, 1997). A graduate textbook with homework problems, covers topics similar to those here, each part written by pioneers or experts in the field.

- A. Bunde and S. Halvin (eds.), *Fractals in Science* (Springer, New York, 1994). Collection of survey articles written by leading scientists; covers self-organized criticality, fractals in biology and medicine, interfaces, polymers and chemistry; a chapter on random walks; another on computer simulations, accompanied by a PC or Macintosh diskette.

- H. G. Schuster, *Deterministic Chaos: An Introduction* (VCH, New York, 1995). Up-to-date accounts of all aspects of chaos, leans on theory, with physical examples.

- G. Weisbuch, *Complex Systems Dynamics* (Addison-Wesley, Menlo Park, 1991). Covers cellular automata, neural networks, random Boolean networks, genotypes and phenotypes.

Popular

- J. Briggs and F. D. Peat, *Turbulent Mirror: An Illustrated Guide to Chaos Theory and the Science of Wholeness* (Harper & Row, San Francisco, 1990). Concise, enjoyable accounts of fractals, chaos, solitons, feedbacks and complex systems; connection to so-called Chinese legends is shaky.

- M. Schroeder, *Fractals, Chaos, Power Laws: Minutes from an Inifinite Paradise* (Freeman, New York, 1991). A must, more mathematical sections can be skipped on first reading.

- J. Gleick, *Chaos: Making a New Science* (Viking, New York, 1987). Written by a non-scientist; exciting stories and clear explanations about chaos.

- W. Poundstone, *The Recursive Universe: Cosmic Complexity and the Limits of Scientific Knowledge* (Contemporary Books, Chicago, 1985). A detailed exposition on "game of life," with PC programs in BASIC and assembly language; very interesting discussion on self-reproducing life and other profound questions connected with cellular automata.

- M. M. Waldrop, *Complexity: The Emerging Science at the Edge of Order and Chaos* (Simon & Schuster, New York, 1992). Captures the excitement of studying complex systems by people at or associated with the Santa Fe Institute, their lives are described and works explained.

- P. Bak, *How Nature Works: The Science of Self-Organized Criticality* (Copernicus, New York, 1996). A personal account of the discovery, concept and application of self-organized criticality, with selected references.

PART II

REPRINTS

Most of the papers reprinted here are reviews or essays written by pioneers. They are selected for their pedagogical values and their accessibility to undergraduates. The others are research papers divided between experimental (Sections 9.2, 10.2, 11.2, 12.3 and 12.4) and theoretical/computational (Sections 9.3, 9.6, 10.4, 11.3, 12.2 and 13.5). With perhaps the exception of one or two in the latter category, all these papers are reasonably easy to read.

It is our experience and conviction that for a good education in any subject, nonlinear physics in particular, there is no substitute for reading articles by the original contributors. However, like looking at an abstract painting, it is not always necessary to understand the content in order to be inspired by it.

Fractal growth processes

Leonard M. Sander

Physics Department, University of Michigan, Ann Arbor, Michigan 48109, USA

The methods of fractal geometry allow the classification of non-equilibrium growth processes according to their scaling properties. This classification and computer simulations give insight into a great variety of complex structures.

ALMOST every theoretical tool of the condensed-matter scientist uses the assumption that the system considered is of high symmetry and is in equilibrium. These assumptions have led to enormous progress; however, to much, if not most, of the natural world such tools cannot be applied. Many systems that we would like to understand are very far indeed from perfectly ordered symmetry and are not even in local equilibrium. Perhaps the most extreme example is disorderly irreversible growth. We mean by this the sort of process which is very familiar in the formation of dust, soot, colloids, cell colonies and many other examples; roughly speaking, things often stick together and do not become unstuck. For example, a particle of soot grows by adding bits of carbon and coagulating with other particles in a random way. A possible result is shown in Fig. 1. We are thinking about cases which are, in some sense, as far from equilibrium as possible, and which have no obvious order.

It is remarkable that the introduction of simplified models has led to quite a good understanding of the morphology of such growth, despite the inapplicability of our usual modes of thinking. Here I will discuss this progress, drawing examples mostly from subjects which have traditionally interested physicists and chemists. However, disorderly growth is ubiquitous in the world around us, and is certainly not limited to inanimate matter. For example, some of the ideas which I will discuss, such as anomalous scaling in kinetic processes, will be useful to biologists. The purpose of the review is to introduce ideas from the area which may be of general use.

The key to our recent progress is the recognition that the most 'interesting' non-equilibrium structures (say, from a visual point of view) are not merely amorphous blobs; they still have a symmetry, despite their random growth habit, albeit a different one than they might have had, had they grown near equilibrium. For example, consider the soot of Fig. 1, or the electrolytic deposit of zinc shown in Fig. 2. Many people will be familiar with branched deposits such as this, and with similar looking objects which form on automobile windshields on cold mornings. In all these cases the structure is disordered, but it is not random. A manifestation of this is that each section of the picture contains holes in the structure comparable in size with that of the section itself. This can only occur if there are long-range correlations in the pattern; particles 'know' about each other over distances far in excess of the range of the forces between them. A truly random pattern, such as that of salt scattered on a table top, shows no such scaling of holes, and correlations are of short range only.

Studies of fractal growth have focused on two questions: how can we characterize and quantify the hidden order in complex patterns of this type, and when and how do such correlations arise? The answer to the first question is now relatively clear, and lies in an application of the fractal geometry of Mandelbrot[1]. The next section gives a brief review of relevant aspects of this subject. The second question has received a partial answer in the formulation and analysis of models suitable for computer

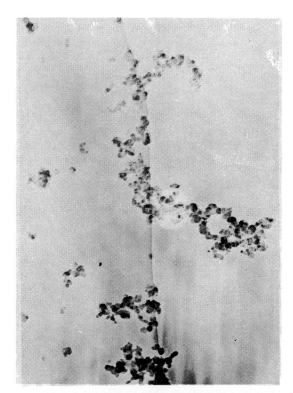

Fig. 1 Electron micrograph of soot. (Supplied by G. Smith, General Motors.)

simulation, which will also be reviewed. For more extensive treatments see refs 2-4.

Fractals and scale invariance

In pure mathematics, it has long been common to study certain 'pathological' geometric shapes that elude ordinary notions such as those of length and area. Figure 3 shows a famous example, which has, in some sense, infinite length, but zero area. It falls between our usual notions of line and solid. Mandelbrot[1] systematized and organized mathematical ideas concerning such objects due to Hausdorff, Besicovitch and others. But, more importantly, he pointed out that such patterns share a central property with complex natural objects such as trees, coastlines, patterns of stars and (as was later discovered) the non-equilibrium growths of Figs 1 and 2. This property is a symmetry which may be called scale invariance. These objects are invariant under a transformation which replaces a small part by a bigger part, that is, under a change in scale of the picture. Scale-invariant structures are called fractals.

There are a number of related properties which follow from the assumption of scale invariance. Consider, for example, the

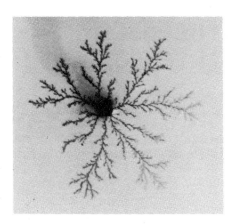

Fig. 2 Zinc electrodeposit produced in a thin cell under conditions of low $ZnSO_4$ concentration (0.01 mol l^{-1}). The outer electrode (not shown) is in the form of a ring 6.3 cm in radius. (Supplied by D. Grier, University of Michigan.)

Fig. 3 Four stages in the growth of an exact fractal, the Koch curve. This and many other examples are discussed in ref. 1. The fractal dimension may be deduced by thinking of each picture as a part of the picture above, with a change of scale. For each scale change by three, we need four such parts. Thus, according to equation (2), $D = \log 4/\log 3 = 1.26$.

density correlation function $c(r)$, of a fractal. This is defined as the average density of the object at distance r from a point on the object, and is a measure of the average environment of a particle. Clearly, $c(r)$ must reflect the scale invariance. It is easy to show that the only way that c may vary is as a power law in r; any other function would have an intrinsic scale. It is convenient to write c in the following form:

$$c(r) = kr^{-(d-D)} \quad (1)$$

Here, k is a constant, and the exponent is written in terms of the dimension of space, d, and a new quantity, D, the fractal dimension. The reason for this terminology will become evident in a moment. As the objects we are dealing with are tenuous, $c(r)$ is a decreasing function of r: the average density decreases as the object becomes larger. Now consider how the total mass of the object, M, scales with the mean radius, R. We can estimate this by multiplying a typical density, from above, by the volume:

$$M(R) = KR^{D-d}R^d = KR^D \quad (2)$$

Here, K is another constant. We can now see why D is called a dimension. For an ordinary curve, $D = 1$: twice the length gives twice the mass. For a disk, $D = 2$. For simple objects D coincides with the usual notion of dimension. But in the cases we are discussing D is not an integer; it has been measured to be ~ 1.7 for the deposit in Fig. 2, and is 1.26 for the fractal of Fig. 3.

This anomalous scaling with radius, measured by D, is a very useful means of characterization because the fractal dimension is a 'robust' quantity. Like the famous scaling exponents of phase-transition physics, it has to do with long-range properties, indeed, with the relationship between properties at different scales. Thus we can expect it to be universal in the sense that it should be independent of the details of the interactions between the objects which stick together during the growth, of their detailed composition, and so forth. But, as we will see, the mechanism of growth does affect D.

Growth models

How might one visualize the growth of an object such as the electrolytic deposit in Fig. 2? As we are interested in long-range properties we can ignore the complications of electrochemistry and simply imagine that ions wander randomly in solution (in many cases the electric field is screened out so that this is a good approximation) and stick to the deposit when they happen to get near it.

To make a computer model which is a literal translation of this process we start with a centre. Then we liberate a diffusing particle, a 'random walker', and let it wander freely until it is within a fixed distance of the centre, where it sticks. Then we liberate another particle and let it walk until it sticks to the centre or the first particle, and so on. We may, for our purposes here, idealize the process of formation as being completely irreversible: we ignore the possibility that the particles rearrange after sticking to find a more energetically favourable location. This is the diffusion-limited aggregation (DLA) model of Witten and Sander[5,6]. The application of DLA to electrodeposition is due to Brady and Ball[7] and Matsushita et al.[8].

Figure 4 shows the result of an extensive simulation according to the DLA rules; its resemblance to Fig. 2 is evident. Measurements of DLA clusters have shown them to scale according to the relations quoted above, with $D = 1.7$ for $d = 2$, and $D = 2.4$ for $d = 3$. Note that the structure is tenuous and open because holes are formed and not filled up. Filling up the holes would require wandering down one of the channels in the cluster without getting stuck on the sides; a random walker cannot do this.

There are several features of the DLA model which should be mentioned. Although it is simple to describe, no progress has been made towards 'solving' it. That is, although we suspect, on the basis of simulations, that DLA clusters are fractals, we cannot prove it. And we have no method of calculating D (or any other property): we must measure it. There are several reasons for this (I will mention a rather technical one below), the primary one being that DLA presents us with a situation in which our experience in equilibrium systems doesn't seem to help. Note that D, along with other scaling properties, arises in a non-trivial way from the kinetics of growth: there is no simple geometric argument with which to predict them.

The DLA model can be generalized in various ways, for example, to describe deposition on a surface[9] rather than a point. A more profound generalization is to use the model to describe systems which apparently have nothing to do with particle aggregation, but which share the same universal properties. We may see how one is led to do this by observing[5,10] that the probability, u, of finding a random walker at some point on its way to the aggregate has the following well-known properties: the flux of walkers; $\mathbf{v}$, is proportional to the gradient of u, and, because walkers are absorbed only on the aggregate, this flux

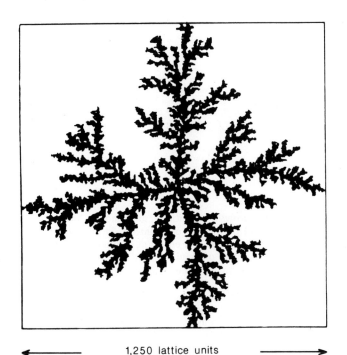

1,250 lattice units

Fig. 4 A large DLA cluster (~50,000 particles) grown on a square lattice. Note the resemblance to Fig. 2, and the beginning of distortion towards a dendritic outline, as discussed in the text. (Supplied by P. Meakin, Dupont.)

has no divergence:

$$\mathbf{v} \propto \nabla u \qquad (3)$$

$$\nabla \cdot \mathbf{v} = \nabla^2 u = 0 \qquad (4)$$

As walkers are not allowed to escape from the aggregate, we set $u = 0$ on the surface. The growth of the aggregate is given by the flux at its surface, that is, by ∇u.

As Niemeyer et al.[10] pointed out, a set of equations of identical form govern dielectric breakdown of a solid if we ignore many short-range details. As we are looking for universal features, making such simple, indeed, crude approximations is justified. If we think of u as the electrostatic potential in a solid about to be destroyed by a discharge, its negative gradient is, of course, the electric field. But u then obeys the Laplace equation of electrostatics, which is of the same form as the steady-state diffusion equation, equation (4), above. The breakdown channel will grow in a way determined by the electric field, that is, the gradient of u, on its surface. If the growth rate is linear in the field, we expect to have exactly the same situation as in DLA, and indeed, direct solutions of the equations, as well as measurements of photographs of real discharges, give the same fractal dimension as DLA. Non-linear breakdowns (lightning in the atmosphere is probably an example) give rise to patterns with different values of D.

Paterson[11] noticed an even more remarkable manifestation of the wide applicability of the model. When a fluid flows under conditions of large friction, inertial effects are negligible and the flow rate can be taken to be proportional to the hydrostatic force, that is, to the gradient of the pressure: this is known as D'Arcy's law. The situation is commonly realized in the laboratory by letting fluid flow between thinly spaced plates, a so-called Hele-Shaw cell. In nature, the flow of crude oil through the porous rock in which it is found is an example of quite serious interest. Suppose we try to force such flow by blowing a bubble of air or another low-viscosity substance into the cell (or by pumping water into an oil field—a scheme known as enhanced recovery). It has long been known that the air will not uniformly displace the fluid; instead it will break up into a complex

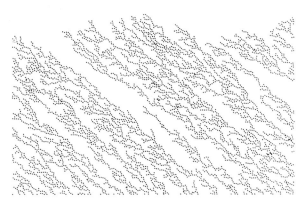

Fig. 5 Columnar microstructure in ballistic aggregation. Particles stick to the substrate and to each other after raining onto the structure in parallel trajectories at an angle to the vertical somewhat smaller than that of the columns. The fluctuations of the upper surface scale with the height for small height, and with the total width for large height. (Simulation performed by P. Ramanlal, University of Michigan.)

structure with many arms[12], which are called 'viscous fingers'. This phenomenon has an obvious detrimental effect on enhanced recovery.

Paterson's[11] speculation was that the pattern of the viscous fingering would scale like DLA. His reasoning was as above: the pressure in an incompressible fluid obeys equation (4), with u now standing for pressure, because fluid, like particles, is conserved. D'Arcy's law is of the same form as equation (3). Once more, many details have been ignored. In particular, the role of surface tension in this and similar situations will be discussed below.

The reasoning has been verified most directly by Chen and Wilkinson[13], who introduced discrete randomness into a Hele-Shaw cell—the effect should be that of the random arrivals of particles. Their patterns look almost exactly like Figs 2 and 4. Another experiment, by Nittmann et al.[14], used the clever trick of eliminating surface effects by taking for the two fluids water and an aqueous polymer solution; the fluids are miscible but mix slowly. Once more the pattern of fingering resembled the simulations. There seems to be a source of randomness in this experiment, probably arising from the non-newtonian flow characteristics of the polymer solution; such shear thinning could amplify noise. Even more startling is the experiment of Ben-Jacob et al.[15], who used a smooth Hele-Shaw cell, and one with a periodic pattern, with newtonian fluids. In some conditions they observed DLA-like scaling without an evident source of randomness, and without discrete 'particles'.

Experts will notice that equations (3) and (4) are of the same form (except for surface effects) as the description of solidification when the limiting factor in growth is diffusion of latent heat away from the surface of the growing crystallite. Why, then, does a snowflake (unlike the crystalline deposit of Fig. 2) not look like DLA, but is instead dominated by the crystal symmetry? I will return to this aspect of growth in the final section.

If particle aggregation doesn't need particles, what does it need? More generally, we can ask what different types of model give rise to scaling objects. For example, it is often the case that aggregates are formed by adding particles with a long mean free path, for example, in the formation of thin films by vapour-phase deposition[16]. In this case we may assume that the paths of the particles are straight lines. This model has become known as ballistic aggregation, and it has a number of very curious features. It is now known that the deposit itself is not a tenuous object but achieves a constant density[17,18]. (In contrast, diffusion-limited growth on a surface[9] yields an open deposit whose average density decreases with height.) It is of great interest to understand the upper surface of the film, which is a

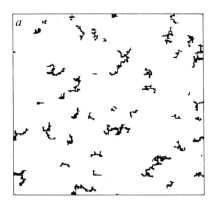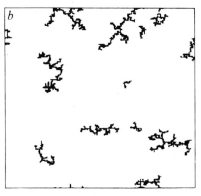

Fig. 6 Two stages in the formation of cluster–cluster aggregates: a, t = 3,669; b, t = 17,409. Note the resemblance of the clusters in b to the soot particle of Fig. 1. (Supplied by P. Meakin, Dupont.)

model of a random rough surface. It has been shown numerically[19] that for normal incidence of the depositing particles this surface also has scaling properties: for example, the fluctuation of the height scales with a non-integral power of the height, for small height. This surface is probably not an ordinary fractal curve, like Fig. 2, but is probably an example of a self-affine fractal[1,20]. 'Self-affine' means that the scaling in two different directions (width and height in this example) is different.

For non-normal incidence another effect appears, which is well-known in thin-film technology[16]. This is the columnar microstructure: the film spontaneously forms as a set of nearly parallel columns as it grows (see Fig. 5). The beginnings of a theory of this effect exist[21], but it is not known what, if any, relationship these giant fluctuations have with the scaling fluctuations at normal incidence.

The simplest aggregation process of all was introduced into mathematical biology by Eden[22]. This is a model for the growth of a cell colony: a cluster is grown by adding particles at random to perimeter sites. Once again the object is compact, but the surface has interesting scaling properties which seem to be the same as for ballistic aggregates with normal incidence[23]. Scaling is ubiquitous, and tends to have common features despite widely different details of growth.

We still have not described how soot forms. The structure of Fig. 1 is far more open than a DLA cluster: its fractal dimension is ~1.8 (DLA in three dimensions has $D = 2.4$). Extensive measurements of soot[24], colloids[25] and other similar objects leads one to suspect that a different class of clusters is involved. In fact, we have omitted a central feature of the formation process of clusters which can coagulate, namely the aggregation of clusters with each other[26,27]. Figure 6 shows two stages of a simulation of this process in two dimensions. We start with a vapour of freely moving particles which stick together whenever they come into contact, and then allow the clusters to continue to move with, perhaps, a smaller diffusion constant. The large fractals which are eventually formed have $D = 1.4$. The corresponding simulations in three dimensions give $D = 1.8$ and yield the open structure of real colloids and aerosols. At each stage of the process almost all of the clusters are of roughly the same size.

The open structure and low fractal dimension which characterize cluster–cluster aggregation are relatively easy to understand. It is difficult for a random-walking particle to penetrate a significant fraction of the radius of a growing cluster for particle aggregation; it is even more difficult for an aggregate of comparable size to do so. Thus, as aggregation proceeds, open, fluffy structures are produced.

One variant of this model which is worth mentioning is reaction-limited (chemically limited) aggregation[28]. In many cases, because of the details of the growth process, the sticking is inefficient, and many attempts are required to form a new cluster. In the limit of a very large number of attempts, the fractal dimension increases from 1.8 to ~2. Reaction-limited aggregation was probably discovered experimentally[29], before the simulations were done. Later experiments[30] have carefully controlled the growth conditions and shown both growth mechanisms, and both types of geometry, in the same system for different growth rates. The encoding of kinetics in the scaling in a form independent of details should be a powerful tool for identifying growth mechanisms.

Attempts at theory

There is no general theory of irreversible growth. The descriptions given in the previous section must be regarded as a kind of phenomenology, albeit a useful one. We can point to situations in which there is scaling, but we are compelled to do experiments, either in the laboratory or on the computer to calculate anything. We do have a few analytical results, but they give only partial information.

The best understood type of aggregation is the cluster–cluster process. Suppose we assume, as stated above, that the dominant cluster–cluster collision is between clusters of similar mass. If we make the masses strictly equal we have a hierarchical model[31]. It is easy to believe then that we do have a fractal: agglomerating parts in this way is exactly how the artificial fractal of Fig. 2 was made. (Note that particle aggregation is not hierarchical, but it seems to be fractal nonetheless.) The specification of the size distribution of clusters in the vapour, and the verification of the hierarchical assumption, have been the objects of detailed studies[32] which have shown that, indeed, the most common collision is between clusters of nearly equal mass. Some of these investigations use the techniques of colloid chemistry, in particular the Schmoluchowski kinetic equations, as well as computer simulations.

There remains the problem of finding the geometry of the clusters formed. Some progress has been made here because of a detail which allows one of the favourite tricks of the theoretical physicist to be applied. The real difficulty in visualizing the process is excluded volume, that is, the tendency of clusters to get in each other's way because they can attach to each other only on the outside. If they could attach anywhere it would be much simpler to sort out what is going on. It is quite obvious that excluded volume problems become less serious in high spatial dimensions: there are more ways into a three-dimensional cluster than into a two-dimensional one. Often, in equilibrium studies, it is found that for sufficiently large dimension of space, d, excluded volume is no problem at all: essentially any part of a cluster is accessible from outside. The dimension at which this starts to happen is called the upper critical dimension.

Above the upper critical dimension, calculations are simple, anomalous scaling is independent of d, and there exist methods (for equilibrium problems) which allow us to extrapolate to the physical world of $d = 3$. For cluster–cluster processes, this is exactly what happens[33]. The fractal dimension, D, for a cluster–cluster aggregate cannot grow above ~3.4 and it attains this value at about $d = 7$. This is rather far from the real world, of course, and no one has yet figured out how to extrapolate.

The situation for particle aggregation is very different. Suppose that the entire cluster were to become accessible to added particles for a large enough value of d. Then the mass in the interior would grow without adding to the volume. The cluster would quickly become so dense that it would no longer be accessible. Thus, there is no upper critical dimension for DLA. In fact, careful considerations of this sort can be turned into a bound[33] on D:

$$d - 1 \leq D \leq d \tag{5}$$

The fractal dimension is never independent of the spatial dimension, and the standard technique cannot be applied.

Some progress has been made in the study of particle aggregation by exploiting the similarity of the process to the famous 'snowflake' problem, that is, the study of dendritic crystallization[34]. We can see, for example, why tenuous structures are likely to arise in DLA and not in ballistic aggregation by noting that in the DLA case we have a growth instability of exactly the same form as the well-known Mullins–Sekerka[35] instability of crystal growth. The reasoning[5,35] goes as follows: suppose we start with a smooth aggregate and ask why it grows sharp tips. If we start with a tiny bump on the surface it will be magnified into a tip by the fact that the bump will grow faster than the rest of the surface: it will catch random walkers more efficiently than the flat portions of the surface, and certainly much more efficiently than the holes in the aggregate. The analogous dielectric breakdown case will make this even clearer: recall that the growth rate of any point on the surface of the structure is proportional to the electric field there. Sharp tips have large electric fields (the lightning rod effect). They grow ever sharper and dominate the growth. In the viscous fingering problem the same instability arises because it is easy for viscous fluid to flow away from a growing tip. It is even possible to specify a relationship between D and the characteristic opening angle of the tips[36] by using the mathematical theory of lightning rods. Unfortunately, no one knows how to calculate these angles. In fact, recent work indicates that there is an array of sharp tips on the surface of the fractal DLA cluster whose distribution is itself fractal[37].

For ballistic aggregates or for the Eden model there is no growth instability: it is easy to see that a bump on the surface neither grows or shrinks, but just adds a uniform skin, and tips do not grow. The bulk of the material remains compact.

Fractals and snowflakes

In the last section we noted the usefulness of the analogy of DLA with the kind of solidification most familiar (at least to those in cold climates) in the formation of snowflakes, that is, branched (dendritic) crystals. But particle aggregates do not look like snowflakes. To be precise, in a typical dendrite, a growing tip forms by the Mullins–Sekerka instability but then stabilizes. It retains its shape and continues in a definite direction, although it may spawn side-branches as it grows. In DLA (and in, for example, the zinc deposit of Fig. 2) the tips repeatedly split and wander.

There are three obvious differences between DLA clusters and dendritic crystals: DLA has essentially zero surface tension, it has a significant source of noise in the discrete arrivals of the particles, and it has (at least in some versions of the model) no analogue of crystal anisotropy. Sorting out how these affect the process is a subject of current controversy and great intrinsic interest.

Surface effects can be added in various ways to DLA simulations[5,38,39]; the result is to thicken the branches of the aggregate, but the scaling is unaffected for large sizes. Nor do surface effects, by themselves, make the equations of crystallization give rise to snowflakes[40,41]. Instead, something unexpected happens: a growing tip with surface tension does not stabilize, but undergoes repeated splittings, which are caused by the surface tension itself. This is because surface tension slows the growth of sharply curved surfaces and the end of the tip is the most sharply curved. In order to make real dendrites, anisotropy arising from the crystal structure must be introduced. The relationship of anisotropy to tip-splitting was verified experimentally[15] using fluid flow in a Hele–Shaw cell with a lattice of grooves.

How does this relate to DLA? It is common to do DLA simulations on a lattice (for convenience). Will the same thing happen here as in the noise-free case; that is, will stable tips form because of lattice anisotropy? It seems that the answer is yes[42,43]: sufficiently large clusters on a lattice have the outline of a crystallite, with tip splitting only on a small scale. But why, without surface tension, do we ever get tip splitting? This is because noise due to the discreteness of the arriving particles can split the tips. This can be verified in various ways, for example, by experiments and calculations which vary the noise[13] at fixed anisotropy. In cases where tip splitting is mainly due to surface tension rather than noise, will we get an object which scales? The answer to this question is not yet clear, but there are indications[15,44] that there is scaling, and that it is close to that of DLA.

These considerations are of more than technical interest, because they show how small effects (such as anisotropy) can make qualitative changes in growth habit. A series of recent experiments[45,46] have shown, for example, how changes in growth conditions, such as an increase in voltage in electrodeposition, can change a fractal pattern like Fig. 2 into an ordered dendritic crystal by increasing the effective anisotropy. This is a fascinating example of the competition between scaling symmetry and ordinary spatial symmetry.

This work was supported in part by NSF grant DMR 85-05474. I thank M. Sander and R. Merlin for comments on the manuscript, P. Meakin, G. Smith, D. Grier and P. Ramanlal for illustrations.

Note added in proof: There have been two interesting recent attempts[47,48] to explicate the competition between anisotropy and noise.

1. Mandelbrot, B. *The Fractal Geometry of Nature* (Freeman, San Francisco, 1982).
2. Family, F. & Landau, D. (eds) *Kinetics of Aggregation and Gelation* (North-Holland, New York, 1984).
3. Pynn, R. & Skjeltorp, A. (eds) *Scaling Phenomena in Disordered Systems* (Plenum, New York, 1985).
4. Stanley, H. & Ostrowsky, N. (eds) *On Growth and Form* (Nijhoff, The Hague, 1985).
5. Witten, T. & Sander, L. *Phys. Rev. Lett.* **47**, 1400–1403 (1981).
6. Witten, T. & Sander, L. *Phys. Rev. B* **27** 5686–5697 (1983).
7. Brady, R. & Ball, R. *Nature* **309**, 225–229 (1984).
8. Matsushita, M., Sano, M. Hayakawa, Y., Honjo, H. & Sawada, Y. *Phys. Rev. Lett.* **53**, 286–289 (1984).
9. Meakin, P. *Phys. Rev. A* **27**, 2616–2623 (1983).
10. Niemeyer, L., Pietronero, L. & Wiesmann, H. *Phys. Rev. Lett.* **52**, 1033–1036 (1984).
11. Paterson, L. *Phys. Rev. Lett.* **52**, 1621–1624 (1984).
12. Saffman, P. & Taylor, G. *Proc. R. Soc.* **A245**, 312–329 (1958).
13. Chen, J. & Wilkinson, D. *Phys. Rev. Lett.* **55**, 1892–1895 (1985).
14. Nittman, J., Daccord, G. & Stanley, H. *Nature* **314**, 141–144 (1985).
15. Ben-Jacob, E. *et al. Phys. Rev. Lett.* **55**, 1315–1318 (1985).
16. Leamy, H., Gilmer, G. & Dirks, A. in *Current Topics in Materials Science* Vol. 6 (ed. Kaldis, E.) (North-Holland, New York, 1980).
17. Bensimon, D., Shraiman, B. & Liang, S. *Phys. Lett.* **102A**, 238–240 (1984).
18. Meakin, P. *Phys. Rev. B* **28**, 5221–5224 (1983).
19. Family, F. & Viscek, T. *J. Phys. A* **18**, L75–L81 (1985).
20. Voss, R., in *Scaling Phenomena in Disordered Systems* (eds Pynn, R. & Skjeltorp, A.) 1–11 (Plenum, New York, 1985).
21. Ramanlal, P. & Sander, L. *Phys. Rev. Lett.* **54**, 1828–1831 (1985).
22. Peters, H. P., Stauffer, D., Holters, H. P. & Loewenich, K. *Z. Phys.* **B34**, 399–408 (1979).
23. Racz, Z. & Plischke, M. *Phys. Rev.* **A31**, 985–993 (1985).
24. Richter, R., Sander, L. & Cheng, Z. *J. Coll. Interf. Sci.* **100**, 203–209 (1984).
25. Weitz, D. & Oliveria, M. *Phys. Rev. Lett.* **52**, 1433–1436 (1984).
26. Meakin, P. *Phys. Rev. Lett.* **51**, 1119–1122 (1983).
27. Kolb, M., Botet, R. & Jullien, R. *Phys. Rev. Lett.* **51**, 1123–1126 (1983).
28. Jullien, R. & Kolb, M. *J. Phys. A* **17**, L639–L643 (1984).
29. Schaefer, D., Martin, J. Wiltzius, P. & Cannell, D. *Phys. Rev. Lett.* **52**, 2371–2374 (1984).
30. Weitz, D., Huang, J., Lin, M. & Sung, J. *Phys. Rev. Lett.* **54**, 1416–1419 (1985).
31. Botet, R. Jullien, R. & Kolb, M. *J. Phys. A* **17**, L75–L79 (1984).
32. Botet, R. & Jullien, R. *J. Phys. A* **17**, 2517–2530 (1984).
33. Ball, R. & Witten, T. *J. statist. Phys.* **36**, 863–866 (1984).
34. Langer, J. *Rev. mod. Phys.* **52**, 1–28 (1980).
35. Mullins, W. & Sekerka, R. *J. appl. Phys.* **34**, 323–329 (1963).
36. Turkevich, L. & Scher, H. *Phys. Rev. Lett.* **55**, 1026–1031 (1985).
37. Halsey, T., Meakin, P. & Procaccia, I. *Phys. Rev. Lett.* **56**, 854–857 (1986).
38. Kadanoff, L. *J. statist. Phys.* **39**, 267–283 (1985).
39. Viscek, T. *Phys. Rev. Lett.* **53**, 2281–2284 (1984).
40. Brower, R., Kessler, D., Koplik, J. & Levine, H. *Phys. Rev.* **A29**, 1335–1342 (1984).
41. Ben-Jacob, E., Goldenfeld, N., Langer, J. & Schon, G. *Phys. Rev.* **A29**, 330–340 (1984).
42. Ball, R. & Brady, R. *J. Phys. A* **18**, L809–L813 (1985).
43. Ball, R., Brady, R., Rossi, G. & Thompson, B. *Phys. Rev. Lett.* **55**, 1406–1409 (1985).
44. Sander, L., Ramanlal, P. & Ben-Jacob, E. *Phys. Rev.* **A32**, 3160–3163 (1985).
45. Grier, D., Ben-Jacob, E., Clarke, R. & Sander, L. *Phys. Rev. Lett.* **56**, 1264–1267 (1986).
46. Sawada, Y., Doughtery, A. & Gollub, J. *Phys. Rev. Lett.* **56**, 1260–1263 (1986).
47. Kertesz, J. & Viscek, T. *J. Phys. A* **19**, L257–L262 (1986).
48. Nittmann, J. & Stanley, H. E. *Nature* **321**, 663–668 (1986).

Fractal geometry in crumpled paper balls

M. A. F. Gomes
Departamento de Física, Universidade Federal de Pernambuco, 50.000 Recife, PE, Brazil

(Received 11 December 1985; accepted for publication 6 October 1986)

The geometry of crumpled paper balls is examined. The analysis stresses some physical, mathematical, and intuitive aspects of the problem, introducing the concept of fractal dimension which underlies many areas of modern physics.

Fractals are now a topic of wide interest[1] and here we describe an interesting example of fractal dimension defined via the mass-size exponent. In what follows, we discuss this example which has been used with success in the last two years in the freshman course Experimental Physics 1 at UFPE.

With two sheets of paper divided in the way indicated in Fig. 1 we construct after crumpling $(n+1)$ handmade balls of different sizes and masses (Fig. 2). We assign mass $M=1$ to the smallest ball and mass $M=2^n$ for the $(n+1)$th ball, with n increasing with the size of the paper (see Fig. 1). In our case $1 \leqslant M \leqslant 2^{13}$, with the largest mass corresponding to a sheet of paper of 98 cm $\times$ 65 cm. Typical log-log graphs of the average diameter (L) versus mass (M) for such crumpled paper balls are quite well described by $L = kM^{1/D}$. D is interpreted as the fractal dimension of the balls and $k \sim (1/\rho)^{1/D}$ is a measure of the average mass-density ρ on these fractal structures. The values obtained for D and k were $D = 2.51 \pm 0.19$, $k = 5.75 \pm 0.71$ for writing paper of surface density $\sigma \sim 80$ g/m^2. The fractal dimension D in this case tells about the complexity or degree of contortion of the area, since a fixed measure of rounded smooth area can enclose a larger volume than a complicated one can. The values of D and k are statistically independent of the students' weight and height. It is experimentally evident that k (lacunarity[1]) has a percent mean square deviation approximately two times larger than that of D, and $(\Delta \rho / \rho) = D(\Delta k / k) \simeq 0.31$. These values show that D is much less affected by the way of crumpling (pressure applied, haste or not, etc.) than the density is. The topological dimension of these balls is $D_T = 2$, since they are made of sheets of paper, which conform with $D_T = 2$. On the other hand, they are embedded in the Euclidean three-dimensional space $(E = 3)$, so their fractal D satisfies $D_T = 2 \leqslant D \leqslant E = 3$. In Fig. 3 we give a diagram indicating the frequency distribution of the Ds obtained in the last semester with 89 students. From the geometrical point of view our crumpled paper balls are self-avoiding surfaces. In this case the Flory argument[2] predicts that the size L obeys $L \sim l^\nu$, where $\nu = 4/(E+2)$, l is the linear size of the uncrumpled paper, and E is the dimension of the imbedding space. Since the mass scales with l according to $M \sim l^2$ we obtain $L \sim M^{\nu/2} = M^{2/(E+2)} \equiv M^{1/D}$, with $D = [(E+2)/2] = 2.5$, for $E = 3$, as experimentally obtained.

On the other hand, the experimental dependence of D with the surface density σ(g/m^2) of the paper is shown in

Fig. 2. A typical set of crumpled paper balls with masses 1,2,4,...,64.

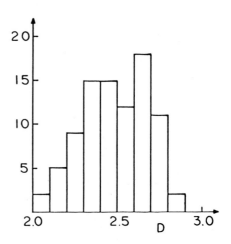

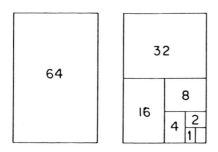

Fig. 1. Dividing two sheets of paper to generate the crumpled paper balls shown in Fig. 2. In this case $n = 6$ (see text).

Fig. 3. The frequency distribution of the fractal dimension D in crumpled paper balls (surface density of the paper $\sigma \sim 80$ g/m^2) from data obtained by 89 students. $\bar{D} = 2.51 \pm 0.19$.

9.2 Fractal Geometry in Crumpled Paper Balls

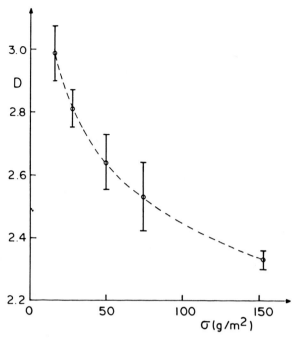

Fig. 4. Experimental dependence of the fractal dimension D with the surface density σ. $\circ$: experimental points; broken line: equation $D = 4.06\sigma^{-0.11}$.

Fig. 4. Using these data we find that D varies according to $D = c\sigma^{\alpha}$, with $c = 4.06$ and $\alpha = -0.11 \simeq -1/9$. The last equation says that for $\sigma = \sigma_3 \simeq 15.7$ g/m^2, $D = 3$ and for $\sigma = \sigma_2 \simeq 624$ g/m^2, $D = 2$. Evidently we have also $D = 2$ for $\sigma > \sigma_2$ and $D = 3$ for $\sigma < \sigma_3$. Deviations from the value $D = 5/2$ observed for $\sigma \lesssim 55$ g/m^2 and $\sigma \gtrsim 120$ g/m^2 (Fig. 4) are indicative that (i) the crumpled surface for low density paper ($\sigma \lesssim 55$ g/m^2) is certainly not self-avoiding (at least in some regions) and (ii) a high density paper ($\sigma \gtrsim 120$ g/m^2) cannot be considered as having $D_T = 2$ to our ends.

ACKNOWLEDGMENTS

I acknowledge Professor B. B. Mandelbrot for helpful criticism and encouragement. Work supported in part by FINEP and CNPq (Brazilian Agencies).

[1] B. B. Mandelbrot, *The Fractal Geometry of Nature* (Freeman, New York, 1983).
[2] P.-G. de Gennes, *Scaling Concepts in Polymer Physics* (Cornell U. P., Ithaca, 1979); see also the work by Y. Kantor, M. Kardar, and D. R. Nelson, Phys. Rev. Lett **57**, 791 (1986).

FRACTAL OF LARGE SCALE STRUCTURE IN THE UNIVERSE

L. Z. FANG

Center for Astrophysics, University of Science and Technology of China, Hefei, Anhui, People's Republic of China

Received 9 October 1986

> The systematic properties of the large scale structure of the universe can be explained in the string model with fractal dimension taken to be 1.4.

Several systematic properties among various distributions of objects, such as galaxies, clusters and superclusters, have recently been found by means of two-point correlation function:[1]

(i) all correlation functions can be fitted by the power law $\xi(r) = \alpha r^{-l}$, the index $l = 1.8$ is the same for all types of objects;[2]

(ii) the correlation strength α increases with the richness of the system $\langle N \rangle$ as $\alpha = \beta \langle N \rangle^m$, $m = 0.7$; 3, α should also increase with the mean separation d of objects in the system being considered by a power law $\alpha = \gamma d^n$.

Here, it will be shown that the above-mentioned results can be explained in the string model of the large scale structure formation in the universe. In particular, the indexes l, m and n are completely consistent with the predictions of the fractal in string model. The fractal dimension is found to be 1.4.

String model[3] pictures the formation of objects as an accretion process by density perturbation seeds of cosmic strings, which are topologically stable, macroscopic defects formed during phase transition of grand unified symmetry breaking in the early universe. At the time of formation, the system of strings consists of Brownian closed loops and infinite Brownian strings. The evolution of strings depends on the intercommuting processes when strings cross each other. The re-connection of intersecting strings will lead to the formation of loops with the rate of about one or few per horizon volume per expansion time. Since the number density of such loops is not high enough to have frequent collision among themselves, the following evolution of loops are dominated by self-intersection which produces new closed loops with smaller size called "daughter" loops. This "daughter" generative processor is self-similar. In this case, the system of loops can be described by a fractal dimensional D[4]. Namely, the number density of loops with size larger than R is given by

$$n(>R) \sim R^{-D}, \qquad (1)$$

9.3 Fractal of Large Scale Structure in the Universe

or the number density of loops with size R to $R + dR$ is

$$n(R)\,dR \sim R^{-D-1}\,dR. \qquad (2)$$

The length L of the loops also depends on the length R to be used as a unit in the measurement, i.e., the L per unit volume is

$$L \sim R^{1-D}. \qquad (3)$$

The expansion of the universe stretches all configuration of the strings in a conformal manner. Therefore, the large scale structure in the distribution of objects formed along the string loops by accretion should have about the same statistical properties shown in Eqs. (2) and (3).

The correlation function of objects in a system with length scale R is mainly determined by loops with radius of the order of R. It is because the number of loops with radius larger than R drops rapidly with R and all small scale loops are smoothed out under the scale of R. The distribution of daughter loops are random along its parent loop. In this case the correlation function of objects in a system of scale R depends only on the number density of loops and the configuration of loops themself.

If all R loops are circular with radius R, it can be shown that

$$\xi(r) = \frac{1}{4\pi^2 n(R) R^3} \frac{1}{\left(\frac{r}{R}\right)^2 \sqrt{1 - \left(\frac{r}{2R}\right)^2}}. \qquad (4)$$

From (4) one can find that the dependance of $\xi(r)$ with r is given by the term $f\left(\frac{r}{R}\right) \equiv 1 / \left(\frac{r}{R}\right)^2 \sqrt{1 - \left(\frac{r}{2R}\right)^2}$, which can be fitted by a power law $\left(\frac{r}{R}\right)^{-1}$ with index $l \sim 1.8$ in the range $r < R$. In fact, $l \sim 2$ at $r \sim 0$ and $l \sim 1.7$ at $r \sim R$. This means that $\xi(r)$ should have a slight flattening from $r \sim 0$ to $r \sim R$. Therefore, it is important to do a further analysis on the confidence of flattening, to be shown in the Bahcall-Soneira statistics.[2] The singularity of $r = 2R$ in $f\left(\frac{r}{R}\right)$ is obviously due to the assumption that all loops are of radius R. It is not important, because when $r \sim 2R$, $\xi(r)$ is already decreased to noise level. For instance, the available ranges of observed correlation functions for galaxies and clusters are, $r < 20h^{-1}$ Mpc and $r \leq 100h^{-1}$ Mpc ($h = H_0/100$), respectively, i.e., all cases being $r \leq (2\text{-}4)d$. Then the scale R in the scale invariant function $f\left(\frac{r}{R}\right)$ can be taken as the object's mean separation d in the system being considered.

Eq. (4) gives the dimensional correlation strength as follows

$$\alpha = \frac{1}{4\pi^2 n(R)R}. \tag{5}$$

From Eq. (2) one finds $\alpha \sim R^D$ or

$$\alpha \sim d^D. \tag{6}$$

On the other hand, the richness is the mean number of objects belonging to R system in volume R^3. Since the length of strings in a system depends on R by Eq. (3), the linear density of objects, such as galaxies, should be $\sigma \sim L^{-1} \sim R^{D-1}$. The number of objects on a R loop is then $\sigma R \sim R^D$. The number of loops in volume R^3 in $n(R)R^3 \sim R^{-D+2}$. The richness is then given by

$$\langle N \rangle \sim \sigma R n(R) R^3 \sim R^2. \tag{7}$$

From Eqs. (5) and (7) we have the relation of α and $\langle N \rangle$ as

$$\alpha \sim \langle N \rangle^{D/2}. \tag{8}$$

Eqs. (7) and (8) show that the indexes m and n should satisfy

$$m = \frac{1}{2}n. \tag{9}$$

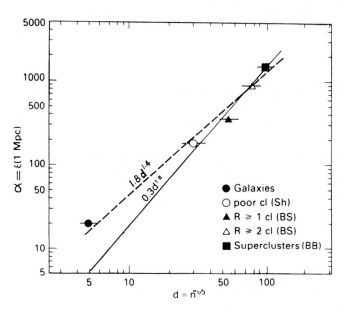

Fig. 1. The dependence of the correlation strength α on the mean separation d of objects in the system. Dash line represents a $d^{1.4}$ dependence. Solid line is the $d^{1.8}$ dependence, only clusters being considered.[1]

This prediction is remarkably consistent with observed results which are (Fig. 1)

$$n = 1.4, \quad m = 0.7. \tag{10}$$

This means the fractal dimension of cosmic string is $D = 1.4$. It should be noted that the index $n = 1.8$, found by Bahcall-Burgett[1], is obtained when only clusters are considered (Fig. 1), while $n = 1.4$ when galaxies are also taken into account.

The dimensionless correlation strength can also be found from Eq. (4) to be

$$\alpha' = \frac{1}{4\pi^2 n(R) R^3}. \tag{11}$$

Szalay-Schramm,[5] Turok[6] and Bahcall-Burgett[1] noted that α' is constant among the samples of clusters (poor, rich $R \geq 1$ and $R \geq 2$). Equivalently one can also assume that the fractal dimension $D = 2$. They explained the difference in α' values of galaxies and clusters by gravitational enhancement. However, this explanation runs into difficulty, when answering the following question: Why is there such a difference shown in the relation of α and $\langle N \rangle$? In fact, both relations of $\alpha - \langle N \rangle$ and $\alpha - d$ should be explained in the same way. Result (10) implies that gravitational enhancement may not be important in the formation of structure with scale larger than a few Mpc.

Finally, it should be stressed that the explanation for the difference in α' between galaxies and cluster can also be given by the model developed here. The key parameter used in phenomenological model[7] is the fraction F of galaxies associated with clusters. Since the number density of objects associated with the system of richness $\langle N \rangle$ is given by Eq. (7) as

$$\langle N \rangle / R^3 \sim 1/R, \tag{12}$$

the fraction of objects with scale R' that participate in the $\langle N \rangle$ system is then

$$F \sim R'/R. \tag{13}$$

Using $d' \sim 5$ Mpc for galaxies and $d \sim 50$ Mpc for clusters, one finds the fraction of galaxies associated with clusters is ten percent, which is about the same as the parameter used in Bahcall's model.

Acknowledgments

The author thanks N. Bahcall for useful discussion. This work has been done in part at the Institute for Advanced Study of Princeton and supported by CSCPRC under the US-China exchange program.

References
1. N. A. Bahcall and W. S. Burgett, *Astrophys. J.* **300** (1986) L35.
2. See, for example: N. A. Bahcall and R. A. Soneira, *Astrophys. J.* **270** (1983) 20; A. A. Klypin

and A. I. Kopylov, *Sov. Astr. Lett.* **9** (1983) 41; S. Schectman, *Astrophys. J. Suppl.* **57** (1985) 77.
3. A. Vilenkin, *Phys. Reports* **121** (1985) 263.
4. B. B. Mandelbrot, *The Fractal Geometry of Nature* (Freeman, Francisco, 1982).
5. A. Szalay and D. Schramm, *Nature* **314** (1985) 718.
6. N. Turok, *Phys. Rev. Lett.* **55** (1985) 1801.
6. N. A. Bahcall, *Astrophys. J.* **302** (1986) L41.

The Devil's staircase

'**The swing,**' a painting by Nicholas Lancret, 1690–1743. The swing and attendant illustrate phase locking in systems containing two competing frequencies. Figure 1

When the interaction between an oscillator and its driver is strong enough, the oscillator will resonate at, or "lock" onto, an infinity of driving frequencies, giving rise to steps with a fractal dimension between 0 and 1.

Per Bak

In the 17th century the Dutch physicist Christian Huyghens observed[1] that two clocks hanging back to back on the wall tend to synchronize their motion. This phenomenon is known as phase locking, frequency locking or resonance, and is generally present in dynamical systems with two competing frequencies. The two frequencies may arise dynamically within the system, as with Huyghens's coupled clocks, or through the coupling of an oscillator to an external periodic force, as with the swing and attendant shown in figure 1. If some parameter is varied—the length of a pendulum or the frequency of the force that drives it, for instance—the system will pass through regimes that are phase locked and regimes that are not. When systems are phase locked the ratio between their frequencies is a rational number. For weak coupling the phase-locked intervals are narrow, so that even if there is an infinity of intervals, the motion is quasiperiodic for most driving frequencies; that is, the ratio between the two frequencies is more likely to be irrational. When the coupling increases, the phase-locked portions increase, and it becomes less likely that the motion is quasiperiodic. This is a unique situation, where it makes sense, despite experimental uncertainty, to ask whether a physical quantity is rational or irrational.

We shall see that if one plots the frequency of the oscillator against the frequency of the applied force the resulting curve may consist of an infinity of steps—the Devil's staircase. In this article I discuss the conditions under which such a staircase appears and how one can understand it through the modern theory of dynamical systems. The Devil's staircase emerges not only in dynamical systems but also in long-range spatially periodic solid structures, so many of my examples will come from condensed-matter physics.

Origins of staircases

Figure 2 shows dynamical behavior as a function of a frequency for a few systems of very different physical nature. The curves all show a characteristic staircase structure where the plateaus of the curves indicate locking at various rational frequency ratios. The current-driven Josephson junction (figure 2a) obeys an equation very similar to that of a damped driven pendulum; the voltage across the junction is a direct measure of the frequency, so the plot essentially shows[2] the current as a function of frequency. Figure 2b shows the frequency of oscillations in a complex *chemical* reaction, the Belusov–Zabotinsky reaction, measured[3] at the University of Texas by Jerzy Maselko and Harry Swinney. Figure 2c shows the frequency of voltage oscillations in the ionic conductor barium sodium niobate, as measured[4] at Frankfurt University by Samuel Martin and Werner Martienssen; in such a conductor the current is carried by ions rather than electrons. Other examples range from Rayleigh–Bénard hydrodynamic convection systems[5,6] and charge-density-wave systems[7] to periodically forced embryonic chicken hearts[8] and firing neurons subjected to external electrical pulses.[9]

As the interaction between two competing frequencies increases, the oscillations eventually begin to interfere with each other, and there is a transition to a state that features chaotic motion in addition to the periodic and quasiperiodic motion. Mogens Jensen, Tomas Bohr and I, working at the University of Copenhagen, recently investigated phase locking in great detail. We found[10] that at the transition to chaos the motion is always locked: As one changes the frequency of either oscillator—for example, by changing the length of a pendulum or the frequency of a driving force—the ratio between the two frequencies locks onto every single rational value p/q. If one subjects a pendulum to a fixed driving frequency and plots the actual frequency of the pendulum against the natural frequency or length of the pendulum, one obtains a curve consisting of an infinity of steps.

Stated slightly differently, a simple pendulum that for weak coupling almost never locks onto the driving frequency (because the resonances are extremely narrow) will for strong enough coupling always lock onto one of the infinity of resonant frequencies. If one then slowly changes the driving frequency, the pendulum will lock onto each resonant frequency, jumping from one to the next, forming an infinite series of steps.

Between any two steps there is an infinity of steps, because between any two rational numbers there is an infinity of rational numbers. It is this property of the curve that has given rise to the name "the Devil's staircase." If part of the curve is blown up, the

---

Per Bak leads the condensed-matter theory group at Brookhaven National Laboratory, in Upton, New York.

9.4 The Devil's Staircase

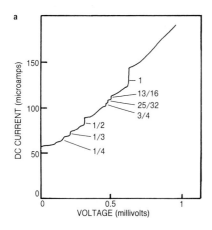

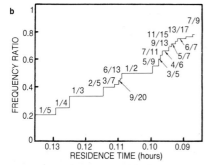

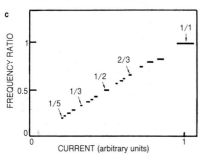

Dynamical-system staircases. The fractions are the ratios of competing frequencies that have locked to form the plateaus. **a:** Current–voltage characteristics of a niobium Josephson junction driven by a 295-GHz microwave current. The voltage is a measure of the frequency. (From reference 2.) **b:** Frequency of oscillations in a complex chemical reaction as a function of the reaction rate. (From reference 3.) **c:** Frequency of voltage oscillations in a superionic conductor, barium sodium niobate, driven by a direct current. (From reference 4.) Figure 2

resulting curve looks very much like the original curve. One can use a scaling index d to describe this self-similarity under magnification. The most striking property of the staircase is that this scaling index is "universal"—the same for all dynamical systems undergoing a mode-locking transition to chaos. The index d is thus a constant of nature. This view has been confirmed for several of the systems mentioned above.

The scaling index d has an interesting mathematical interpretation. Consider the horizontal frequency axis, and remove all the intervals where the frequency is locked. What remains is a set of points. This set of points, called a Cantor set after Georg Cantor, the mathematician who first constructed such sets, has measure zero because the frequency-locked intervals fill the entire axis. The total width of all the points is zero. However, the dimension of the set is not 0 as it would be for a countable set of points, nor is it 1 as it would be for a line segment or collection of line segments. In fact, the scaling index d is approximately 0.87 and can be interpreted as the Cantor set's dimension—in this case a "fractal" dimension between 0 and 1.

The fractal dimension is a generalization of our usual concept of dimension, and can be explained in the following way. Consider a circle or sphere of radius r around a point belonging to the Cantor set of points. If the number of points within the sphere scales as r^d, then d is the fractal dimension of the set. Clearly, for a line the number of points is directly proportional to the radius, so the fractal dimension d is 1, and for a plane the number of points goes as r^2, so d is 2.

The Cantor set has traditionally been thought of as an artificial mathematical construction with no physical application. It is thus quite fascinating that one can relate it directly to a system's mode locking and that one can measure its characteristic scaling dimension directly in rather simple experiments despite its zero measure. I return to the theory and the numerous experimental realizations below.

Solids. The Devil's staircase also shows up in an entirely different context—periodic structures with long spatial periods. Several intermetallic compounds, such as Ag_3Mg, $CuAu$, Cu_3Pt, Au_3Mn, Au_3Cd and Au_3Zn, form[11] crystals with extremely long periodicities along a unique direction. The long-range structures can be built from blocks of the form XY_3 by putting them together with or without stacking faults, as figure 3a indicates. The figure shows the structure 11111..., which has distance 1 between stacking faults, and the structure 22222..., which has distance 2 between stacking faults. The composition of the compound clearly depends on the density of stacking faults because each fault removes a layer of the majority Y atoms. If a crystal structure with period q has p stacking faults over a distance of q unit lengths, then the structure's average periodicity M is defined as q/p.

Figure 3a also shows a plot of the average periodicity M as a function of temperature for the compounds $Ti_{1+x}Al_{3-x}$, as measured[11] by a French group. Note again the staircaselike dependence where M assumes only rational values. The stairs appear even though the quantity M plotted here is the ratio between two *spatial* periodicities—that of the ordered structure and that of the lattice—in contrast to the dynamical systems such as the pendulum, where we plotted the ratio between two *temporal* periods or frequencies. The analog in a structurally modulated system of the frequency in a dynamical system is the wavevector.

This phenomenon of compounds crystallizing in a multitude of long-range periodic structures is known as polytypism. Structures arising from the stacking of individual atomic layers that have hexagonal or triangular symmetry are another example of polytypism. Hexagonal close-packed structures can be formed by stacking hexagonal layers in the pattern ABABAB...; face-centered-cubic phases have the pattern ABCABC.... Using a notation where h means that near neighbors are identically stacked and c means opposite stacking, these structures can be written hhhhhh... and cccccc.... Certain magnesium-based ternary alloys crystallize in long-range periodic structures with a mixed stacking pattern hchchc.... Figure 3b shows the length of the period divided by the number of c's, or stacking faults, as a function of composition in these

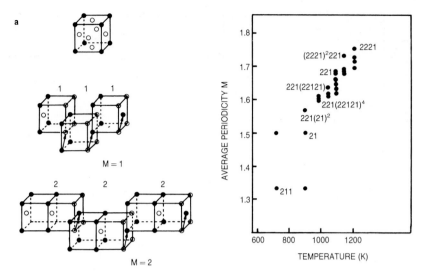

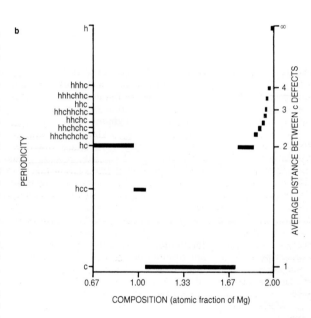

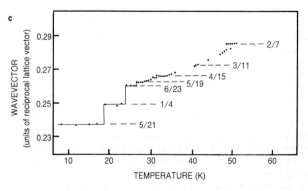

alloys, as measured[12] at the University of Hiroshima, Japan.

Figure 3c shows the periodicity of the magnetic structure of the rare-earth element erbium as a function of temperature. Physicists at Brookhaven measured[13] this curve by the novel technique of magnetic x-ray scattering. Erbium has a modulated magnetic structure with aligned spins. The various spin configurations are formed by periodic sequences of layers of up spins and layers of down spins. The structure with period 4 (and wavevector $1/4$), for example, has two layers of up spins followed by two layers of down spins and so on. Similar structures have been found[14] in the rare-earth element holmium and in cerium antimonide. Other systems exhibiting[15] structural staircases are ferroelectrics and the stacking structures of graphite intercalation compounds.

All these structures arise from competition between spatial periodicities. Despite the seemingly enormous complexity of these structures and the wide range of physics that underlies them, two simple models explain the main features of them all. The first[16] is a simple Ising system with long-range repulsive interactions; such a system has a complete Devil's staircase with all possible rational periodicities. The second is the "axial next-nearest-neighbor Ising model," which exhibits[17] the most spectacular phase diagram of any model studied so far; shown in figure 4, the diagram is a "Devil's flower" consisting of an infinity of leaves that represent periodic phases and that spring from a "multiphase point."

Experiments with dynamical systems

Theoretical analysis of dynamical systems leads to the conjecture (see the box on page 44) that the Devil's staircase is universal at the transition to chaos. It is interesting to examine in this light the real physical systems considered above and illustrated in figure 2.

For the driven pendulum, one can show[10] numerically that (in the termin-

Structural staircases in periodic lattice systems. **a:** Building blocks for long-range periodic crystal structures in $Ti_{1+x}Al_{3-x}$, and a plot of the average periodicity as a function of temperature for an alloy with 72% aluminum. Open circles represent aluminum atoms; closed circles, titanium. (From reference 11.) **b:** Periodicity as a function of composition for ternary magnesium-based compounds. The h phase is a hexagonal close-packed structure; the c phase is a face-centered cubic structure. (From reference 12.) **c:** Modulation wavevector as a function of temperature for the magnetic structure of erbium. (From reference 13.) Figure 3

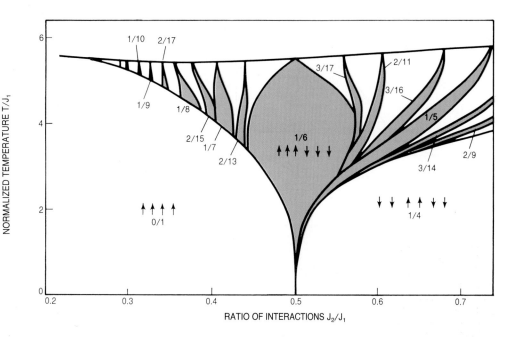

'Devil's flower.' In this phase diagram for the axial next-nearest-neighbor Ising model, the horizontal axis is the ratio of antiferromagnetic interactions to ferromagnetic interactions. The white areas contain yet more leaves, representing infinities of periodic phases with incommensurate phases between. Arrows indicate the sequences of ferromagnetic layers. (From ref. 17.) **Figure 4**

ology of the theoretical analysis given in the box) the return map indeed develops an inflection point at a critical surface in the parameter space. Numerical simulations by groups at Stony Brook[18] and the University of Copenhagen[19] confirm the existence of a complete Devil's staircase along a line where the map has an inflection point. Preben Alstrom and Mogens Levinsen at the University of Copenhagen found that the fractal dimension of the staircase is about 0.87, in agreement with prediction. Stewart Brown, George Mozurkewich and George Grüner at UCLA also found[7] indications of a complete Devil's staircase, with dimension about 0.92, for charge-density waves in niobium triselenide driven by the combination of a constant current and an oscillating current, but they made little attempt to verify that the system was indeed at the transition line. The most accurate experiments are probably the Rayleigh–Bénard experiment[6] by Albert Libchaber's group at the University of Chicago and the experiments[4] on the ionic conductor BSN by Martin and Martienssen at Frankfurt University (figure 2c).

The Chicago group used a cell filled with mercury and heated from below. At a critical point above the onset of convection there is an oscillatory instability into a time-dependent state involving an ac vertical vorticity in the fluid. The period of this oscillation defines one frequency. The experimenters generated the second oscillator by applying a small horizontal dc magnetic field parallel to the axis of the convection cells in the fluid and by applying a vertical sheet of current through the fluid. The Lorenz force induces an ac vorticity in the fluid's velocity. This is the second oscillator. By varying the amplitude and frequency of the current, they were able to scan a very large range of winding numbers and amplitudes, resulting in a phase diagram with numerous Arnold tongues, similar to the one shown for a circle map in figure 5b. They identified the critical curve by tuning the experiment to quasiperiodic states and looking to see when broadband noise, indicating chaos, appears in the spectrum. They determined the fractal dimension by looking at the scaling of the mode-locked steps at the critical curve. Indeed, they verified that the mode-locked steps form a complete Devil's staircase, and found the fractal dimension to be 0.86 ± 3%, in agreement with theoretical predictions. The Chicago experiment is a spectacular example of the physical relevance of the Cantor set. The universal scaling behavior allows one to extrapolate to the limiting set in a situation where one can only measure a finite number of steps.

The ionic conductor BSN is unique in that a constant driving current gives rise to an oscillating voltage. In the experiment by Martin and Martienssen the periodic voltage oscillations define one frequency. An additional ac current defines the second frequency, and again one can scan a large range of winding numbers and amplitudes by varying the frequency and strength of the ac current. The measured return map indeed appears to develop an inflection point at a transition to chaotic voltage oscillations. Thus the theory described in the box applies to this system. Martin and Martienssen studied the scaling behavior of the staircase formed by the mode-locked steps (figure 2c) following our procedures described in the box. They confirmed that the staircase is complete and has fractal dimension 0.93, in reasonable agreement with theory.

Long-range periodic structures

We can understand the essential features of the magnetic structures, ferroelectric structures and crystal structures, whether long-range periodic or incommensurate, in terms of two very simple models. The first[16] is the one-dimensional Ising model with long-range repulsive interactions whose energy E is given by

$$E = -\sum_i H S_i + \sum_{\langle i,j \rangle} J(i-j)(S_i + \tfrac{1}{2})(S_j + \tfrac{1}{2}) \quad (1)$$

Here S_i are the spins, which are $\pm \tfrac{1}{2}$, H is the magnetic field and the function $J(i-j)$ is the antiferromagnetic interaction between up spins at sites i and j. The antiferromagnetic interaction could, for instance, decay as a power law or exponentially.

It is easy to relate the spins and their interactions to physical quantities for the systems discussed at the beginning of this article. For long-range periodic crystal structures such as $Ti_{1+x}Al_{3-x}$, the spin-up state represents[11] the presence of a defect (figure 3a) and the spin-down state the absence of a defect. In this case, the interaction $J(i-j)$ is the interaction between defects, and the background H a "chemical potential"

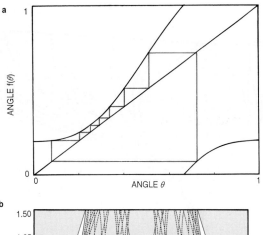
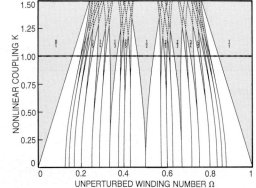
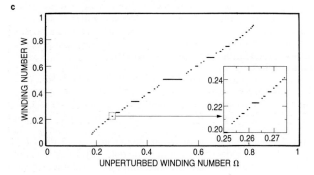

Circle map, phase diagram and self-similar staircase. **a:** Sine circle map with periodic orbit (color) for nonlinear coupling equal to 1—the critical point for the transition to chaos. The curve appears broken because it is plotted modulo 1. **b:** Phase diagram for the circle map. Note the "Arnold tongues" (color) where the winding number assumes rational values. **c:** Winding number at the transition to chaos. Note the self-similarity of the curve under magnification. Figure 5

for defects that depends on the temperature and composition. For the magnesium-based ternary alloys[12] (figure 3b), which crystallize in long-range periodic structures with mixed stacking hhchchc..., spin $-\frac{1}{2}$ could represent h stacking, so that a chain of down spins would represent a hexagonal close-packed structure, and spin $+\frac{1}{2}$ could represent c stacking, so that a chain of up spins would represent a face-centered cubic structure. The interaction $J(i-j)$ would be the interaction between the c stacking "defects." Robijn Bruinsma at UCLA and Andy Zangwill at the Polytechnic Institute of New York have constructed[20] a microscopic theory for the origin of these interactions based on the electronic structure and elastic properties of the compounds. An interpretation similar to the one for long-range periodic structures would apply to the stacking sequences in graphite intercalation compounds; Roy Clarke and M. J. Winokur have studied[15] these sequences at the University of Michigan.

Bruinsma and I originally constructed[16] the Ising model with long-range interactions (equation 1) to study neutral–ionic transitions in organic charge-transfer compounds, where the parameter H represents the electron ionization energy and J the interaction between ionized layers. For a given fraction q of up spins, the position x_i of the ith up spin in the stable configuration is given by the simple formula

$$x_i = \text{integer}(i/q) \qquad (2)$$

Hence, for $q = \frac{1}{2}$ the structure is

$$- + - + - + \ldots$$

For $q = \frac{10}{23}$ the ground state is

$$- + - + - + - - + - + - + - - + - + - \ldots$$

Trivially, these structures translate into 222 and 2223223223, respectively, for the ternary alloys, and hchc and hchchchhchchchhchchchhc for the TiAl$_3$ structures. All the structures that have been found experimentally can be expressed in terms of equation 2. When the fraction q of up spins is irrational—for instance, for q equal to the golden mean, $(\sqrt{5}-1)/2$—the sequence given by equation 2 is not periodic, and the resulting crystal structures have been called[21] "quasicrystalline." The Ising model of equation 1 thus is an extremely simple microscopic model for quasicrystals.

It turns out that for every rational concentration q equal to m/n there exists an interval $\Delta H(m/n)$ in which the structure is stable, and these intervals fill the entire H interval. Hence the concentration of up spins plotted against the parameter H constitutes a complete Devil's staircase. Figure 6 shows the staircase for an interaction of the form $J(n) = 1/n^2$. Note the self-similarity of the staircase under magnification. The dimension d for the Cantor set can easily be determined

9.4 The Devil's Staircase

Theory for dynamical systems

Consider first the simple pendulum driven by a periodic force. A second-order differential equation describes the variation of the angle θ as a function of time t:

$$\alpha\ddot{\theta} + \beta\dot{\theta} + \gamma \sin\theta = A + B\cos(2\pi t)$$

Here α is the inertia, β the damping, γ the gravitation, A the amplitude of a constant torque and B the amplitude of an external periodic force. Despite its simplicity, this equation cannot be solved analytically; it has a richness of periodic, quasiperiodic and chaotic solutions. For the other systems mentioned in the article, the situation is even worse: We do not even know with any confidence the equations that govern their behavior. However, we shall see that this ignorance does not prevent us from making quantitative predictions about mode locking in the systems.

Let us look at the pendulum with a stroboscopic light that flashes at the discrete times $t = n$, where n is an integer. We are in a sense using the external force as a clock. Let θ_n be the phase, or angle, of the pendulum, and let $\dot{\theta}_n$ be its derivative at time $t = n$. Because the equation is of second order the phase at time $n + 1$ is an (unknown) function h of the phase and its derivative at time n:

$$\theta_{n+1} = h(\theta_n, \dot{\theta}_n)$$

The function h is called a Poincaré map, or return map, of the system. Damping may cause the derivative of the phase to become a "slave" of the phase after a transient period: $\dot{\theta}_n = g(\theta_n)$. Then

$$\theta_{n+1} = h(\theta_n, g(\theta_n)) = f(\theta_n)$$

The two-dimensional map h has thus collapsed into a one-dimensional map, which is called a "circle map" because it maps one point θ_n on the circle $0 < \theta < 2\pi$ onto another point θ_{n+1} on the circle.

For given values of the parameters we do not know whether or not this dimensional reduction actually takes place. The best we can do is to generate the return map either by solving the differential equation numerically or by measuring it experimentally. Figure 5a shows an example of a one-dimensional map $f(\theta)$ where we have chosen the period to be 1. Numerical calculations show[10] that the return map for the damped driven pendulum indeed reduces to such a one-dimensional map. Measurements[4] on the ionic conductor BSN, a system for which the underlying equations are not very well known, have also yielded a one-dimensional map. The advantage of studying simple maps of this form is obvious. It is much easier to identify periodic, quasiperiodic and chaotic solutions by iterating the map than by a cumbersome numerical integration of the underlying differential equation. We shall see that the qualitative behavior of the system does not depend on the details of the map as long as it is a circle map. This universality allows us to make predictions with confidence even for systems where we do not know the underlying equations.

Let us therefore study the simple sine circle map

$$\theta_{n+1} = f_\Omega(\theta_n) = \theta_n + \Omega + (K/2\pi)\sin(2\pi\theta_n)$$

The map has a linear term θ_n and a bias term Ω representing the frequency of the system in the absence of nonlinear coupling K. (The factor 2π is used so that the transition to chaos occurs at $K = 1$.) To study the mode locking in the circle map we consider iterations of the map f: θ, $f(\theta)$, $f^2(\theta)$..., or $\theta_1, \theta_2, \theta_3$.... The frequency of the dynamical system is given by the mapping's "winding number" W, which is the average phase increase per unit time:

$$W = \lim_{n\to\infty} (f^n(\theta_1) - \theta_1)/n$$

In the absence of nonlinear coupling, the winding number W is equal to the bias Ω, so that Ω is the frequency of the unperturbed system. Under iteration the variable θ_n may converge to a series that is periodic, quasiperiodic or chaotic. If the series is periodic, then $\theta_{n+q} = \theta_n + p$ and the frequency of the system is given by a rational winding number p/q. If the series is quasiperiodic, the winding number is irrational. If the series is chaotic, the winding number is not defined. When the nonlinear coupling K is less than 1, the map is strictly monotonic and has only periodic and quasiperiodic winding numbers. When the nonlinear coupling is 1 (figure 5a), the map develops a cubic inflection point at $\theta = 0$. And when the nonlinear coupling is greater than 1, the map has a local maximum.

The transition to chaos takes place precisely at $K = 1$. Figure 5a shows a periodic orbit with a winding number W of $1/8$, a nonlinear coupling K of 1 and a bias Ω of 0.2. Figure 5b shows the phase diagram for the circle map. The regimes where the winding number W assumes rational values are called "Arnold tongues" after the Soviet mathematician V. I. Arnold. When the nonlinear coupling K is close to zero, all intervals of resonance are quite small, so the probability that the winding number is rational is almost zero (as for the Huyghens clocks), and the probability of hitting an irrational winding number is almost 1. However, with increasing nonlinear coupling the widths of all intervals increase. Clearly, the widths of the resonances cannot grow indefinitely; at some point they will interact and overlap.

Avoiding unending calculations. One might speculate that when the nonlinear coupling is 1 there is a resonance for all values of the unperturbed frequency Ω. How is it possible to check this without calculating an infinity of intervals? To do so, we calculated[10] the widths $\Delta(p/q)$ of about 1400 steps for a nonlinear coupling of 1. Figure 5c shows the "staircase" formed by plotting the winding number W against the bias Ω. As one includes more and more steps the Ω axis becomes more and more filled. To investigate whether or not the mode-locked steps will eventually cover the entire Ω axis, we calculated the total width $S(r)$ of all steps wider than a given scale r. The quantity of interest is the space between the steps, $1 - S(r)$, which eventually shrinks to a Cantor set. We measured this quantity on the scale r to find the number N of "holes," given by $N(r) = (1 - S(r))/r$. The points in a log–log plot of $N(r)$ fall excellently on a straight line, indicating a power law

$$N(r) \approx (1/r)^d$$

From the slope of the straight line one finds the exponent d to be 0.8700 ± 0.0004. This result means that the space between the steps vanishes as r^{1-d} as the scale r goes to 0. Thus there is no room for quasiperiodic motion and the Devil's staircase is complete. We can interpret the exponent d as the dimension of the Cantor set that is the complementary set to the mode-locked intervals on the Ω axis.

When the nonlinear coupling K passes beyond 1, the widths of the steps continue to increase. Because they fill the entire Ω axis at $K = 1$, they must necessarily overlap for $K > 1$, as figure 5b indicates. The transition to chaos is basically caused by the overlap of resonances, and one can visualize the chaotic motion as an erratic jumping between resonances. Most nonlinear systems perturbed by an external periodic field will probably exhibit a transition to chaos caused by the overlap of resonances that follow the Devil's staircase as described here.

An important question is whether or not the critical behavior at the transition is universal, that is, whether or not it depends on the specific function f defined above. If the behavior is not universal, we cannot predict with confidence how any specific system will behave in an experiment because the function f is generally not known. To check universality, we studied a broad class of circle maps with more complicated nonlinear terms. Generally the details are different from those of the staircase in figure 5c. Some steps become narrower, some wider. The scaling, however, remains the same, with the dimension about 0.870.

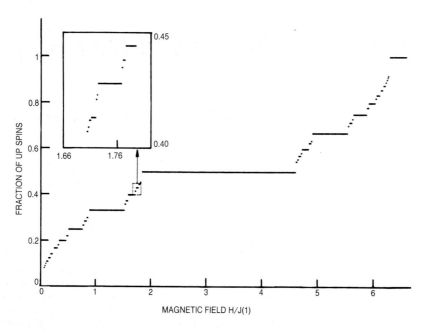

Density of up spins as a function of magnetic field for an Ising model with long-range interactions. (From reference 16.)
Figure 6

analytically. For power-law interactions of the form $n^{-\alpha}$, the dimension d is $2/(1+\alpha)$. This dimension is obviously not universal because it depends on the exponent α, and it is not possible to predict the dimension from a simple general theory.

For the magnetic structures of cerium antimonide and erbium (figure 3c), and for many of the ferroelectric structures with staircase behavior, the temperature and hence the entropy play an essential role in determining the sequence of the periodic phases. Also, there often appear to be incommensurate phases between the high-order periodic ones. It is essential to construct a real three-dimensional thermodynamic model for these systems. To accomplish this, Juhani von Boehm and I, then at the Bohr Institute in Copenhagen, constructed[17] a simple Ising model with competing nearest- and next-nearest-neighbor interactions. The model was later named the axial next-nearest-neighbor Ising model. Within each layer the model has nearest-neighbor ferromagnetic interactions J_1. In the axial direction the model has ferromagnetic nearest-neighbor interactions J_1, but antiferromagnetic next-nearest-neighbor interactions J_2. When the ratio J_2/J_1 is small, it is energetically favorable for the system to order ferromagnetically into a structure of the form + + + +, where the pluses indicate a sequence of up layers. When the ratio J_2/J_1 is large, it is energetically favorable for the system to order into a structure of period 4, or wavevector $1/4$, of the form + + − − + + − − + + − −.

Figure 4 shows the resulting phase diagram of the model. The diagram is possibly the most spectacular found for any statistical-mechanical model whatsoever. At zero temperature only the + + + and + + − − + + phases can exist. At finite temperatures there is an infinity of periodic phases with wavevectors between 0 and $1/4$; these spring out of a multiphase point at $T=0$. It can be shown that near the transition line to the paramagnetic phase all possible rational phases become stable, so there is a Devil's staircase. The white areas between the leaves of the Devil's flower in figure 4 indicate regimes with yet more leaves, leaves of infinities of long-range periodic phases with incommensurate phases between.

* * *

I would like to thank Juhani von Boehm, Mogens Høgh Jensen, Tomas Bohr and Robijn Bruinsma for a very stimulating collaboration on studies involved in this devilish enterprise. Our work is supported by the Department of Energy, Division of Materials Sciences, under contract DE-AC02-76CH00016.

References

1. C. Huyghens, letter to his father, dated 26 February 1665. *Oeuvres completes des Christian Huyghens*, M. Nijhoff, ed., Societé Hollandaise des Sciences, The Hague, The Netherlands (1893), vol. 5, p. 243. I am grateful to Carson Jeffries and Paul Bryant for bringing this reference to my attention.
2. V. N. Belykh, N. F. Pedersen, O. H. Sorensen, Phys. Rev. B **16**, 4860 (1977).
3. J. Maselko, H. L. Swinney, Phys. Scr. **T9**, 35 (1985).
4. S. Martin, W. Martienssen, Phys. Rev. Lett. **56**, 1522 (1986).
5. A. P. Fein, M. S. Heutmacher, J. P. Gollub, Phys. Scr. **T9**, 79 (1985).
6. J. Stavans, F. Heslot, A. Libchaber, Phys. Rev. Lett. **55**, 596 (1985).
7. S. E. Brown, G. Mozurkewich, G. Grüner, Phys. Rev. Lett. **52**, 2277 (1984).
8. M. R. Guevara, L. Glass, A. Shrier, Science **214**, 1350 (1980).
9. L. D. Harmon, Kybernetik **1**, 89 (1961). T. Allen, Physica D **6**, 305 (1983).
10. M. H. Jensen, P. Bak, T. Bohr, Phys. Rev. Lett. **50**, 1637 (1983); Phys. Rev. A **30**, 1960, 1970 (1984). P. Bak, T. Bohr, M. H. Jensen, Phys. Scr. **T9**, 50 (1985). See also P. Cvitanovic, M. H. Jensen, L. P. Kadanoff, I. Procaccia, Phys. Rev. Lett. **55**, 343 (1985).
11. A. Loiseau, G. Van Tendeloo, R. Portier, F. Ducastelle, J. Phys. (Paris) **46**, 595 (1985).
12. Y. Komura, Y. Kitano, Acta Crystallogr. Sect. B **33**, 2496 (1977).
13. D. Gibbs, D. E. Moncton, K. L. D'Amico, J. Bohr, B. H. Grier, Phys. Rev. Lett. **55**, 234 (1985).
14. P. Fischer, B. Lebech, G. Meier, B. D. Rainford, O. Vogt, J. Phys. C **11**, 346 (1977). J. Rossat-Mignod, P. Burlet, J. Villain, H. Bartholin, W. Tcheng-Si, D. Florence, O. Vogt, Phys. Rev. B **16**, 440 (1977).
15. M. J. Winokur, R. Clarke, Phys. Rev. Lett. **56**, 2072 (1986).
16. P. Bak, R. Bruinsma, Phys. Rev. Lett. **49**, 249 (1982); Phys. Rev. B **27**, 5824 (1983).
17. P. Bak, J. von Boehm, Phys. Rev. Lett. **42**, 122 (1978); Phys. Rev. B **21**, 5297 (1980).
18. W. J. Yeh, D.-R. He, Y. H. Kao, Phys. Rev. Lett. **52**, 480 (1984).
19. P. Alstrom, M. T. Levinsen, Phys. Rev. B **31**, 2753 (1985).
20. R. Bruinsma, A. Zangwill, Phys. Rev. Lett. **55**, 214 (1985).
21. D. Levine, P. Steinhardt, Phys. Rev. Lett. **53**, 2477 (1984). □

Multifractal phenomena in physics and chemistry

H. Eugene Stanley* & Paul Meakin†

*Center for Polymer Studies and Department of Physics, Boston University, Boston, Massachusetts 02215, USA
†Central Research and Development Department, E. I. du Pont de Nemours and Company, Experimental Station, Wilmington, Delaware 19898, USA

The neologism 'multifractal phenomena' describes the concept that different regions of an object have different fractal properties. Multifractal scaling provides a quantitative description of a broad range of heterogeneous phenomena.

A WIDE range of complex structures of interest to physicists and chemists have in recent years been quantitatively characterized using the idea of a fractal dimension; a dimension that corresponds in a unique fashion to the geometrical shape under study and that often is not an integer[1-7]. The key to this progress has been the recognition that many objects with random structure possess a scale symmetry. Scale symmetry implies that objects look the same on many different scales of observation.

To be more specific, consider an object with fractal dimension d_f. Imagine that we digitize the object by representing it by the pixels of a computer. If a unit mass is associated with each pixel, the total mass M of the object corresponds to its volume and its 'density' $\rho \equiv M/L^d$ is a measure of the fraction of d-dimensional space occupied by the object. Here L is a characteristic diameter, such as the caliper diameter or radius of gyration. If, however, $d_f < d$, the mass increases more slowly than L^d as the size of the object increases; for example, if we double L ($L \to L' = 2L$) then M increases by a power less than 2^d (that is, $M \to M' = 2^{d_f} M < 2^d M$). Thus the density decreases ($\rho \to \rho' = 2^{d_f-d} \rho < \rho$).

As long as a unit mass is associated with each pixel, a single scaling exponent d_f characterizes the structure of the object. In recent years, however, very interesting phenomena have been studied which seem to require not one but an infinite number of exponents for their description. Such multifractal phenomena have recently become an extremely active area of investigation and here we provide the non-specialist with a brief introduction to them. There are many types of multifractal phenomena but we shall concentrate on two examples, the behaviour of complex surfaces and interfaces[8], and fluid flow in porous media.

Complex surfaces

Figure 1a shows an object formed by a process called diffusion-limited aggregation (DLA)[9,10]; such structures arise naturally in many processes currently of interest to physicists and chemists, ranging from electrochemical deposition[11,12], thin-film

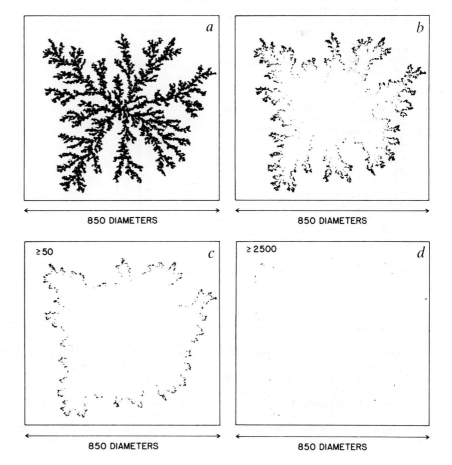

Fig. 1 The harmonic measure for a 50,000 particle off-lattice two-dimensional DLA aggregate. *a*, The cluster; *b*, all 6,803 perimeter sites that have been contacted by at least one out of 10^6 random walkers, following off-lattice trajectories. *c*, All the perimeter sites that have been contacted 50 or more times. *d*, The sites that have been contacted 2,500 or more times. The maximum number of contacts for any perimeter site 8,197, so that $p_{max} = 8.2 \times 10^{-3}$ (after ref. 27).

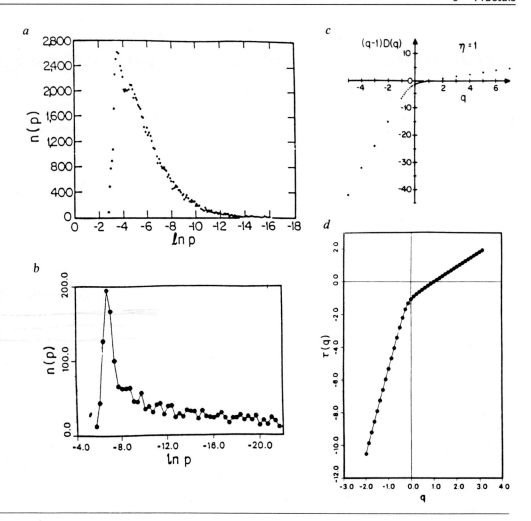

Fig. 2 Comparison between: *a*, *b*, the distribution functions $n(p)$ and *c*, *d*, the critical exponents $\tau(q) = (q-1)D(q)$ for viscous fingering patterns. *a*, Theoretical and experimental values for $n(p)$, where $n(p)\delta p$ is the number of perimeter sites with growth probabilities in the range $[p, p+\delta p]$. The simulated patterns and their growth probabilities were obtained[31] using the dielectric breakdown model. The growth probabilities for the experimental patterns were obtained by numerically solving Laplace's equation in the vicinity of a digitized representation of the pattern with absorbing boundary conditions on the sites occupied by the pattern. Similar results were obtained for small α by directly subtracting two successive experimental patterns. *c*, Theoretical and (*d*) experimental values for $\tau(q)$ in both cases was obtained numerically as in *a* and *b* (refs 31 and 32).

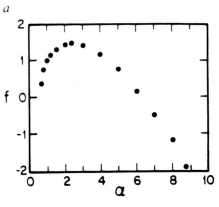
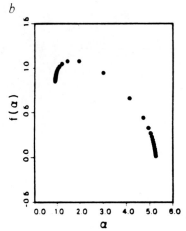

Fig. 3 Comparison between theoretical (*a*) and experimental (*b*) plots of the function $f(\alpha)$ (see Box 1) (refs 31 and 32).

morphology[13] and dendritic solidification[14–17] to various 'breakdown phenomena' such as dielectric breakdown[18,19] viscous fingering[20–24], and chemical dissolution[25,26]. If patterns such as the one shown in Fig. 1*a* are digitized, the fractal dimension $d_f \approx 1.7$ is obtained. Thus $d_f - d = -0.3$, and ρ decreases as the object grows due to the presence of 'fjords' whose size increases as the DLA cluster grows.

What is meant by the surface of the DLA cluster in Fig. 1*a* depends on the way it is to be measured. The exposed tips define one surface that is most likely to matter when diffusion is the essential feature, for example, if the surface is probed by particles undergoing random-walk motion. Figure 1*b–d* shows where the same object has been 'hit' by 10^6 random walkers[27], highlighting the surface sites touched by one, 50 and 2,500 random walkers respectively. Any of these three figures could be used to define the accessible surface, but the actual surfaces shown in Fig. 1*b–d* differ a great deal from one another. This example emphasizes that there is no unambiguous definition of the surface of this object.

An unambiguous quantity is the 'hit probability' p_i, defined as the probability that surface site i is the next to be hit. Operationally, we calculate $p_i \equiv N_i/N_T$, where N_i is the number

9.5 Multifractal Phenomena in Physics and Chemistry

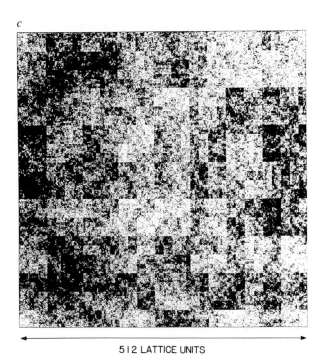

Fig. 4 A multifractal lattice: a, generator; b, second stage of construction; c, results after 8 generations. Lattice shown contains 2^8 pixels along each edge and shading is proportional to the value of π_i for pixel i. Here π_i is the product of the 8 probabilities assigned to that pixel as a result of the 8 generations (ref. 61).

of trajectories that hit site i and $N_T \equiv \sum_i N_i$ is the total number of trajectories (for the example of Fig. 1 $N_T = 10^6$). The set of numbers p_i may be used to form a probability distribution function $n(p)$, where $n(p)\delta p$ is the number of p_i in the range $[p, p+\delta p]$, as shown in Fig. 2. This probability distribution, like all probability distributions, is characterized by its moments

$$Z(q) \equiv \sum_p n(p) p^q \tag{1}$$

A central dogma of critical point phenomena has been the statement that the probability distributions that arise are characterized by only two independent exponents[28]. This means that we obtain no more information about the system by studying increasingly higher moments: moment $q+1$ is described by an exponent related to that of moment q by a simple gap exponent Δ. In general, one finds that

$$Z(q) \sim L^{-\tau(q)} \tag{2}$$

If the central dogma were correct, then $\tau(q)$ would be a linear function of q so that only *two* independent exponents would be needed to specify $\tau(q)$. It was discovered recently that this idea fails for the probability distribution $n(p)$ for DLA: both simulation[27,29-31], (Fig. 2c) and experiments[32] (Fig. 2d) show that $\tau(q)$ is a continuous curve. An infinite hierarchy of independent exponents is required. The Legendre transform $f(\alpha)$ of the function $\tau(q)$ contains the same information as $\tau(q)$ itself (see Fig. 3 and Box A), and is the characteristic customarily studied when dealing with multifractals[27,30-42].

Fluid flow in complex media

Consider a second example, fluid flow in random porous media. Such flow is customarily represented by considering an idealized network of bonds, a fraction p of which are open, and the remaining fraction $1-p$ of which are blocked[43]. For p lower than a critical value p_c, termed the percolation threshold, no fluid passes across a macroscopic system. At p_c a single macroscopic cluster appears, called the incipient infinite cluster, which carries fluid across the entire system. The singly connected bonds

---

A. Analogies of multifractals with thermodynamics and multifractal scaling

Consider the sum in equation (1) in the form

$$Z(q) = \sum_p e^{F(p)} \tag{A1}$$

where

$$F(p) = \log n(p) + q \log p \tag{A2}$$

The sum in (A1) is dominated by some value $p = p^*$, where p^* is the value of p that maximizes $F(p)$. Thus

$$Z(q) \sim e^{F(p^*)} = n(p^*)(p^*)^q \tag{A3}$$

For fixed q, p^* and $(n(p^*)$ both depend on the system size L, leading one to define the new q-dependent exponents α and f by

$$p^* \sim L^{-\alpha}; \qquad n(p^*) \sim L^f \tag{A4}$$

Substituting (A4) into (A3) gives

$$Z(q) \sim L^{f-\alpha q} \tag{A5}$$

Comparing (A5) with equation (2), we find the desired result

$$\tau(q) = q\alpha(q) - f(q). \tag{A6}$$

From (A2) it follows that

$$\frac{d}{dq}\tau(q) = \alpha(q). \tag{A7}$$

Hence we can interpret $f(\alpha)$ as the negative of the Legendre transform of the function $\tau(q)$

$$f(\alpha) = -(\tau(q) - q\alpha) \tag{A8}$$

where $\alpha \equiv d\tau/dq$. The function $Z(q)$ is formally analogous to the partition function $Z(\beta)$ in thermodynamics, so that $\tau(\beta)$ is like the free energy. The Legendre transform $f(\alpha)$ is thus the analogue of the entropy, with α being the analogue of the energy E. Indeed, the characteristic shape of plots of $f(\alpha)$ against α (compare Fig. 3) are reminiscent of plots of the dependence on E of the entropy for a thermodynamic system.

B. Random multiplicative processes
[A cautionary note for random sampling algorithms]

Multifractal phenomena seem to be associated with systems where the underlying physics is governed by a random multiplicative process. A simple random additive process might be the sum of 8 numbers, each number being chosen to be either $a-1$ or $a+1$ (this has a geometrical interpretation as an 8-step random walk on a one-dimensional lattice). Similarly, a simple random multiplicative process could be the product of 8 numbers, each number randomly chosen to be either a $1/2$ or $a/2$ (S. Redner, personal communication). The results of simulations of such a process are shown in Fig. 5. The y-axis is the running average of the product after R realizations and the x-axis is the number of realizations R.

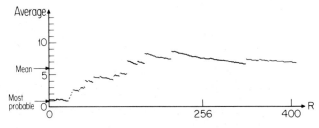

Fig. 5 A computer simulation of a random multiplicative process in which a string of 8 numbers is multiplied together. Each number is chosen with equal probability to be either 2 or $1/2$. The limiting or asymptotic value of the product is $(5/4)^8 = 5.96$. The simulations do not give this value unless the number of realizations R is approximately the same as the total number of configurations of this product, $2^8 = 256$. Simulation provided by R. Selinger.

In total there are 2^8, or 256, possible configurations of such random products. Normally, random sampling procedures give approximately correct answers when only a small fraction of the possible 256 configurations has been realized. Here, however, one sees from Fig. 5 that the correct asymptotic value of the product is attained only after ~256 realizations (S. Redner, personal communication). Monte Carlo sampling of only a small fraction of the 256 configurations is doomed to failure because a rare few configurations—consisting of, say, all 2s or seven 2s and a single 1—bias the average significantly and give rise to the upward steps in the running average shown in Fig. 5.

A simple random multiplicative process that gives rise to multifractal phenomena is found in the simple hierarchical model of the percolation backbone shown in Fig. 6. If the potential

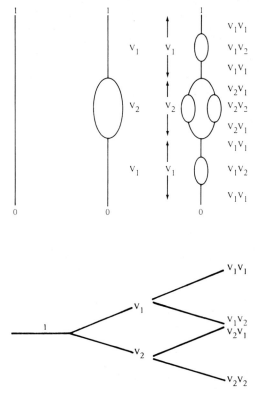

Fig. 6 A hierarchical model of the percolation backbone which displays multifractal scaling of the potential distribution $n(V)$. A unit potential is applied across the extremities of the cluster. The potentials shown are given by $V_1 = 2/5$ and $V_2 = 1/5$ (after ref. 37).

drop across the singly connected links is V_1 and that across the multiply-connected links is V_2, then when this structure is iterated, the potential drops across each of the bonds are products of the potential drops of the original structure. For this hierarchical structure $Z(q) = (V_1^q + V_2^q)^N$, where N is the number of iterations carried out. It turns out that $Z(q)$ obeys a power-law relation of the form of equation (2), with an infinite hierarchy of exponents given by $\tau(q) = 1 + \log[V_1^q + V_2^q]/\log 2$. To obtain this result, the relation $N_1 = L^{3/4}$ must be used, where N_1 is the number of singly connected bonds[54-57].

of this incipient infinite cluster carry the entire current and so sustain the largest potential drops, whereas the multiply connected bonds partition the current among them and so have smaller potential drops. The analogue of the hit-probability distribution $n(p)$ for DLA is the potential-drop distribution $n(V)$, where $n(V) dV$ is the number of bonds whose current lies in the interval $[V, V+dV]^{44-53}$. The distribution $n(V)$ is characterized by its moments and the analogue of (1) is $Z(q) = \sum n(V) V^q$. If the size of the random network is doubled, the distribution function $n(V)$ changes and so do the moments $Z(q)$. An infinite hierarchy of exponents $\tau(q)$ is found and again $\tau(q)$ is not linear in q.

It is natural to ask why the constant gap or single exponent idea breaks down when studying the surface of a DLA cluster. The key idea is that the underlying probability distribution $n(p)$ develops a long 'tail' extending to extremely small values of the variable p (Fig. 2). For DLA, this tail arises from the presence of extremely deep cluster 'fjords'. As the cluster grows, the hit probability p_i for all cluster sites decreases. The hit probability p_i for a site deep in a fjord, however, decreases much faster than p_i for a site on the exposed tips. Thus the long tail of the $n(p)$ distribution shifts its relative weight to smaller and smaller values of p as the cluster grows.

Similarly for flow in porous media, the distribution function $n(V)$ has a long tail extending to extremely small values of the potential V. This tail arises from the presence of large multiply connected regions of the network, termed blobs[54-57]. When $L \to L' = 2L$, the characteristic size of the largest blobs increases from M to $M' = 2^{1.62} M$ (ref. 58). Thus the minimum potential drops decrease dramatically and the distribution function $n(V)$ has a larger fraction of its weight from very small values of V.

Generality

It is not necessary to have a fractal structure to find multifractal phenomena. For example, consider the electric field E_i at every point i on the surface of a charged needle—a non-fractal object of dimension one. The set $\{E_i\}$ of electric-field values is formally analogous to the set $\{p_i\}$ of DLA growth probabilities, and indeed one finds that the $\{E_i\}$ also form a multifractal set[27,35]. A question that arises naturally concerns the conditions under which multifractal phenomena can be expected. As the example of the needle suggests, it is necessary to define a 'measure' on an object such that this measure has a different fractal dimension in different regions of the object[59]. Thus when the length of a needle is doubled, the electric field near its tip changes by a factor which differs from the factor by which the electric field near the centre changes. Similarly, when the mass of DLA is doubled, the growth probability near the tips of a DLA structure changes by a factor different from the growth probability deep in the fjords. This is because the screening in the deep fjords increases dramatically with increasing cluster size.

Multifractal theory permits the characterization of complex phenomena in a fully quantitative fashion. Just as completely random phenomena in nature may generate shapes that are fractal, phenomena with spatial correlations are sometimes multifractals. For example, randomly porous media are traditionally modelled by the random-resistor network of percolation theory: the resistance of each element corresponds to the permeability in a suitable digitization of the porous medium Although this model captures much of the essential physics, it neglects the phenomenon of spatial correlation that in turn leads to both short- and long-range heterogeneities in the porous medium. For this reason, it has been recently proposed that atmospheric turbulence and porous media should be modelled by a multifractal lattice[60-64]. This is obtained by a random multiplicative process (see Fig. 4 and Box B). Transport in such a lattice can be anomalously slow, just as it is in the random-resistor network model. But the exponent d_w describing the anomaly can be continuously tuned; in fact the slowing down can, under suitable conditions, become large without limit. A similar model has been solved analytically in one dimension[76].

Analogous phenomena are found for a wide variety of systems; in fact, multifractal phenomena were first found in studies of fluid turbulence[65,66], and in analysis of non-linear dynamical systems[67-69]. Recently it has been demonstrated[33] that experimental data concerning the onset of turbulence can be analysed using a method derived from multifractal theory. Various multifractal sets have been mapped on to the thermodynamics of one-dimensional spin models[70]. In a study of the depletion of a diffusion substance near an absorbing polymer it has been found that the scaling with distance r of each moment of the Laplace field is governed by an independent exponent[71].

Several authors have examined the multifractal properties of random walks[72-74]. In particular, it has been shown[72-74] that the fractal dimension d_f is a member of a continuous set of scaling exponents; consideration of the entire hierarchy of scaling exponents provides a more complete description of random walks than was possible previously. The main idea is to characterize an infinite walk with an exponent α that measures how fast its total probability decays to zero with increasing mass of the walk. The analogue of the function $f(\alpha)$ discussed above is the growth rate $z(\alpha)$ for the subset of walks with decay rate α. The analogue of $\tau(q)$ is the Legendre transform of $z(\alpha)$. A log-normal distribution has been found for the first-passage time in percolation[77], and some understanding of the conditions under which such a log-normal distribution will occur has also developed recently[37,75].

We thank our collaborators P. Alstrøm, A. Coniglio, G. Daccord, T. C. Halsey, J. Nittmann, I. Procaccia, D. Stassinopoulos, E. Touboul and T. A. Witten. We are also grateful to C. Amitrano, L. de Arcangelis, F. di Liberto, L. Pietronero and S. Redner for helpful discussions. A. Bunde, P. Grassberger and S. Havlin commented on the first draft of this article. The work at Boston University was supported by grants from NSF and ONR.

1. Mandelbrot, B. B. *The Fractal Geometry of Nature* (Freeman, San Francisco, 1982).
2. Family, F. & Landau, D. P. (eds) *Kinetics of Aggregation and Gelation* (Elsevier, Amsterdam, 1984).
3. Stanley, H. E. & Ostrowsky, N. (eds) *On Growth and Form: Fractal and Non-Fractal Patterns in Physics* (Nijhoff, Dordrecht, 1985).
4. Boccara, N. & Daoud, M. (eds) *Physics of Finely Divided Matter* (Springer, Heidelberg, 1985).
5. Pynn, R. & Skjeltorp, A. (eds) *Scaling Phenomena in Disordered Systems* (Plenum, New York, 1986).
6. Pietronero, L. & Tosatti, E. (eds) *Fractals in Physics* (North-Holland, Amsterdam, 1986).
7. Stanley, H. E. *Introduction to Fractal Phenomena* (Oxford University Press, in the press).
8. Meakin, P. *CRC Critical Rev. in Solid State and Materials Sciences* **13**, 143-189 (1987).
9. Witten, T. A. & Sander, L. M. *Phys. Rev. Lett.* **47**, 1400-1403 (1981); *Phys. Rev. B* **27**, 5686-5697 (1983).
10. Sander, L. M. *Nature* **332**, 789-793 (1986); *Scient. Am.* **256**, 94-100 (1986).
11. Brady, R. M. & Ball, R. C. *Nature* **309**, 225-229 (1984).
12. Matsushita, M., Sano, M., Hakayawa, Y., Honjo, H. & Sawada, Y. *Phys. Rev. Lett.* **53**, 286-289 (1984).
13. Elam, W. T. et al. *Phys. Rev. Lett.* **54**, 701 (1985).
14. Ben-Jacob, E. et al. *Phys. Rev. Lett.* **55**, 1315-1318 (1985).
15. Buka, A., Kertèsz, J. & Vicsek, T. *Nature* **323**, 424-425 (1986).
16. Couder, Y., Cardoso, O., Dupuy, D., Tavernier, P. & Thom, W. *Europhys. Lett.* **2**, 437-443 (1986).
17. Nittmann, J. N. & Stanley, H. E. *Nature* **321**, 663-668 (1986).
18. Niemeyer, L., Pietronero, L. & Wiesmann, H. J. *Phys. Rev. Lett.* **52**, 1033-1036 (1984).
19. Pietronero, L. & Wiesmann, H. J. *J. statist. Phys.* **36**, 909-916 (1984).
20. Nittmann, J., Daccord, G. & Stanley, H. E. *Nature* **314**, 141-144 (1985).
21. Daccord, G., Nittmann, J. & Stanley, H. E. *Phys. Rev. Lett.* **56**, 336-339 (1986).
22. Chen, J. D. & Wilkinson, D. *Phys. Rev. Lett.* **55**, 1892-1895 (1985).
23. Måløy, K. J., Feder, J. & Jøssang, T. *Phys. Rev. Lett.* **55**, 2688-2691 (1985).
24. Van Damme, H., Obrecht, F., Levitz, P., Gatineau, L. & Laroche, C. *Nature* **320**, 731-733 (1986).
25. Daccord, G. *Phys. Rev. Lett.* **58**, 479-482 (1987).
26. Daccord, G. & Lenormand, R. *Nature* **325**, 41-43 (1987).
27. Meakin, P., Coniglio, A., Stanley, H. E. & Witten, T. A. *Phys. Rev. A* **34**, 3325-3340 (1986).
28. Stanley, H. E. *Introduction to Phase Transitions and Critical Phenomena* (Oxford University Press, 1971).
29. Meakin, P., Stanley, H. E., Coniglio, A. & Witten, T. A. *Phys. Rev. A* **32**, 2364-2369 (1985).
30. Halsey, T. C., Meakin, P. & Procaccia, I. *Phys. Rev. Lett.* **56**, 854-857 (1986).
31. Amitrano, C., Coniglio, A. & di Liberto, F. *Phys. Rev. Lett.* **57**, 1016-1019 (1987).
32. Nittmann, J., Stanley, H. E., Touboul, E. & Daccord, G. *Phys. Rev. Lett.* **58**, 619 (1987).
33. Jensen, M. H., Kadanoff, L. P., Libchaber, A., Procaccia, I. & Stavans, J. *Phys. Rev. Lett.* **55**, 2798-2801 (1985).
34. Coniglio, A. in *On Growth and Form: Fractal and Non-Fractal Patterns in Physics* (eds Stanley, H. E. & Ostrowsky, N.) 101 (Nijhoff, Doredrecht, 1985).
35. Halsey, T. C., Jensen, M. H., Kadanoff, L. P., Procaccia, I. & Shraiman, B. *Phys. Rev. A* **33**, 1141-1151 (1986).
36. Coniglio, A. in *Fractals in Physics* (eds Pietronero, L. & Tosatti, E.) (North-Holland, Amsterdam, 1986).
37. Coniglio, A. *Physica A* **140**, 51-61 (1986).
38. Meakin, P. *Phys. Rev. A* **34**, 710-713 (1986).
39. Grassberger, P. & Procaccia, I. *Physica* **13D**, 34-54 (1984).
40. Grassberger, P. *Phys. Lett.* **107A**, 101-105 (1985).
41. Benzi, R., Paladin, G., Parisi, G. & Vulpiani, A. *J. Phys. A* **17**, 3521-3531 (1984); *Ibid.* **18**, 2157-2165 (1985).
42. Badii, R. & Politi, A. *J. statist. Phys.* **40**, 725-750 (1985).
43. Stauffer, D. *Introduction to Percolation Theory* (Taylor and Francis, Philadelphia, 1985).
44. Rammal, R., Tannous, C., Breton, P. & Tremblay, A. M. S. *Phys. Rev. Lett.* **54**, 1718-1721 (1985).
45. de Arcangelis, L., Redner, S. & Coniglio, A. *Phys. Rev. B* **31**, 4725-4727 (1985); *Phys. Rev. B* **34**, 4656-4673 (1986).
46. Rammal, R., Tannous, C. & Tremblay, A. M. S. *Phys. Rev. A* **31**, 2662-2671 (1985).
47. Rammal, R. *J. Phys., Paris* **46**, L129 (1985); *Phys. Rev. Lett.* **55**, 1428 (1985).
48. Meir, Y., Blumenfeld, R., Aharony, A. & Harris, A. B. *Phys. Rev. B* **34**, 3424-3428 (1986).
49. Blumenfeld, R., Meir, Y., Harris, A. B. & Aharony, A. *J. Phys. A* **19**, L791-L796 (1986).
50. Blumenfeld, R., Meir, Y., Aharony, A. & Harris, A. B. *Phys. Rev. B* **35**, 3524-3535 (1987).
51. Meir, Y. & Aharony, A. *Phys. Rev. A* **37**, 596-600 (1988).
52. Rammal, R. & Tremblay, A.-M. S. *Phys. Rev. Lett.* **58**, 415-418 (1987).
53. Fourcade, B., Breton, P. & Tremblay, A.-M. S. *Phys. Rev. Lett.* (in the press).
54. Stanley, H. E. *J. Phys. A* **10**, L211-L220 (1977).
55. Coniglio, A. *Phys. Rev. Lett.* **46**, 250-253 (1981).
56. Coniglio, A. *J. Phys. A* **15**, 3829-2844 (1981).
57. Pike, R. & Stanley, H. E. *J. Phys. A* **14**, L169-L177 (1981).
58. Nittmann, H. J. & Stanley, H. E. *Phys. Rev. Lett.* **53**, 1121-1124 (1984).
59. Alstrøm, P. in *Time-Dependent Effects in Disordered Materials* (eds Pynn, R. & Riste, T.) 185-193 (Plenum, New York, 1987); *Phys. Rev. A* **37**, 1378-1380 (1988); *Phys. Rev. A* **36**, 827-833 (1987).
60. Frisch, U., Sulem, P. & Nelkin, M. *J. Fluid Mech.* **87**, 719-736 (1978).
61. Meakin, P. *Phys. Rev. A* **36**, 2833-2837; *J. Phys. A* **20**, L779-L784 (1987).
62. Lovejoy, S. & Schertzer, D. *Bull. Am. met. Soc.* **67**, 221 (1986).
63. Lovejoy, S., Schertzer, D. & Tsonis, A. A. *Science* **231**, 1036-1038 (1987).
64. Schertzer, D. & Lovejoy, S. in *IUTAM Symposium on Turbulence and Chaotic Phenomena in Fluids, Kyoto, Japan* 141-144 (1983).
65. Mandelbrot, B. B. *J. Fluid Mech.* **62**, 331-358 (1974).
66. Mandelbrot, B. B. in *Proc. 13th IUPAP Conference on Statistical Physics* (eds Cabib, E., Kuper, C. G. & Reiss, I.) (Hilger, Bristol, 1978).
67. Grassberger, P. *Phys. Lett.* **A 97**, 277-230 (1983).
68. Hentschel, H. G. E. & Procaccia, I. *Physica* **8D**, 435-444 (1983).
69. Frisch, U. & Parisi, G. in *Turbulence and Predictability in Geophysical Fluid Dynamics and Climate Dynamics* Proc. Int. School of Physics Enrico Fermi, Course LXXXVIII (eds Ghil, M., Benzi, R. & Parisi G.) (North-Holland, Amsterdam, 1985).
70. Katzen, D. & Procaccia, I. *Phys. Rev. Lett.* **58**, 1169-1172 (1987).
71. Cates, M. E. & Witten, T. A. *Phys. Rev. Lett.* **56**, 2497-2500 (1986); *Phys. Rev. A* **35**, 1809-1824 (1987).
72. Evertsz, C. & Lyklema, J. W. *Phys. Rev. Lett.* **58**, 397-400 (1987).
73. Lyklema, J. W., Evertsz, C. & Pietronero, L. *Europhys. Lett.* **2**, 77-82 (1986).
74. Lyklema, J. W. & Evertsz, C. *Phys. Rev. A* **19**, L895-L900 (1986).
75. Pietronero, L. & Siebesma, A. P. *Phys. Rev. Lett.* **57**, 1098-1101 (1986).
76. Waissmann, H. & Havlin, S. *Phys. Rev. B* **37**, 5994-5997 (1988).
77. Trus, B., Havlin, S & Stauffer, D. *J. Phys. A* **20**, 6627-6630 (1987).

Simple Multifractals with Sierpinski Gasket Supports

L. Lam, R.D. Freimuth and J.L. Drake

*Nonlinear Physics Group, Department of Physics,
San Jose State University, San Jose, CA 95192-0106*

Abstract. Simple multifractals with Sierpinski gasket supports are constructed. The construction is based on the process of multiplicative cascade. Properties of these multifractals are given analytically. Generalizations to random multifractals embedded in a space of dimension other than two are given. The case of negative fractal dimension is dicussed. These multifractals can be generated by a computer with very brief algorithms. Two methods are shown: The first one is based on the use of logic function, the second is the generalized chaos game.

I. INTRODUCTION

Multifractals [1,2] have been studied intensely in the natural sciences in recent years [3-6]. A multifractal may be considered as an object consisting of subsets of varying fractal dimensions. The spatial distribution of the constituents of the object may be deterministic or random.

In many cases, mathematical multifractals are constructed for which exact calculations can be made. In addition to elucidate the basic properties of multifractals in general, these mathematical multifractals serve as the supports on which random walks and growth models are studied. Examples include the Cantor type multifractals which are embedded in one-dimensional space with either fractal or nonfractal support [5], and the multifractal filling up a two-dimensional square on which growth processes are studied [7]. (Note that in the latter, the support is two-dimensional and is nonfractal in nature.)

Here, we are interested in the construction of *simple* multifractals with *fractal* supports. Similar to the examples mentioned above, these multifractals are constructed by a recursive process and possess analytic results (Sec. II). They can be considered as generalizations of the Cantor types [2] to a two-dimensional embedding space, or the multifractals on a two-dimensional square to similar ones on a Sierpinski gasket. Generalizations are presented in Sec. III. The most simple of these multifractals can be generated with a computer by a novel use of the logic function AND (or OR) in an extremely brief algorithm. The more general ones can be generated with a generalized "chaos game" (Sec. IV).

These multifractals are simple and analytical, and can be easily generated with a computer. They are interesting for their pedagogical value as introductory examples to multifractals. These simple multifractals may also be used as the substrates on which additional properties can be studied.

II. A SIMPLE MULTIFRACTAL

The construction of the multifractal can be described by a curdling process starting with a square. The simplest case is introduced in this section.

Let us start with a square piece of paper. Without loss of generality, the linear size and the total mass are each taken to be unity. Divide the square into four equal smaller squares; cut and stack the upper left square onto the lower left square. The mass of each square is now changed (Fig. 1). This process of cutting and stacking is repeated *ad infinitum*. The end product is our multifractal.

This construction process may be described equivalently as follows. At the first step (n = 1), divide the square into four equal squares; redistribute the total mass of the original square into the three smaller squares at the right and at the

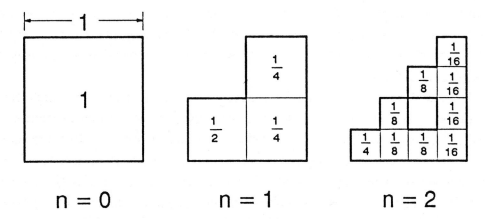

FIGURE 1. The "cut and stack" process in constructing the simple multifractal. Numbers in each square indicate the mass contained in that square. Note that the total mass at each step is conserved and is equal to one.

bottom (Fig. 2) according to the fractional distribution of p_1 and p_2 such that $p_1 + 2p_2 = 1$, $0 < p_j < 1$, $j = 1$ and 2, where $p_1 = 1/2$ and $p_2 = 1/4$. Repeat this process for each of the small squares. The case of $n = 2$ is shown in Fig. 2. It can be shown that at the n-th step, the linear size of each square is $\epsilon = 2^{-n}$, and there are $(n+1)$ different types of squares when classified according to their masses. Each type can be designated by an integer k ($k = 0, 1, ..., n$) such that the mass of each of the k-type squares is give by $\mu_k = p_1^{n-k} p_2^k$. The number of the k-type squares N_k is given by $N_k = 2^k \binom{n}{k}$.

These results can be easily seen by observing that the different masses in the n-th step can be obtained in a multiplicative process. At the n-th step, the possible masses in the squares are given by all the terms in the expression of $(p_1 + 2p_2)^n$, which is equivalent to

$$\sum_{k=0}^{n} \binom{n}{k} p_1^{n-k} (2p_2)^k = \sum_{k=0}^{n} 2^k \binom{n}{k} p_1^{n-k} p_2^k \qquad (1)$$

A. Mass Exponents $\tau(q)$

The fraction of mass in the i-th cell μ_i can be taken to be the fractal measure. The sequence of mass exponents $\tau(q)$ is defined by [5]

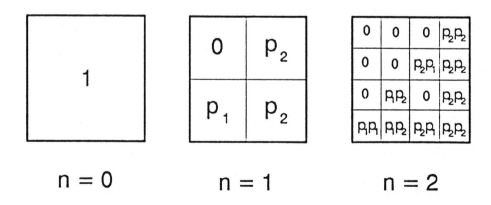

FIGURE 2. An equivalent construction process of the simple multifractal. The quantity in each square indicates the mass contained there. The case in Fig. 1 corresponds to $p_1 = 1/2$ and $p_2 = 1/4$.

9.6 Simple Multifractals with Sierpinski Gasket Supports

$$N(q,\epsilon) = \sum_i \mu_i^q \sim \epsilon^{-\tau(q)} \qquad (2)$$

as $\epsilon \to 0$, where ϵ is the size of each cell. Specialized to our case here, one obtains

$$N(q,\epsilon) = \sum_{k=0}^{n} N_k \mu_k^q$$

$$= \sum_k 2^k \binom{n}{k} (p_1^{n-k} p_2^k)^q$$

$$= \sum_k \binom{n}{k} (p_1^q)^{n-k} (2p_2^q)^k$$

$$= (p_1^q + 2p_2^q)^n \qquad (3)$$

By (1) and (2), we have

$$\tau(q) = [\ln(p_1^q + p_2^q)]/\ln 2 \qquad (4)$$

Numerical plot of $\tau(q)$ is shown in Fig. 3.

B. The f(α) Curve

Using the transformations [5],

$$\begin{cases} \alpha(q) = -d\tau(q)/dq \\ f(\alpha(q)) = q\alpha(q) + \tau(q) \end{cases} \qquad (5)$$

one obtains the f(α) curve,

$$f(\alpha) = q\alpha + [\ln(p_1^q + p_2^q)]/\ln 2 \qquad (6)$$

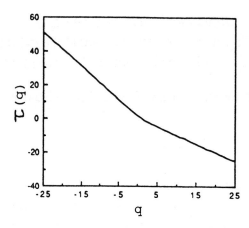

FIGURE 3. The $\tau(q)$ curve corresponding to the multifractal of Fig. 2. $p_1 = 1/2$ and $p_2 = 1/4$.

where $q = q(\alpha)$ and is given by

$$q = \frac{\ln[2(\ln p_2 + \alpha \ln 2)/(-\ln p_1 - \alpha \ln 2)]}{\ln(p_1/p_2)} \qquad (7)$$

The $f(\alpha)$ curve corresponding to the $\tau(q)$ of Fig. 3 is plotted in Fig. 4. Note that the fractal dimension of the support is given by $\tau(q=0) = \ln 3/\ln 2$, which, as expected, is that of the Sierpinski gasket.

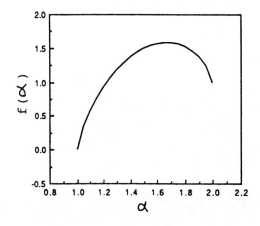

FIGURE 4. The $f(\alpha)$ curve corresponding to the multifractal of Fig. 2. $p_1 = 1/2$ and $p_2 = 1/4$.

In Fig. 4, α ranges from 1 to 2; $f(1) = 0$ and $f(2) = 1$. The subset in the multifractal represented by $\alpha = 1$ is a single point at the extreme left end; the one represented by $\alpha = 2$ is a solid vertical line at the extreme right (see Fig. 5 in Sec. IV).

C. Mass Density Distributions

In the n-th step, the position of each square may be denoted by the pair of numbers (i,j), $i,j = 0,1,2,\ldots,2^n-1$. The position of the lower left corner of the square is given by $(x,y) = (i\epsilon, j\epsilon)$, where $\epsilon = 2^{-n}$, the linear size of the square. Since $0 \le x, y < 1$, one can use the binary fraction representation with $x = 0.x_1 x_2 \ldots x_n$, where $x_\nu = 0$ or 1, $\nu = 1, 2, \ldots, n$; similarly for y. By definition,

$$x = \sum_{\nu=1}^{n} 2^{-\nu} x_\nu, \quad y = \sum_{\nu=1}^{n} 2^{-\nu} y_\nu, \tag{8}$$

Given (i,j) and hence the two sets of numbers $\{x_\nu\}$ and $\{y_\nu\}$, it is easy to show that the mass density at the square (x,y) is given by

$$\rho_n(x,y) = \prod_{\nu=1}^{n} r_\nu \tag{9}$$

where $r_\nu = g(x_\nu, y_\nu)$ such that $g(0,0) = 2$, $g(0,1) = 0$, $g(1,0) = 1$ and $g(1,1) = 1$. The proof of (9) is given in the Appendix. We thus see that all the properties of the multifractal are given analytically.

III. GENERALIZATIONS

The multifractal in Sec II can be easily generalized in several different directions. For example, one may redistribute the mass of a square in the n-th step into the three smaller ones in the next step with the fractions p_1, p_2 and p_3. Similar to the derivations in Sec. II, Eq. (1) now becomes

$$(p_1 + p_2 + p_3)^n = \sum_{k=0}^{n} \sum_{s=0}^{n-k} \binom{n}{k} \binom{n-k}{s} p_1^{n-k-s} p_2^{s} p_3^{k} \tag{10}$$

leading to $\mu_{ks} = p_1^{n-k-s} p_2^s p_3^k$ and $N_{ks} = \binom{n}{k}\binom{n-k}{s}$, where each type of square is characterized by the pair (k,s). Similar to (3), one has

$$N(q,\epsilon) = (p_1^q + p_2^q + p_3^q)^n \tag{11}$$

and

$$\tau(q) = [\ln(p_1^q + p_2^q + p_3^q)]/\ln 2 \tag{12}$$

Note that the constraint $p_1 + p_2 + p_3 = 1$ is not used in obtaining these results and can thus be discarded. (The multifractal of Sec. II is a special case with $p_1 = 1/2$ and $p_2 = p_3 = 1/4$.)

In fact, the derivation of (11) remains valid if we take an arbitrary object of any shape of Euclidean spatial dimension d; shrink it in size by a fraction x; make three identical copies of this shrunken object; distribute these three copies in either orderly or random fashion in space without overlapping with each other (even though touching contact is allowed); then redistribute the mass of the original object into these three smaller replicas with fractional distribution p_1, p_2 and p_3. Now shrink each of the three objects by the same fraction ξ and repeat the above process for each one of them. The location and orientation of each shrunken object, and the placement of the three distributions in each set of three replicas may or may not be the same at each step, or between different steps. Consequently, both ordered or random multifractals [including the two cases of (i) random subsets on ordered substrate, and (ii) random subsets on random substrate] may be generated.

In any case, Eq. (12) is now replaced by

$$\tau(q) = [\ln(p_1^q + p_2^q + p_3^q)]/(-\ln \xi) \tag{13}$$

If any two of the three p_i's are equal to each other, say, $p_2 = p_3$, then $f(\alpha)$ is given by (5) and (13), with

$$q = \frac{\ln[2(\ln p_2 - \alpha \ln \xi)/(\alpha \ln \xi - \ln p_1)]}{\ln(p_1/p_2)} \tag{14}$$

Another generalization is to replace the number of replicas, three, by any positive integer m. Equation (13) becomes

$$\tau(q) = [\ln(p_1^q + p_2^q + ... + p_m^q)]/(-\ln \xi) \tag{15}$$

From (15) one sees that the fractal dimension of the support, $\tau(0) = (-\ln m)/(\ln \xi)$, can be negative when $\xi > 1$. It means that at each stage of generation of the multifractal, the size of the replica is larger than the original. If one looks at this as a growth process with a source (or several sources) being added continuously, this can be a real physical process. Examples include the formation of galaxies or structures in space, the formation of clusters in colloidal aggregates [6] and the growth of droplets from saturated vapors [9]. (Of course, the negative "fractal dimension" so defined differs from the usual definition of a positive fractal dimension. The latter refers to the scaling during the growth process of a set.) Since growth processes are quite common in nature, such a concept of negative fractal dimension should be of some value in their descriptions.

IV. COMPUTER GENERATION OF MULTIFRACTALS

A. Method with Logic Function

A very brief algorithm can be constructed to generate with a computer the simple multifractal of Sec. II. The trick is to take advantage of the logic function AND (denoted by $\wedge$ here) which is built into many computer languages such as BASIC and C.

Given (x_ν, y_ν) one can solve the following equation,

$$y_\nu = x_\nu \wedge \lambda_\nu, \quad \nu = 1, 2, \ldots, n \tag{16}$$

to obtain $\{\lambda_\nu\}$. The $\wedge$ operation is standard and is defined as follows: $y_\nu = 1$ if $x_\nu = 1$ and $\lambda_\nu = 1$; $y_\nu = 0$ otherwise. Solutions of (16) are given in Table I. From Table I one sees that, given (x_ν, y_ν), the number of possible solutions

TABLE I. Solution of Eq. (16). Here s_ν is the number of solutions of λ_ν, given (x_ν, y_ν).

x	y	λ_ν	s_ν
0	0	0,1	2
0	1	*	0
1	0	0	1
1	1	1	1

* Does not exist.

of λ_y, s_y, has the same properties as r_y of (9). One may thus make the identification of $r_y = s_y$, in (9).

A computer algorithm written in QUICK BASIC consisting of 8 lines is given as follows.

```
n = ?
DO
  x = INT(RND * 2^n)
  lambda = INT(RND * 2^n)
  y = (x AND lambda)/2^n
  x = x/2^n
  PSET(x,y)
LOOP
```

The multifractal generated with n = 15, using this algorithm, is shown in Fig. 5a. It this algorithm, (i) x and λ are generated randomly, where $\lambda = 0.\lambda_1\lambda_2...\lambda_n$. (ii) Equation (16) is used to find the possible y's. (iii) The computer automatically stores the cell location (x,y) in the binary representation during the calculation. (iv) The key point is that the random sampling reflects directly the density of points in the (x,y) plane.

One may replace the AND by another logic function, e.g., OR. One still gets a multifractal (see figure 5b). Note that if any two quadrants at the n = 1 step have zero mass, one still has a multifractal with a solid straight line as support. When the square is divided into m equal parts, the case covered by (15), one may not be able to find a suitable logic function to generate the multifractal in the fashion described here. However, in this case, it is always possible to write an algorithm to generate the multifractal by using a look-up table similar to that in Table I.

B. The Generalized Chaos Game

Another method for producing the multifractal of Sec. II, as well as its generalizations described by (10)-(12), is to proceed as follows. One starts with three points (denoted by i = 1, 2, and 3) on a plane which are the vertices of an arbitrary triangle. A random point is chosen as the starting point of an iterative process. The i-th vertex is chosen randomly with a weight p_i, where $p_1 + p_2 + p_3 = 1$. The midpoint of the line formed by connecting the starting point and the chosen vertex is marked and used as the starting point for the next step. The set of "initial" points, with the exception of the first few, form the multifractal described by (10)-(12), the support of which is a Sierpinski gasket.

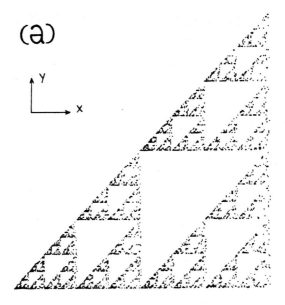

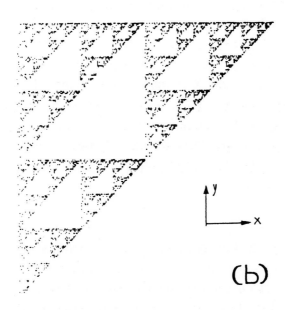

FIGURE 5. Simple multifractals generated by the algorithm of Sec. IVA. n = 15; 5000 iterations. (a) With the use of the AND logic function; (b) with AND replaced by OR, another logic function.

This repetitive algorithm is easily programmed in almost all computer languages. An example of the algorithm is shown below in C.

```c
int Select_Direction (void)
{
    int a_adjust, b_adjust;
int number;
int direction;
number = random (10000);
a_adjust =
(int)((p_1_weight/(p_1_weight+p_2_weight+p_3_weitht))
    *10000);
b_adjust =
(int)((p_2_weight/(p_1_weight+p_2_weight+p_3_weitht))
    *10000);
if (number <= a_adjust)      direction = 1;
   else
      if (number <= a_adjust + b_adjust)   direction = 2
      else              direction = 3;
      return direction;
}

for (count = 0; count <= max_count; count++)
{
  direction = Select_Direction ();
  *x = *x + ((vertex[direction][0] - *x)/2);
  *y = *y + ((vertex[direction][0] - *y)/2);
Plot_Point (x,y);
}
```

Here, 'Vertex' is a 3 x 2 array containing the coordinates of the vertices of the triangle. "Select_Direction" is a function that will randomly pick which vertex is to be used for the calculation, taking into account specified weightings of the vertices. The function "Plot_Point" is self-explanatory.

The density of points generated by the process is determined by the weights (see Fig. 6). When $p_1 = p_2 = p_3$, this algorithm becomes the simple "chaos game" [9]. The result is a simple fractal—the Sierpinski gasket. For the case in Sec. II, the appropriate weights are $p_1 = 2$, $p_2 = 1$ and $p_3 = 1$. Therefore this method for generating a multifractal covers the results of the previous method as a special case.

9.6 Simple Multifractals with Sierpinski Gasket Supports

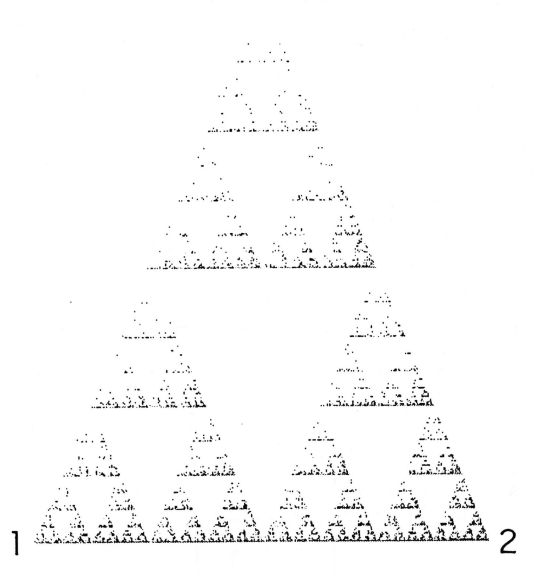

FIGURE 6. The multifractal generated by the algorithm of Sec. IVB. $p_1 = 3$, $p_2 = 5$ and $p_3 = 1$. 10,000 iterations.

ACKNOWLEDGMENTS

This work is supported by the NSF-REU program and a Syntex Corporation grant from the Research Corporation.

APPENDIX: DERIVATION OF EQUATION (9)

The proof of (9) is by induction. For n = 1, the locations of the four squares are represented by the four points a, a_1, a_2 and a_3, respectively (Fig. 7). In the binary fraction representation, each point is given by (x,y) = $(0.\bar{x}_1, 0.\bar{y}_1)$ with $(\bar{x}_1, \bar{y}_1)$ = (0,0), (1,0), (0,1) and (1,1) for a, a_1, a_2 and a_3, respectively. According to the definition of the function g, defined after (9), we see immediately that $r_1 [= g(\bar{x}_1, \bar{y}_1) \equiv \bar{r}_1]$ assumes the value for each of the four quadrants as shown in Fig. 7. From the n = 1 diagram in Fig. 1, we see that the mass densities ρ_1 (obtained by dividing the mass in each quadrant there by 1/4, the area of each quadrant) coincides with the numbers shown in Fig. 7. In other words, Eq. (9) is proved for n = 1.

Now assume that at the n-th step, S_n, the mass density of an arbitrary square with position represented by the point b (Fig. 8), is given correctly by (9), i.e., $S_n = \prod_{\nu=1}^{n} r_\nu$, with $r_\nu = g(x_\nu, y_\nu)$. Here, b is given by $(0.x_1 x_2 \ldots x_n, 0.y_1 y_2 \ldots y_n)$ in the binary fraction representation. According to the construction of the multifractal, the (n+1)-th step is obtained by dividing this square into four equal quadrants with mass density in each quadrant as shown in Fig. 8, resulting in $\rho_{n+1} = S_n \times \bar{r}_1$. The crucial point is to note that the four points b, b_1, b_2 and b_3, representing the locations of these four quadrants, are given by $(0.x_1 x_2 \ldots x_n x_{n+1}, 0.y_1 y_2 \ldots y_n y_{n+1})$ with (x_{n+1}, y_{n+1}) = (0,0), (1,0), (0,1) and (1,1) for b, b_1, b_2 and b_3, respectively, while x_ν and y_ν with $\nu = 1, \ldots, n$ are the same as those in the n-th step for all these four points. Comparing with the

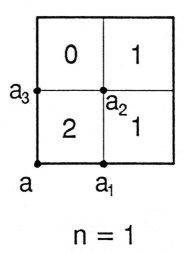

FIGURE 7. The mass density ρ_1 for each of the four quadrants at the n = 1 step.

9.6 Simple Multifractals with Sierpinski Gasket Supports

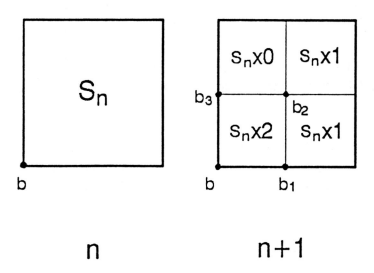

FIGURE 8. The mass densities ρ_n and ρ_{n+1} at the n-th and (n+1)-th steps, respectively.

n = 1 case above, since $(x_{n+1}, y_{n+1}) = (\bar{x}_1, \bar{y}_1)$ we see that $r_{n+1} = \bar{r}_1$, where $r_{n+1} = g(x_{n+1}, y_{n+1})$. Therefore, $\rho_{n+1} = S_n \times r_{n+1} = \prod_{\nu=1}^{n+1} r_\nu$. This completes the proof of (9) for the n+1 case. By induction, Eq. (9) is valid for an arbitrary n (≥ 1).

It is easy to see that the above proof remains valid if the binary representation (instead of the binary fraction representation) is used in denoting all the a and b points. In this case, b is given by $(x_n...x_2 x_1, y_n...y_2 y_1)$ at the n-th step. In fact, binary representation is the one used in the algorithm of Sec IVA. In a computer with k-bit word, say, b is actually represented by $(00...0 x_n...x_2 x_1, 00...0 y_n...y_2 y_1)$ where there are (k-n) zeros preceding x_n or y_n. Since $g(0,0) = 2$, the density of points in the multifractal generated by the algorithm is actually equal to the actual value multiplied by 2^{k-n}. However, this will not affect the overall qualitative appearance of the pictures in Fig. 5. The reason is that due to the finite number of iterations and the finite resolution of the printer, the densities of points in these pictures should not be understood to represent the real values anyway.

REFERENCES

[1] B.B. Mandelbrot, in *Random Fluctuations and Pattern Growth*, edited by H.E. Stanley and N. Ostrowsky (Kluwer, dordrecht, 1988).

[2] T.C. Halsey, M.H. Jensen, L.P. Kadanoff, I. Procaccia, and B.I. Shraiman, "Fractal measures and their singularities: The characterization of strange sets," Phys. Rev. A **33**, 1141-1151 (1986).

[3] H.E. Stanley and P. Meakin, "Multifractal phenomena in physics and chemistry," Nature **335**, 405-409 (1988).

[4] H. gould and J. Tobochnik, "More on fractals and chaos: Multifractals," Comput. Phys. **4**(2), 202-207 (1990).

[5] J. Feder, *Fractals* (Plenum, New York, 1988).

[6] T. Vicsek, *Fractal Growth Phenomena* (World Scientific, Teaneck, 1989); *Fractals and Disordered Systems* edited by A. Bunde and S. Havlin (Spriner, New York, 1991).

[7] P. Meakin, "Diffusion limited aggregation on multifractal lattices: A model for fluid-fluid displacement in porous media," Phys. Rev. A **36**, 2833-2837 (1987); "Random walk on multifractal lattices," J. Phys. A **20**, L771-L777 (1987).

[8] F. Family, in *Random Fluctuations and Pattern Growth*, edited by H.E. Stanley and N. Ostrowsky (Kluwer, dordrecht, 1988).

[9] M.F. Barnsley, "Fractal modelling of real world images," in *The Science of Fractal Images*, edited by H.-O. Peitgen and D. Saupe (Springer, New York, 1988).

Chaos

There is order in chaos: randomness has an underlying geometric form. Chaos imposes fundamental limits on prediction, but it also suggests causal relationships where none were previously suspected

by James P. Crutchfield, J. Doyne Farmer, Norman H. Packard and Robert S. Shaw

The great power of science lies in the ability to relate cause and effect. On the basis of the laws of gravitation, for example, eclipses can be predicted thousands of years in advance. There are other natural phenomena that are not as predictable. Although the movements of the atmosphere obey the laws of physics just as much as the movements of the planets do, weather forecasts are still stated in terms of probabilities. The weather, the flow of a mountain stream, the roll of the dice all have unpredictable aspects. Since there is no clear relation between cause and effect, such phenomena are said to have random elements. Yet until recently there was little reason to doubt that precise predictability could in principle be achieved. It was assumed that it was only necessary to gather and process a sufficient amount of information.

Such a viewpoint has been altered by a striking discovery: simple deterministic systems with only a few elements can generate random behavior. The randomness is fundamental; gathering more information does not make it go away. Randomness generated in this way has come to be called chaos.

A seeming paradox is that chaos is deterministic, generated by fixed rules that do not themselves involve any elements of chance. In principle the future is completely determined by the past, but in practice small uncertainties are amplified, so that even though the behavior is predictable in the short term, it is unpredictable in the long term. There is order in chaos: underlying chaotic behavior there are elegant geometric forms that create randomness in the same way as a card dealer shuffles a deck of cards or a blender mixes cake batter.

The discovery of chaos has created a new paradigm in scientific modeling. On one hand, it implies new fundamental limits on the ability to make predictions. On the other hand, the determinism inherent in chaos implies that many random phenomena are more predictable than had been thought. Random-looking information gathered in the past—and shelved because it was assumed to be too complicated—can now be explained in terms of simple laws. Chaos allows order to be found in such diverse systems as the atmosphere, dripping faucets and the heart. The result is a revolution that is affecting many different branches of science.

What are the origins of random behavior? Brownian motion provides a classic example of randomness. A speck of dust observed through a microscope is seen to move in a continuous and erratic jiggle. This is owing to the bombardment of the dust particle by the surrounding water molecules in thermal motion. Because the water molecules are unseen and exist in great number, the detailed motion of the dust particle is thoroughly unpredictable. Here the web of causal influences among the subunits can become so tangled that the resulting pattern of behavior becomes quite random.

The chaos to be discussed here requires no large number of subunits or unseen influences. The existence of random behavior in very simple systems motivates a reexamination of the sources of randomness even in large systems such as weather.

What makes the motion of the atmosphere so much harder to anticipate than the motion of the solar system? Both are made up of many parts, and both are governed by Newton's second law, $F = ma$, which can be viewed as a simple prescription for predicting the future. If the forces F acting on a given mass m are known, then so is the acceleration a. It then follows from the rules of calculus that if the position and velocity of an object can be measured at a given instant, they are determined forever. This is such a powerful idea that the 18th-century French mathematician Pierre Simon de Laplace once boasted that given the position and velocity of every particle in the universe, he could predict the future for the rest of time. Although there are several obvious practical difficulties to achieving Laplace's goal, for more than 100 years there seemed to be no reason for his not being right, at least in principle. The literal application of Laplace's dictum to human behavior led to the philosophical conclusion that human behavior

CHAOS results from the geometric operation of stretching. The effect is illustrated for a painting of the French mathematician Henri Poincaré, the originator of dynamical systems theory. The initial image (*top left*) was digitized so that a computer could perform the stretching operation. A simple mathematical transformation stretches the image diagonally as though it were painted on a sheet of rubber. Where the sheet leaves the box it is cut and reinserted on the other side, as is shown in panel *1*. (The number above each panel indicates how many times the transformation has been made.) Applying the transformation repeatedly has the effect of scrambling the face (*panels 2–4*). The net effect is a random combination of colors, producing a homogeneous field of green (*panels 10 and 18*). Sometimes it happens that some of the points come back near their initial locations, causing a brief appearance of the original image (*panels 47–48, 239–241*). The transformation shown here is special in that the phenomenon of "Poincaré recurrence" (as it is called in statistical mechanics) happens much more often than usual; in a typical chaotic transformation recurrence is exceedingly rare, occurring perhaps only once in the lifetime of the universe. In the presence of any amount of background fluctuations the time between recurrences is usually so long that all information about the original image is lost.

46

10.1 Chaos

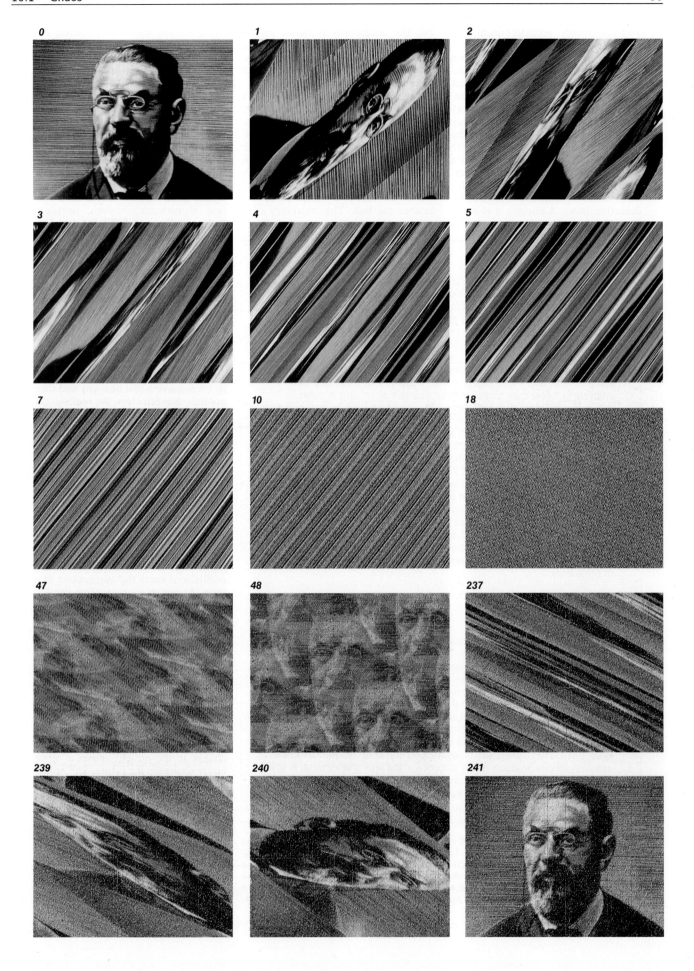

was completely predetermined: free will did not exist.

Twentieth-century science has seen the downfall of Laplacian determinism, for two very different reasons. The first reason is quantum mechanics. A central dogma of that theory is the Heisenberg uncertainty principle, which states that there is a fundamental limitation to the accuracy with which the position and velocity of a particle can be measured. Such uncertainty gives a good explanation for some random phenomena, such as radioactive decay. A nucleus is so small that the uncertainty principle puts a fundamental limit on the knowledge of its motion, and so it is impossible to gather enough information to predict when it will disintegrate.

The source of unpredictability on a large scale must be sought elsewhere, however. Some large-scale phenomena are predictable and others are not. The distinction has nothing to do with quantum mechanics. The trajectory of a baseball, for example, is inherently predictable; a fielder intuitively makes use of the fact every time he or she catches the ball. The trajectory of a flying balloon with the air rushing out of it, in contrast, is not predictable; the balloon lurches and turns erratically at times and places that are impossible to predict. The balloon obeys Newton's laws just as much as the baseball does; then why is its behavior so much harder to predict than that of the ball?

The classic example of such a dichotomy is fluid motion. Under some circumstances the motion of a fluid is laminar—even, steady and regular—and easily predicted from equations. Under other circumstances fluid motion is turbulent—uneven, unsteady and irregular—and difficult to predict. The transition from laminar to turbulent behavior is familiar to anyone who has been in an airplane in calm weather and then suddenly encountered a thunderstorm. What causes the essential difference between laminar and turbulent motion?

To understand fully why that is such a riddle, imagine sitting by a mountain stream. The water swirls and splashes as though it had a mind of its own, moving first one way and then another. Nevertheless, the rocks in the stream bed are firmly fixed in place, and the tributaries enter at a nearly constant rate of flow. Where, then, does the random motion of the water come from?

The late Soviet physicist Lev D. Landau is credited with an explanation of random fluid motion that held sway for many years, namely that the motion of a turbulent fluid contains many different, independent oscillations. As the fluid is made to move faster, causing it to become more turbulent, the oscillations enter the motion one at a time. Although each separate oscillation may be simple, the complicated combined motion renders the flow impossible to predict.

Landau's theory has been disproved, however. Random behavior occurs even in very simple systems, without any need for complication or indeterminacy. The French mathematician Henri Poincaré realized this at the turn of the century when he noted that unpredictable, "fortuitous" phenomena may occur in systems where a small change in the present causes a much larger change in the future. The notion is clear if one thinks of a rock poised at the top of a hill. A tiny push one way or another is enough to send it tumbling down widely differing paths. Although the rock is sensitive to small influences only at the top of the hill, chaotic systems are sensitive at every point in their motion.

A simple example serves to illustrate just how sensitive some physical

---

Laplace, 1776

"The present state of the system of nature is evidently a consequence of what it was in the preceding moment, and if we conceive of an intelligence which at a given instant comprehends all the relations of the entities of this universe, it could state the respective positions, motions, and general affects of all these entities at any time in the past or future.

"Physical astronomy, the branch of knowledge which does the greatest honor to the human mind, gives us an idea, albeit imperfect, of what such an intelligence would be. The simplicity of the law by which the celestial bodies move, and the relations of their masses and distances, permit analysis to follow their motions up to a certain point; and in order to determine the state of the system of these great bodies in past or future centuries, it suffices for the mathematician that their position and their velocity be given by observation for any moment in time. Man owes that advantage to the power of the instrument he employs, and to the small number of relations that it embraces in its calculations. But ignorance of the different causes involved in the production of events, as well as their complexity, taken together with the imperfection of analysis, prevents our reaching the same certainty about the vast majority of phenomena. Thus there are things that are uncertain for us, things more or less probable, and we seek to compensate for the impossibility of knowing them by determining their different degrees of likelihood. So it is that we owe to the weakness of the human mind one of the most delicate and ingenious of mathematical theories, the science of chance or probability."

Poincaré, 1903

"A very small cause which escapes our notice determines a considerable effect that we cannot fail to see, and then we say that the effect is due to chance. If we knew exactly the laws of nature and the situation of the universe at the initial moment, we could predict exactly the situation of that same universe at a succeeding moment. But even if it were the case that the natural laws had no longer any secret for us, we could still only know the initial situation *approximately*. If that enabled us to predict the succeeding situation with *the same approximation*, that is all we require, and we should say that the phenomenon had been predicted, that it is governed by laws. But it is not always so; it may happen that small differences in the initial conditions produce very great ones in the final phenomena. A small error in the former will produce an enormous error in the latter. Prediction becomes impossible, and we have the fortuitous phenomenon."

OUTLOOKS OF TWO LUMINARIES on chance and probability are contrasted. The French mathematician Pierre Simon de Laplace proposed that the laws of nature imply strict determinism and complete predictability, although imperfections in observations make the introduction of probabilistic theory necessary. The quotation from Poincaré foreshadows the contemporary view that arbitrarily small uncertainties in the state of a system may be amplified in time and so predictions of the distant future cannot be made.

systems can be to external influences. Imagine a game of billiards, somewhat idealized so that the balls move across the table and collide with a negligible loss of energy. With a single shot the billiard player sends the collection of balls into a protracted sequence of collisions. The player naturally wants to know the effects of the shot. For how long could a player with perfect control over his or her stroke predict the cue ball's trajectory? If the player ignored an effect even as minuscule as the gravitational attraction of an electron at the edge of the galaxy, the prediction would become wrong after one minute!

The large growth in uncertainty comes about because the balls are curved, and small differences at the point of impact are amplified with each collision. The amplification is exponential: it is compounded at every collision, like the successive reproduction of bacteria with unlimited space and food. Any effect, no matter how small, quickly reaches macroscopic proportions. That is one of the basic properties of chaos.

It is the exponential amplification of errors due to chaotic dynamics that provides the second reason for Laplace's undoing. Quantum mechanics implies that initial measurements are always uncertain, and chaos ensures that the uncertainties will quickly overwhelm the ability to make predictions. Without chaos Laplace might have hoped that errors would remain bounded, or at least grow slowly enough to allow him to make predictions over a long period. With chaos, predictions are rapidly doomed to gross inaccuracy.

The larger framework that chaos emerges from is the so-called theory of dynamical systems. A dynamical system consists of two parts: the notions of a state (the essential information about a system) and a dynamic (a rule that describes how the state evolves with time). The evolution can be visualized in a state space, an abstract construct whose coordinates are the components of the state. In general the coordinates of the state space vary with the context; for a mechanical system they might be position and velocity, but for an ecological model they might be the populations of different species.

A good example of a dynamical system is found in the simple pendulum. All that is needed to determine its motion are two variables: position and velocity. The state is thus a point in a plane, whose coordinates are position and velocity. Newton's laws provide

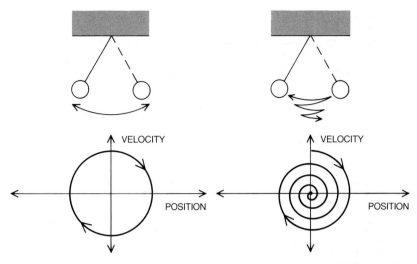

STATE SPACE is a useful concept for visualizing the behavior of a dynamical system. It is an abstract space whose coordinates are the degrees of freedom of the system's motion. The motion of a pendulum (*top*), for example, is completely determined by its initial position and velocity. Its state is thus a point in a plane whose coordinates are position and velocity (*bottom*). As the pendulum swings back and forth it follows an "orbit," or path, through the state space. For an ideal, frictionless pendulum the orbit is a closed curve (*bottom left*); otherwise, with friction, the orbit spirals to a point (*bottom right*).

a rule, expressed mathematically as a differential equation, that describes how the state evolves. As the pendulum swings back and forth the state moves along an "orbit," or path, in the plane. In the ideal case of a frictionless pendulum the orbit is a loop; failing that, the orbit spirals to a point as the pendulum comes to rest.

A dynamical system's temporal evolution may happen in either continuous time or in discrete time. The former is called a flow, the latter a mapping. A pendulum moves continuously from one state to another, and so it is described by a continuous-time flow. The number of insects born each year in a specific area and the time interval between drops from a dripping faucet are more naturally described by a discrete-time mapping.

To find how a system evolves from a given initial state one can employ the dynamic (equations of motion) to move incrementally along an orbit. This method of deducing the system's behavior requires computational effort proportional to the desired length of time to follow the orbit. For simple systems such as a frictionless pendulum the equations of motion may occasionally have a closed-form solution, which is a formula that expresses any future state in terms of the initial state. A closed-form solution provides a short cut, a simpler algorithm that needs only the initial state and the final time to predict the future without stepping through intermediate states. With such a solution the algorithmic effort required to follow the motion of the system is roughly independent of the time desired. Given the equations of planetary and lunar motion and the earth's and moon's positions and velocities, for instance, eclipses may be predicted years in advance.

Success in finding closed-form solutions for a variety of simple systems during the early development of physics led to the hope that such solutions exist for any mechanical system. Unfortunately, it is now known that this is not true in general. The unpredictable behavior of chaotic dynamical systems cannot be expressed in a closed-form solution. Consequently there are no possible short cuts to predicting their behavior.

The state space nonetheless provides a powerful tool for describing the behavior of chaotic systems. The usefulness of the state-space picture lies in the ability to represent behavior in geometric form. For example, a pendulum that moves with friction eventually comes to a halt, which in the state space means the orbit approaches a point. The point does not move—it is a fixed point—and since it attracts nearby orbits, it is known as an attractor. If the pendulum is given a small push, it returns to the same fixed-point attractor. Any system that comes to rest with the passage of time can be characterized by a fixed point in state space. This is an example of a very general phenomenon, where losses due to friction or viscosity, for example, cause orbits to be attracted to a smaller region of the state space with lower dimension. Any such region is called an attractor. Roughly speaking, an attractor is what the behavior of a system settles down to, or is attracted to.

Some systems do not come to rest in the long term but instead cycle periodically through a sequence of states. An example is the pendulum clock, in which energy lost to friction is replaced by a mainspring or weights. The pendulum repeats the same motion over and over again. In the state space such a motion corresponds to a cycle, or periodic orbit. No matter how the pendulum is set swinging, the cycle approached in the long-term limit is the same. Such attractors are therefore called limit cycles. Another familiar system with a limit-cycle attractor is the heart.

A system may have several attractors. If that is the case, different initial conditions may evolve to different attractors. The set of points that evolve to an attractor is called its basin of attraction. The pendulum clock has two such basins: small displacements of the pendulum from its rest position result in a return to rest; with large displacements, however, the clock begins to tick as the pendulum executes a stable oscillation.

The next most complicated form of attractor is a torus, which resembles the surface of a doughnut. This shape describes motion made up of two independent oscillations, sometimes called quasi-periodic motion. (Physical examples can be constructed from driven electrical oscillators.) The orbit winds around the torus in state space, one frequency determined by how fast the orbit circles the doughnut in the short direction, the other regulated by how fast the orbit circles the long way around. Attractors may also be higher-dimensional tori, since they represent the combination of more than two oscillations.

The important feature of quasi-periodic motion is that in spite of its complexity it is predictable. Even though the orbit may never exactly repeat itself, if the frequencies that make up the motion have no common divisor, the motion remains regular. Orbits that start on the torus near one another remain near one another, and long-term predictability is guaranteed.

Until fairly recently, fixed points, limit cycles and tori were the only known attractors. In 1963 Edward N. Lorenz of the Massachusetts Institute

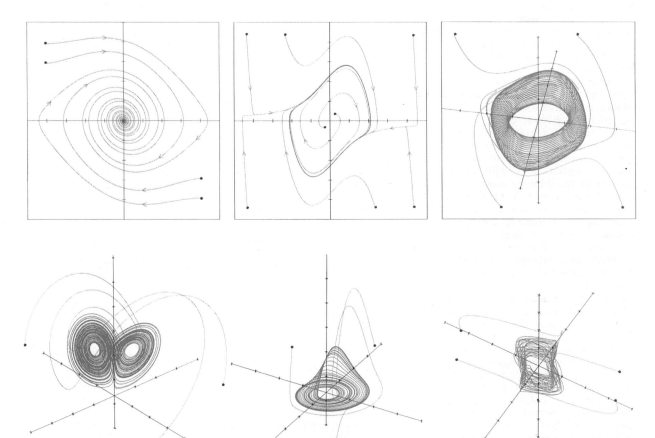

ATTRACTORS are geometric forms that characterize long-term behavior in the state space. Roughly speaking, an attractor is what the behavior of a system settles down to, or is attracted to. Here attractors are shown in blue and initial states in red. Trajectories (*green*) from the initial states eventually approach the attractors. The simplest kind of attractor is a fixed point (*top left*). Such an attractor corresponds to a pendulum subject to friction; the pendulum always comes to the same rest position, regardless of how it is started swinging (*see right half of illustration on preceding page*). The next most complicated attractor is a limit cycle (*top middle*), which forms a closed loop in the state space. A limit cycle describes stable oscillations, such as the motion of a pendulum clock and the beating of a heart. Compound oscillations, or quasi-periodic behavior, correspond to a torus attractor (*top right*). All three attractors are predictable: their behavior can be forecast as accurately as desired. Chaotic attractors, on the other hand, correspond to unpredictable motions and have a more complicated geometric form. Three examples of chaotic attractors are shown in the bottom row; from left to right they are the work of Edward N. Lorenz, Otto E. Rössler and one of the authors (Shaw) respectively. The images were prepared by using simple systems of differential equations having a three-dimensional state space.

of Technology discovered a concrete example of a low-dimensional system that displayed complex behavior. Motivated by the desire to understand the unpredictability of the weather, he began with the equations of motion for fluid flow (the atmosphere can be considered a fluid), and by simplifying them he obtained a system that had just three degrees of freedom. Nevertheless, the system behaved in an apparently random fashion that could not be adequately characterized by any of the three attractors then known. The attractor he observed, which is now known as the Lorenz attractor, was the first example of a chaotic, or strange, attractor.

Employing a digital computer to simulate his simple model, Lorenz elucidated the basic mechanism responsible for the randomness he observed: microscopic perturbations are amplified to affect macroscopic behavior. Two orbits with nearby initial conditions diverge exponentially fast and so stay close together for only a short time. The situation is qualitatively different for nonchaotic attractors. For these, nearby orbits stay close to one another, small errors remain bounded and the behavior is predictable.

The key to understanding chaotic behavior lies in understanding a simple stretching and folding operation, which takes place in the state space. Exponential divergence is a local feature: because attractors have finite size, two orbits on a chaotic attractor cannot diverge exponentially forever. Consequently the attractor must fold over onto itself. Although orbits diverge and follow increasingly different paths, they eventually must pass close to one another again. The orbits on a chaotic attractor are shuffled by this process, much as a deck of cards is shuffled by a dealer. The randomness of the chaotic orbits is the result of the shuffling process. The process of stretching and folding happens repeatedly, creating folds within folds ad infinitum. A chaotic attractor is, in other words, a fractal: an object that reveals more detail as it is increasingly magnified [*see illustration on page 53*].

Chaos mixes the orbits in state space in precisely the same way as a baker mixes bread dough by kneading it. One can imagine what happens to nearby trajectories on a chaotic attractor by placing a drop of blue food coloring in the dough. The kneading is a combination of two actions: rolling out the dough, in which the food coloring is spread out, and folding the dough over. At first the blob of food coloring simply gets longer, but eventually it is folded, and after considerable time the blob is stretched and refolded many

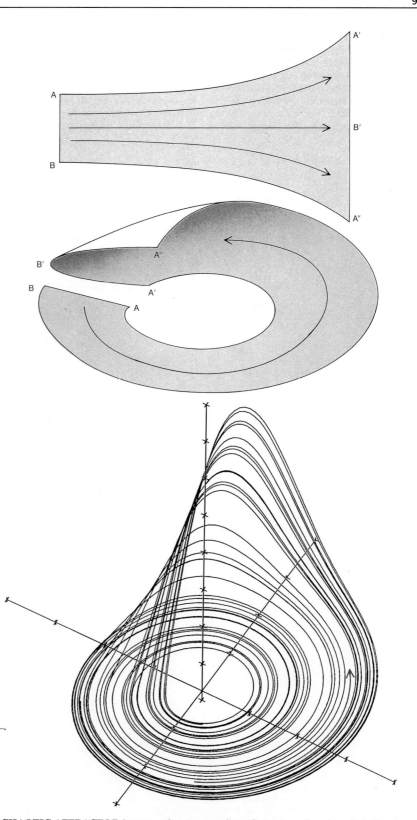

CHAOTIC ATTRACTOR has a much more complicated structure than a predictable attractor such as a point, a limit cycle or a torus. Observed at large scales, a chaotic attractor is not a smooth surface but one with folds in it. The illustration shows the steps in making a chaotic attractor for the simplest case: the Rössler attractor (*bottom*). First, nearby trajectories on the object must "stretch," or diverge, exponentially (*top*); here the distance between neighboring trajectories roughly doubles. Second, to keep the object compact, it must "fold" back onto itself (*middle*): the surface bends onto itself so that the two ends meet. The Rössler attractor has been observed in many systems, from fluid flows to chemical reactions, illustrating Einstein's maxim that nature prefers simple forms.

times. On close inspection the dough consists of many layers of alternating blue and white. After only 20 steps the initial blob has been stretched to more than a million times its original length, and its thickness has shrunk to the molecular level. The blue dye is thoroughly mixed with the dough. Chaos works the same way, except that instead of mixing dough it mixes the state space. Inspired by this picture of mixing, Otto E. Rössler of the University of Tübingen created the simplest example of a chaotic attractor in a flow [*see illustration on preceding page*].

When observations are made on a physical system, it is impossible to specify the state of the system exactly owing to the inevitable errors in measurement. Instead the state of the system is located not at a single point but rather within a small region of state space. Although quantum uncertainty sets the ultimate size of the region, in

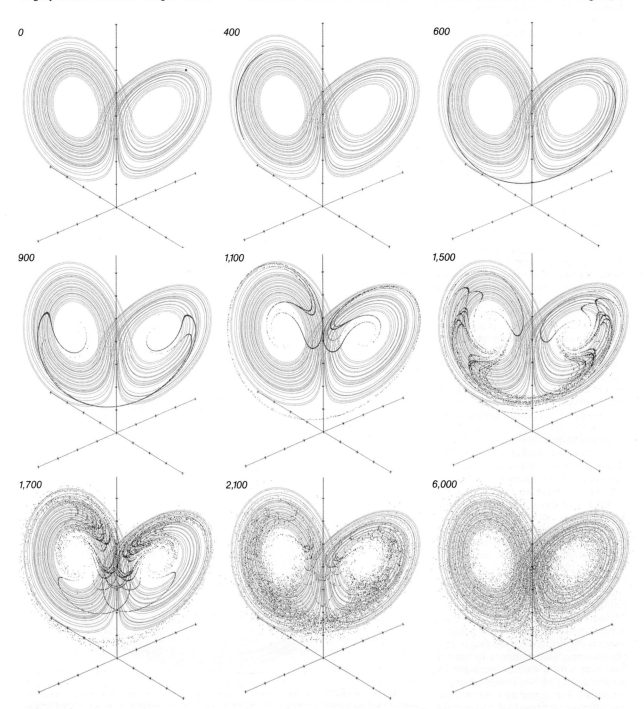

DIVERGENCE of nearby trajectories is the underlying reason chaos leads to unpredictability. A perfect measurement would correspond to a point in the state space, but any real measurement is inaccurate, generating a cloud of uncertainty. The true state might be anywhere inside the cloud. As shown here for the Lorenz attractor, the uncertainty of the initial measurement is represented by 10,000 red dots, initially so close together that they are indistinguishable. As each point moves under the action of the equations, the cloud is stretched into a long, thin thread, which then folds over onto itself many times, until the points are spread over the entire attractor. Prediction has now become impossible: the final state can be anywhere on the attractor. For a predictable attractor, in contrast, all the final states remain close together. The numbers above the illustrations are in units of 1/200 second.

practice different kinds of noise limit measurement precision by introducing substantially larger errors. The small region specified by a measurement is analogous to the blob of blue dye in the dough.

Locating the system in a small region of state space by carrying out a measurement yields a certain amount of information about the system. The more accurate the measurement is, the more knowledge an observer gains about the system's state. Conversely, the larger the region, the more uncertain the observer. Since nearby points in nonchaotic systems stay close as they evolve in time, a measurement provides a certain amount of information that is preserved with time. This is exactly the sense in which such systems are predictable: initial measurements contain information that can be used to predict future behavior. In other words, predictable dynamical systems are not particularly sensitive to measurement errors.

The stretching and folding operation of a chaotic attractor systematically removes the initial information and replaces it with new information: the stretch makes small-scale uncertainties larger, the fold brings widely separated trajectories together and erases large-scale information. Thus chaotic attractors act as a kind of pump bringing microscopic fluctuations up to a macroscopic expression. In this light it is clear that no exact solution, no short cut to tell the future, can exist. After a brief time interval the uncertainty specified by the initial measurement covers the entire attractor and all predictive power is lost: there is simply no causal connection between past and future.

Chaotic attractors function locally as noise amplifiers. A small fluctuation due perhaps to thermal noise will cause a large deflection in the orbit position soon afterward. But there is an important sense in which chaotic attractors differ from simple noise amplifiers. Because the stretching and folding operation is assumed to be repetitive and continuous, any tiny fluctuation will eventually dominate the motion, and the qualitative behavior is independent of noise level. Hence chaotic systems cannot directly be "quieted," by lowering the temperature, for example. Chaotic systems generate randomness on their own without the need for any external random inputs. Random behavior comes from more than just the amplification of errors and the loss of the ability to predict; it is due to the complex orbits generated by stretching and folding.

It should be noted that chaotic as

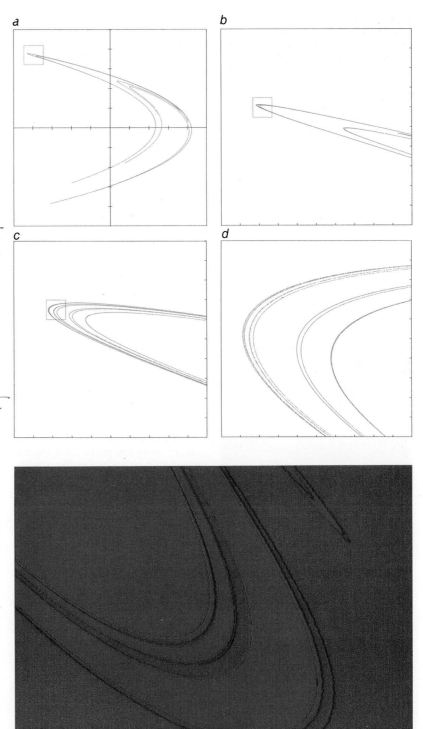

CHAOTIC ATTRACTORS are fractals: objects that reveal more detail as they are increasingly magnified. Chaos naturally produces fractals. As nearby trajectories expand they must eventually fold over close to one another for the motion to remain finite. This is repeated again and again, generating folds within folds, ad infinitum. As a result chaotic attractors have a beautiful microscopic structure. Michel Hénon of the Nice Observatory in France discovered a simple rule that stretches and folds the plane, moving each point to a new location. Starting from a single initial point, each successive point obtained by repeatedly applying Hénon's rule is plotted. The resulting geometric form (*a*) provides a simple example of a chaotic attractor. The small box is magnified by a factor of 10 in *b*. By repeating the process (*c*, *d*) the microscopic structure of the attractor is revealed in detail. The bottom illustration depicts another part of the Hénon attractor.

well as nonchaotic behavior can occur in dissipationless, energy-conserving systems. Here orbits do not relax onto an attractor but instead are confined to an energy surface. Dissipation is, however, important in many if not most real-world systems, and one can expect the concept of attractor to be generally useful.

Low-dimensional chaotic attractors open a new realm of dynamical systems theory, but the question remains of whether they are relevant to randomness observed in physical systems. The first experimental evidence supporting the hypothesis that chaotic attractors underlie random motion in fluid flow was rather indirect. The experiment was done in 1974 by Jerry P. Gollub of Haverford College and Harry L. Swinney of the University of Texas at Austin. The evidence was indirect because the investigators focused not on the attractor itself but rather on statistical properties characterizing the attractor.

The system they examined was a Couette cell, which consists of two concentric cylinders. The space between the cylinders is filled with a fluid, and one or both cylinders are rotated with a fixed angular velocity. As the angular velocity increases, the fluid shows progressively more complex flow patterns, with a complicated time dependence [*see illustration on this page*]. Gollub and Swinney essentially measured the velocity of the fluid at a given spot. As they increased the rotation rate, they observed transitions from a velocity that is constant in time to a periodically varying velocity and finally to an aperiodically varying velocity. The transition to aperiodic motion was the focus of the experiment.

The experiment was designed to distinguish between two theoretical pictures that predicted different scenarios for the behavior of the fluid as the rotation rate of the fluid was varied. The Landau picture of random fluid motion predicted that an ever higher number of independent fluid oscillations should be excited as the rotation rate is increased. The associated attractor would be a high-dimensional torus. The Landau picture had been challenged by David Ruelle of the Institut des Hautes Études Scientifiques near Paris and Floris Takens of the University of Groningen in the Netherlands. They gave mathematical arguments suggesting that the attractor associated with the Landau picture would not be likely to occur in fluid motion. Instead their results suggested that any possible high-dimensional tori might give way to a chaotic attractor, as originally postulated by Lorenz.

Gollub and Swinney found that for low rates of rotation the flow of the fluid did not change in time: the underlying attractor was a fixed point. As the rotation was increased the water began to oscillate with one independent frequency, corresponding to a limit-cycle attractor (a periodic orbit), and as the rotation was increased still further the oscillation took on two independent frequencies, corresponding to a two-dimensional torus attractor. Landau's theory predicted that as the rotation rate was further increased the pattern would continue: more distinct frequencies would gradually appear. Instead, at a critical rotation rate a continuous range of frequencies suddenly appeared. Such an observation was consistent with Lorenz' "deterministic nonperiodic flow," lending credence to his idea that chaotic attractors underlie fluid turbulence.

Although the analysis of Gollub and Swinney bolstered the notion that chaotic attractors might underlie some random motion in fluid flow, their work was by no means conclusive. One would like to explicitly demonstrate the existence in experimental data of a simple chaotic attractor. Typically, however, an experiment does not record all facets of a system but only a few. Gollub and Swinney could not record, for example, the entire Couette flow but only the fluid velocity at a single point. The task of the investigator is to "reconstruct" the attractor from the limited data. Clearly that cannot always be done; if the attractor is too complicated, something will be lost. In some cases, however, it is possible to reconstruct the dynamics on the basis of limited data.

A technique introduced by us and put on a firm mathematical foundation by Takens made it possible to reconstruct a state space and look for chaotic attractors. The basic idea is that the evolution of any single component of a system is determined by the other components with which it interacts. Information about the relevant components is thus implicitly contained in the history of any single component. To reconstruct an "equivalent" state space, one simply looks at a single component and treats the measured values at fixed time delays (one second ago, two seconds ago and so on, for example) as though they were new dimensions.

The delayed values can be viewed as new coordinates, defining a single point in a multidimensional state space. Repeating the procedure and taking delays relative to different times generates many such points. One can then use other techniques to test whether or not these points lie on a

a
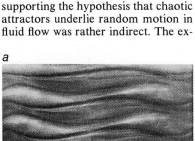

b

c

d

EXPERIMENTAL EVIDENCE supports the hypothesis that chaotic attractors underlie some kinds of random motion in fluid flow. Shown here are successive pictures of water in a Couette cell, which consists of two nested cylinders. The space between the cylinders is filled with water and the inner cylinder is rotated with a certain angular velocity (*a*). As the angular velocity is increased, the fluid shows a progressively more complex flow pattern (*b*), which becomes irregular (*c*) and then chaotic (*d*).

10.1 Chaos

chaotic attractor. Although this representation is in many respects arbitrary, it turns out that the important properties of an attractor are preserved by it and do not depend on the details of how the reconstruction is done.

The example we shall use to illustrate the technique has the advantage of being familiar and accessible to nearly everyone. Most people are aware of the periodic pattern of drops emerging from a dripping faucet. The time between successive drops can be quite regular, and more than one insomniac has been kept awake waiting for the next drop to fall. Less familiar is the behavior of a faucet at a somewhat higher flow rate. One can often find a regime where the drops, while still falling separately, fall in a never repeating patter, like an infinitely inventive drummer. (This is an experiment easily carried out personally; the faucets without the little screens work best.) The changes between periodic and random-seeming patterns are reminiscent of the transition between laminar and turbulent fluid flow. Could a simple chaotic attractor underlie this randomness?

The experimental study of a dripping faucet was done at the University of California at Santa Cruz by one of

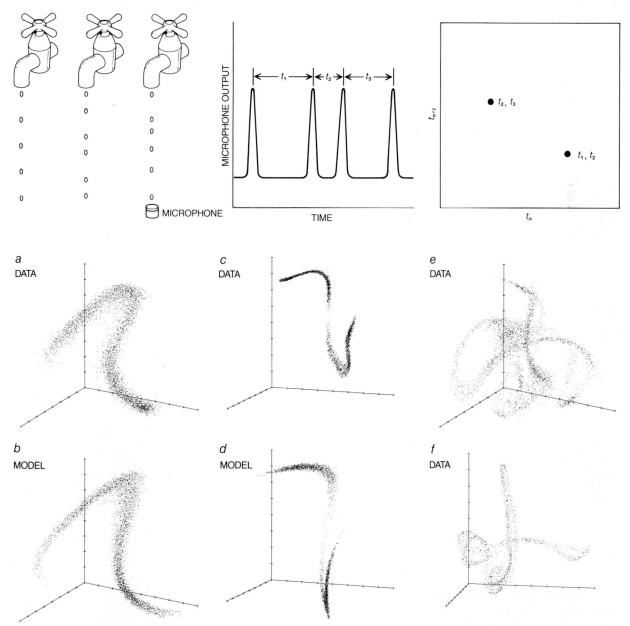

DRIPPING FAUCET is an example of a common system that can undergo a chaotic transition. The underlying attractor is reconstructed by plotting the time intervals between successive drops in pairs, as is shown at the top of the illustration. Attractors reconstructed from an actual dripping faucet (*a, c*) compare favorably with attractors generated by following variants of Hénon's rule (*b, d*). (The entire Hénon attractor is shown on page 53.) Illustrations *e* and *f* were reconstructed from high rates of water flow and presumably represent the cross sections of hitherto unseen chaotic attractors. Time-delay coordinates were employed in each of the plots. The horizontal coordinate is t_n, the time interval between drop *n* and drop $n-1$. The vertical coordinate is the next time interval, t_{n+1}, and the third coordinate, visualized as coming out of the page, is t_{n+2}. Each point is thus determined by a triplex of numbers (t_n, t_{n+1}, t_{n+2}) that have been plotted for a set of 4,094 data samples. Simulated noise was added to illustrations *b* and *d*.

us (Shaw) in collaboration with Peter L. Scott, Stephen C. Pope and Philip J. Martein. The first form of the experiment consisted in allowing the drops from an ordinary faucet to fall on a microphone and measuring the time intervals between the resulting sound pulses. Typical results from a somewhat more refined experiment are shown on the preceding page. By plotting the time intervals between drops in pairs, one effectively takes a cross section of the underlying attractor. In the periodic regime, for example, the meniscus where the drops are detaching is moving in a smooth, repetitive manner, which could be represented by a limit cycle in the state space. But this smooth motion is inaccessible in the actual experiment; all that is recorded is the time intervals between the breaking off of the individual drops. This is like applying a stroboscopic light to regular motion around a loop. If the timing is right, one sees only a fixed point.

The exciting result of the experiment was that chaotic attractors were indeed found in the nonperiodic regime of the dripping faucet. It could have been the case that the randomness of the drops was due to unseen influences, such as small vibrations or air currents. If that was so, there would be no particular relation between one interval and the next, and the plot of the data taken in pairs would have shown only a featureless blob. The fact that any structure at all appears in the plots shows the randomness has a deterministic underpinning. In particular, many data sets show the horseshoelike shape that is the signature of the simple stretching and folding process discussed above. The characteristic shape can be thought of as a "snapshot" of a fold in progress, for example, a cross section partway around the Rössler attractor shown on page 51. Other data sets seem more complicated; these may be cross sections of higher-dimensional attractors. The geometry of attractors above three dimensions is almost completely unknown at this time.

If a system is chaotic, how chaotic is it? A measure of chaos is the "entropy" of the motion, which roughly speaking is the average rate of stretching and folding, or the average rate at which information is produced. Another statistic is the "dimension" of the attractor. If a system is simple, its behavior should be described by a low-dimensional attractor in the state space, such as the examples given in this article. Several numbers may be required to specify the state of a more complicated system, and its corresponding attractor would therefore be higher-dimensional.

The technique of reconstruction, combined with measurements of entropy and dimension, makes it possible to reexamine the fluid flow originally studied by Gollub and Swinney. This was done by members of Swinney's group in collaboration with two of us (Crutchfield and Farmer). The reconstruction technique enabled us to make images of the underlying attractor. The images do not give the striking demonstration of a low-dimensional attractor that studies of other systems, such as the dripping faucet, do. Measurements of the entropy and dimension reveal, however, that irregular fluid motion near the transition in Couette flow can be described by chaotic attractors. As the rotation rate of the Couette cell increases so do the entropy and dimension of the underlying attractors.

In the past few years a growing number of systems have been shown to exhibit randomness due to a simple chaotic attractor. Among them are the convection pattern of fluid heated in a small box, oscillating concentration levels in a stirred-chemical reaction, the beating of chicken-heart cells and a large number of electrical and mechanical oscillators. In addition computer models of phenomena ranging from epidemics to the electrical activity of a nerve cell to stellar oscillations have been shown to possess this simple type of randomness. There are even experiments now under way that are searching for chaos in areas as disparate as brain waves and economics.

It should be emphasized, however, that chaos theory is far from a panacea. Many degrees of freedom can also make for complicated motions that are effectively random. Even though a given system may be known to be chaotic, the fact alone does not reveal very much. A good example is molecules bouncing off one another in a gas. Although such a system is known to be chaotic, that in itself does not make prediction of its behavior easier. So many particles are involved that all that can be hoped for is a statistical description, and the essential statistical properties can be derived without taking chaos into account.

There are other uncharted questions for which the role of chaos is unknown. What of constantly changing patterns that are spatially extended, such as the dunes of the Sahara and fully developed turbulence? It is not clear whether complex spatial patterns can be usefully described by a single attractor in a single state space. Perhaps, though, experience with the simplest attractors can serve as a guide to a more advanced picture, which may involve entire assemblages of spatially mobile deterministic forms akin to chaotic attractors.

The existence of chaos affects the scientific method itself. The classic approach to verifying a theory is to make predictions and test them against experimental data. If the phenomena are chaotic, however, long-term predictions are intrinsically impossible. This has to be taken into account in judging the merits of the theory. The process of verifying a theory thus becomes a much more delicate operation, relying on statistical and geometric properties rather than on detailed prediction.

Chaos brings a new challenge to the reductionist view that a system can be understood by breaking it down and studying each piece. This view has been prevalent in science in part because there are so many systems for which the behavior of the whole is indeed the sum of its parts. Chaos demonstrates, however, that a system can have complicated behavior that emerges as a consequence of simple, nonlinear interaction of only a few components.

The problem is becoming acute in a wide range of scientific disciplines, from describing microscopic physics to modeling macroscopic behavior of biological organisms. The ability to obtain detailed knowledge of a system's structure has undergone a tremendous advance in recent years, but the ability to integrate this knowledge has been stymied by the lack of a proper conceptual framework within which to describe qualitative behavior. For example, even with a complete map of the nervous system of a simple organism, such as the nematode studied by Sidney Brenner of the University of Cambridge, the organism's behavior cannot be deduced. Similarly, the hope that physics could be complete with an increasingly detailed understanding of fundamental physical forces and constituents is unfounded. The interaction of components on one scale can lead to complex global behavior on a larger scale that in general cannot be deduced from knowledge of the individual components.

Chaos is often seen in terms of the limitations it implies, such as lack of predictability. Nature may, however, employ chaos constructively. Through amplification of small fluctuations it can provide natural systems with access to novelty. A prey escaping a predator's attack could use chaotic

10.1 Chaos

flight control as an element of surprise to evade capture. Biological evolution demands genetic variability; chaos provides a means of structuring random changes, thereby providing the possibility of putting variability under evolutionary control.

Even the process of intellectual progress relies on the injection of new ideas and on new ways of connecting old ideas. Innate creativity may have an underlying chaotic process that selectively amplifies small fluctuations and molds them into macroscopic coherent mental states that are experienced as thoughts. In some cases the thoughts may be decisions, or what are perceived to be the exercise of will. In this light, chaos provides a mechanism that allows for free will within a world governed by deterministic laws.

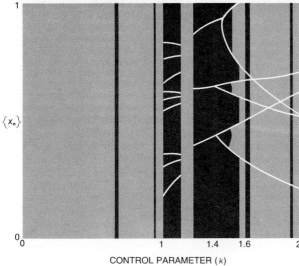

TRANSITION TO CHAOS is depicted schematically by means of a bifurcation diagram: a plot of a family of attractors (*vertical axis*) versus a control parameter (*horizontal axis*). The diagram was generated by a simple dynamical system that maps one number to another. The dynamical system used here is called a circle map, which is specified by the iterative equation $x_{n+1} = \omega + x_n + k/2\pi \cdot \sin(2\pi x_n)$. For each chosen value of the control parameter k a computer plotted the corresponding attractor. The colors encode the probability of finding points on the attractors: red corresponds to regions that are visited frequently, green to regions that are visited less frequently and blue to regions that are rarely visited. As k is increased from 0 to 2 (*see drawing at left*), the diagram shows two paths to chaos: a quasi-periodic route (from $k = 0$ to $k = 1$, which corresponds to the green region above) and a "period doubling" route (from $k = 1.4$ to $k = 2$). The quasi-periodic route is mathematically equivalent to a path that passes through a torus attractor. In the period-doubling route, which is based on the limit-cycle attractor, branches appear in pairs, following the geometric series 2, 4, 8, 16, 32 and so on. The iterates oscillate among the pairs of branches. (At a particular value of k—1.6, for instance—the iterates visit only two values.) Ultimately the branch structure becomes so fine that a continuous band structure emerges: a threshold is reached beyond which chaos appears.

Chaos in a dripping faucet

H N Núñez Yépez[†‡], A L Salas Brito[†‡], C A Vargas[†‡] and L A Vicente[§]

† Departamento de Física, Facultad de Ciencias, Universidad Nacional Autónoma de México, Apartado Postal 21-726, México 04000 D F, Mexico
‡ Departamento de Ciencias Básicas, Universidad Autónoma Metropolitana—Azcapotzalco, México D F, Mexico.
§ Facultad de Química, Universidad Nacional Autónoma de México, México D F, Mexico

Received 6 April 1988, in final form 10 August 1988

Abstract. An advanced undergraduate experiment on the chaotic behaviour of a dripping faucet is presented. The experiment can be used for the demonstration of typical features of chaotic phenomena and also allows the advanced physics student to learn about the use of microcomputers as data-taking devices. For convenience a brief introduction to the basic concepts of non-linear dynamics and to the period-doubling route to chaos are included.

Resumen. Se propone un experimento sobre el comportamiento caótico del goteo en una llave mal cerrada. Este resulta útil para la demostración de características típicas de los fenómenos caóticos, y permite que los estudiantes de física aprendan a usar una microcomputadora para la toma de datos en un experimento. Hemos creído conveniente incluir una breve introducción a la dinámica no lineal y, en particular, a la aparición de caos por sucesivas bifurcaciones subarmónicas.

1. Introduction

Chaodynamics is a recent area of research (Ott 1981, Ford 1983, Bai Lin 1984, Jensen 1987); even its name is recent (Andrey 1986). It concerns the occurrence of complex and seemingly random phenomena in non-linear but otherwise deterministic systems. Common examples of this behaviour include the results of tossing a coin, or the swirling paths of leaves falling from a tree on a windy day. Similar aperiodic phenomena have been observed in an impressive number of experimental systems, even in some previously thought to be very well understood, as is the case of the driven pendulum (Koch et al 1983). Electrical, optical, mechanical, chemical, hydrodynamical and biological systems can all exhibit the kind of dynamical instabilities that produce chaotic behaviour (Jensen 1987 and references therein). Despite this, recent discoveries in the field of non-linear dynamics are still not well known to many undergraduate physics students.

With the above ideas in mind, we have developed an experiment that can be useful for introducing some of the ideas and methods used in the description of non-linear chaotic systems. Our experiment follows the work of Martien et al (1985), based on a suggestion of Rössler (1977), which shows that drops falling from a leaky faucet behave chaotically under appropriate conditions. Other experiments, demonstrations or computer simulations have been recently proposed to introduce students to the field of non-linear phenomena (e.g. Berry 1981, Viet et al 1983, Salas Brito and Vargas 1986, Briggs 1987), but curiously none of them deals with liquids despite the fact that much original work has been done on such systems. In our experiment the students investigate the dripping behaviour of a leaky faucet, a system which remains incompletely understood and hence may still offer some surprises to both teachers and students. In this system, the students can measure the time interval between successive drops, the drip interval—as we, following Martien et al (1985), will call it—as a function of the flow rate of water.

The students become acquainted with the concepts of non-linear dynamics (as deterministic chaos, attractors, subharmonic bifurcations, and the like) by reading the basic literature, paying particular attention to the logistic map (May 1976, Feigenbaum 1980, Hofstadter 1981, Schuster 1984, Jensen 1987). Then, since many aspects of this mapping are common to a large class of dynamical systems showing chaotic behaviour, they are encouraged to explore it on a microcomputer to obtain firsthand experience of the behaviour of a chaotic system, before they begin the experiment.

In the following we summarise the experimental set-up and show the results obtained so far in our laboratories. Since the dripping faucet seems to follow the period-doubling route to chaos (Martien *et al* 1985), after a brief introduction to illustrate the basic concepts of the field, in § 2 we examine in some detail the logistic map, a paradigmatic example of a system following such a route to chaos. In § 3 we describe our experimental device and show the return maps obtained from the data collected. These data confirm the existence of a sequence of period doublings in the system, at least up to period four, before the onset of chaos. Finally, we present our conclusions in § 4.

2. Basic concepts and the period-doubling route to chaos

Let us first introduce some basic notions and terminology of non-linear dynamics. Consider an harmonically driven pendulum: given the frequency and strength of the driving force, the motion of the system is completely determined if the angle θ and angular speed $\dot\theta$ of the pendulum are known. These variables can be used as coordinates in the phase space of the pendulum; as it swings back and forth, the point representing its state moves along an orbit in phase space. For example, if the strength of the driving force vanishes, due to the effect of friction, no matter how we start its motion the pendulum will come to rest at its point of stable equilibrium after a number of oscillations. From the point of view of phase space, the orbit spirals to the fixed point at the origin. The motion is quite different for non-zero values of the driving force; in this case the pendulum settles to a stationary oscillation with the same frequency as the external driving force. These stationary motions in which the system settles after the transients have died out are examples of *attractors*, a term which conveys the idea that many nearby orbits are 'attracted' to them. We have mentioned two types of attractors, a stable fixed point and a stable limit cycle, but there exists a more complicated attractor, the so-called *strange attractors* which only occur in dissipative non-linear systems. They capture the solution of a deterministic system into a perfectly defined region of phase space, but in which there is a very complex structure (these objects are usually fractals) and the motion shows every feature associated with random motion. Such behaviour is a manifestation of the very sensitive dependence on the initial conditions developed by the system (Ruelle 1980). The existence of strange attractors is one of the fingerprints of *chaos* i.e. the loss of long-term predictability in a supposedly deterministic system.

Various attractors may be present in the long-term behaviour of a dynamical system; in most, its presence or absence is governed by the value of a single *control parameter*. For example, the magnitude the driving force determines if the pendulum settles to a point or to a limit cycle (or possibly even to a more complex attractor (D'Humieres *et al* 1982)). In the case of the dripping faucet, it is the flow rate of water which governs its dynamics: for low values of flow, the dripping is simply periodic and the system is attracted to a stable fixed point; but for much larger values of flow, strange attractors can appear. The succession of stationary states which a system follows prior of the onset of chaos, as the control parameter is varied, determines what is called *the route to chaos* followed by the system (Kadanoff 1983). The dripping faucet seems to follow the period-doubling route to chaos (Martien *et al* 1985). We will explain this route in some detail below.

The evolution of a dynamical system can be described in either continuous time (a flow) or in discrete time (a mapping). The pendulum is a good example of a system that may be described by a flow in phase space—although it can also be described by a mapping (Testa *et al* 1982). On the other hand, the sequence of drip intervals in a leaky faucet is naturally described by a discrete map. For any given value of the dripping rate, a plot of the next drip interval versus the previous one can give a clear idea of its dripping behaviour and of the possible existence of attractors. This is the representation we use for the data obtained in the experiment (see figure 5); it is called a *return* or *Poincaré map*.

As an illustration of some of these ideas, and because they offer perhaps the simplest examples of systems undergoing a period-doubling transition to chaos, we shall consider iterative processes of the form

$$x_{n+1} = f(x_n) \qquad (1)$$

where $f(x)$ is a continuous function defined in a suitable one-dimensional interval. Discussing only one-dimensional mappings as (1) is not as restrictive as it may seem at first, since it can be viewed as a discrete time version of a continuous but dissipative dynamical system. The dissipative terms shrink the volume of phase space occupied by the system until it becomes effectively one-dimensional. In this instance it can be modelled, at least in its universal qualitative features, by a simple mapping as (1) (Collet and Eckmann 1980). In fact, such iterations have been advocated frequently as qualitative models for many complex physical systems, from the behaviour of a driven non-linear oscillator (Linsay 1981, Testa *et al* 1982) to the onset of turbulence in the Rayleigh–Benard phenomena (Gollub and Benson 1980). Most of the results here do not

Chaos in a dripping faucet

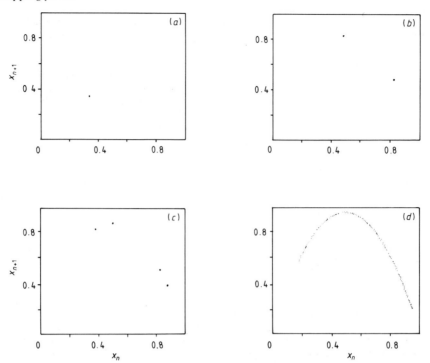

Figure 1. Return maps, i.e. plots of x_{n+1} versus x_n for large values of n, obtained from the logistic map for different values of μ: (a) $\mu = 1.5$; (b) $\mu = 3.3$; (c) $\mu = 3.5$; (d) $\mu = 3.8$. This illustrates the dynamics of the map up to the four cycle as well as the chaotic attractor for $\mu > \mu_\infty$.

depend on the precise form of the function $f(x)$, as long as it has a single quadratic maximum but to be specific we will analyse the dynamics of the logistic map. This mapping is defined by

$$f(x) = \mu x(1-x) \qquad (2)$$

where $0 \leq \mu \leq 4$ is a parameter measuring the strength of the non-linearity. With this choice for $f(x)$, equation (1) describes a non-linear and non-invertible map of the unit interval on itself. The evolution of the sequence of x_n generated by this simple equation exhibits a transformation from periodic to chaotic behaviour as the control parameter μ is increased. Let us see how this occurs. The behaviour of the sequence of iterates is trivial when $\mu = 0$: for every initial value x_0 all the iterates are zero. We can say then that the solution quickly reaches an attractor, the single point $x = 0$; this is called a period-one cycle, orbit or attractor. For values of μ between 0 and 1, the large n behaviour of the x_n is identical; they approach the point $x = 0$ after a certain number of steps. But for larger values of μ the dynamics is much more interesting as can be easily verified using a hand-held calculator. Various types of stationary solutions of the logistic map are exemplified by figures 1 and 2.

Figure 1 shows return maps (plots of x_{n+1} versus x_n) for $\mu = 1.5$, 3.3, 3.5 and 3.8. The successive appearance of attractors of period one, two, four and of a one-dimensional chaotic attractor can be appreciated in these plots. Figure 2 illustrates this kind of behaviour in a different and more global way; it shows a plot of the large n behaviour of the iterates (i.e. the attractors) of the logistic map as a function of the value of μ. This graph gives a 'pictorial meaning' to the way the onset of chaos occurs via a sequence of 'pitchfork' (period-doubling) bifurcations as the value of μ changes. It also shows the critical dependence of the behaviour with the value of this parameter. For values of μ between 1 and 3, and almost all initial values x_0, there is a single point attractor (figure 1(a)). Then, as μ is increased between 3 and 4, the dynamics changes in surprising ways. First, for $3 < \mu \leq (1+\sqrt{6})$ the stationary solution bifurcates to a period-two attractor—the period of the solution has doubled and its frequency halved, hence the names of period-doubling or subharmonic bifurcation given to the phenomena—as can be seen in the bifurcation diagram (figure 2), where the solution hops back and forth between the upper and lower branches of the pitchfork, and in figure 1(b). As μ is increased further, the solution bifurcates again to a period-four attractor, then to a period-eight attractor and so on. This sequence (or cascade) of bifurcations continues indefinitely, but the interval of values of μ in which a given periodic orbit acts as an attractor shrinks very quickly at a rate governed by the universal parameter

$$\delta = \lim_{n \to \infty} \frac{\mu_n - \mu_{n-1}}{\mu_{n+1} - \mu_n} = 4.6692\ldots \quad (3)$$

until a critical value $\mu_\infty = 3.5699\ldots$ is reached (Feigenbaum 1978, 1979). This value marks the beginning of the aperiodic regime: the iterates seem to wander erratically around a subset of the unit interval. If we increase μ further, windows of periodic motion of every integer period reappear. Chaotic or periodic motion can be found for suitable values of $\mu > \mu_\infty$. A complete discussion of the properties of the logistic map can be found in the account given by Feigenbaum (1983). For a more complete discussion of the period doubling as well as other possible routes to chaos in a dynamical system see Kadanoff (1983).

As with many other properties discovered in systems making a period-doubling transition to chaos, the constant δ is universal in the sense that it is found to be valid for a large number of systems and not only for the logistic map. For example, if the dripping faucet effectively follows the period-doubling route to chaos and we were able to calculate δ, we should find a numerical value very close to that given in (3). Now, obviously, not every feature of the logistic map is shared by other systems, for example, the values quoted above for the onset of instabilities in the attractors are not universal—they are specific for the logistic map.

3. The dripping faucet experiment

The apparatus used in the experiment is rather simple and widely available. We use a Commodore 64 microcomputer for data acquisition and subsequent analysis and display. The inclusion of an automatic data-taking procedure is fundamental in an experiment which requires the taking of 2000 data points every time it is run. In fact, this represents an additional advantage, for it allows the students to learn simple interfacing techniques and to work with a microcomputer-assisted experiment.

The basic apparatus is shown schematically in figure 3. It consists of a large reservoir of water (a large Mariotte bottle) kept at a constant pressure with the help of a float valve. The water can flow through a valve to a plastic tube with a nozzle at the end. This valve, as well as the float valve, were obtained from a used automobile carburetor. With its help we can control the dripping rate, which is the control parameter in our experiment. Drops falling from the nozzle pass through an optocoupler (General Electric H23L1, with a Schmidt trigger included at the output) which produces a TTL pulse for each drop. The pulses are sent, via a very simple interface (figure 4), to the user port of the Commodore 64 microcomputer. The computer is used to store the data, to compute the drip interval

Figure 2. A section of the bifurcation diagram of the logistic map. The graph shows the asymptotic behaviour of x_n for values of μ between 2.94 and 4.

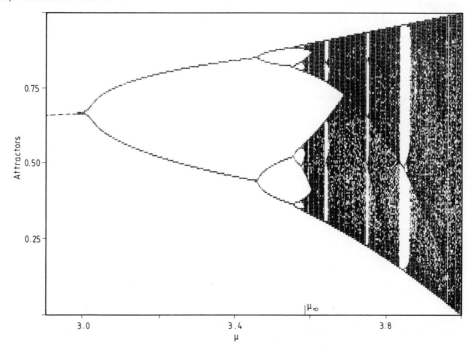

Chaos in a dripping faucet

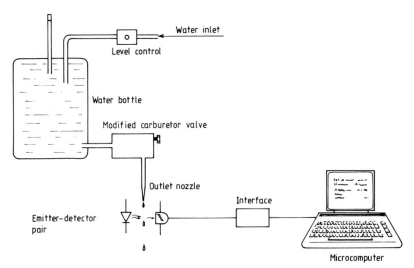

Figure 3. Schematic diagram of the experimental set-up. We use a float valve (marked 'level control' in the diagram) to maintain the water level in an upper reservoir (not shown).

and to display the return maps obtained. With this arrangement students are able to take, store and analyse up to 3072 drips (using 6 Kbyte of memory).

The machine-language subroutine used for acquiring the data and measuring the drip interval T_n is capable of taking data up to a rate of 1.2 kHz, far above the dripping rates occurring in the experiment, and has an estimated resolution of $50 \,\mu s$. This estimation has been tested with good results with the help of a signal generator (Wavetek 181) used as the input of our data-taking device.

The flow rate is controlled by means of the carburetor valve, but we do not measure it directly, preferring instead to use the valve setting as an indicator. The program we use to analyse the data computes a mean dripping rate. The mean dripping rates students are able to investigate under experimental conditions vary from 0.1 to 40 drips/s, a rate at which the drops become a continuous stream of water. In this interval, the system moves from a stable period-one attractor and undergoes period doublings until strange attractors appear for dripping rates greater than 7 drips/s. At such large dripping rates the behaviour is irregular and, surely, is very complex (figures 5 and 6). In fact, much to our surprise the dynamics of the system is very rich

Figure 4. The interface is a single 74LS00 chip. The connections to the microcomputer user's port are shown.

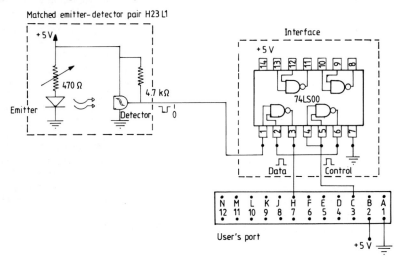

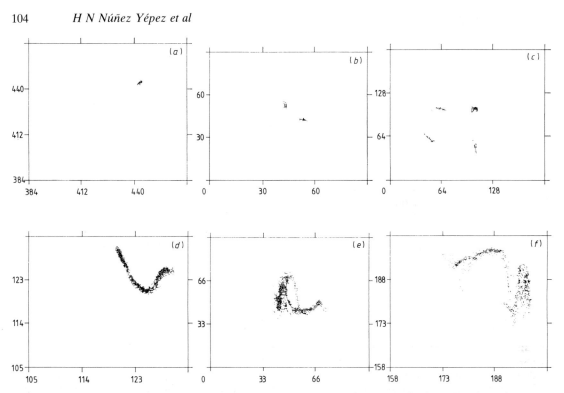

Figure 5. Example of the experimental results shown as T_{n+1} (vertical axes) versus T_n (horizontal axes) graphs redrawn from the printout of our data. Periodic behaviour, (a)–(c); complex chaotic behaviour, (d)–(f). All values of time are in milliseconds.

and shows patterns not discussed in Martien et al. All of this has generated a great deal of interest among our students.

Typical experimental results are shown as T_{n+1} versus T_n plots in figures 5 and 6 (notice the qualitative similarity of figures 5(a)–(c) with figures 1(a)–(c). These are plots of the 2000 typical points taken each time the experiment is run. The beginning of a period-doubling sequence can be appreciated; the dripping behaviour shows attractors of period one, two and four prior to the chaotic regime. With the current experimental arrangement it is not possible to ascertain precisely the ranges of stability of the attractors but, roughly, the students have found the periodic attractors to be present up to 7 drips/s. For greater dripping rates we observe chaotic behaviour, signalled by what seem to be strange attractors; typical examples are shown in figures 5(d)–(f) and in figure 6. This last attractor has been singled out because it illustrates the folding, stretching and fractioning that occur in the attractors in the process of becoming more complex, as a result of increasing the dripping rate. We have not been able to see periodic attractors of period larger than four, due perhaps to the inherent noise in the system or to the somewhat poor control of dripping rates allowed by the carburetor valve. But, occasionally, students were able to observe cycles of period three immersed in the chaotic regime. As these observations are very sensitive to the valve setting and to vibrations produced near the apparatus, we have been unable to reproduce them at will with the current experimental arrangement.

The result of the experiment has been taken as an indication of a period-doubling route to chaos in the system, but to be conclusive further evaluation is needed. For example, it may require the computation of universal parameters like δ. But before we

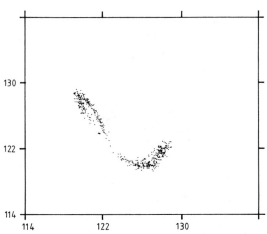

Figure 6. Another example of an attractor in the chaotic region. Note the folding and separation developed as it becomes a more complex attractor. Axes and units as for figure 5.

can determine such parameters we must be able to measure with greater confidence the stability intervals of the attractors and to discern at least a period-eight attractor.

As figures 5 and 6 show, the aperiodic regime exhibits patterns of behaviour which seem to have an underlying one-dimensional structure somewhat blurred by the noise in the system. This quasi-one-dimensional appearance of the attractors is an indicator of chaotic behaviour as a characteristic of the system and not a result of external noise generated, for example, by the carburetor valve or produced by small air currents. On the other hand, these results show that a qualitative model in terms of a one-dimensional mapping may be appropriate. In fact, to analyse the results of their experiment Martien *et al* proposed a very simple one-dimensional analogue model. It is worth mentioning here that the system exhibits hysteresis and the bifurcation points may differ for increasing and decreasing dripping rates. Despite the fact that the system is expected to show hysteresis, we believe our observations to be due mainly to the valve used to control the flow of water. We are now trying to improve the arrangement and to use a good needle valve in order to determine this.

4. Conclusions

In summary, we have presented an experiment in which students can investigate the non-linear behaviour and the route to chaos in a dripping faucet. Students are able to observe a sequence of period doublings preceding chaos and the existence of a chaotic regime with various types of strange attractors. They can also convince themselves that despite the large number of variables involved in the phenomenon it can be qualitatively modelled by a one-dimensional mapping (although we may expect better agreement with a mapping of greater dimensionality).

In view of the above results, and to the relative simplicity of the experimental arrangement, we think that this system is very suitable for introducing the concept of non-linear dynamics and the techniques for its experimental study. The experiment is a very good example of the type of behaviour possible in classical dynamic systems. Our experimental set-up can also be useful as an exhibit or to inform conferences addressing wider audiences. On the other hand, when used in an open-ended investigation, it has allowed our students to explore the many features of the transition to chaos at their own level and interest. Another useful feature of the experiment is that it allows advanced physics students to learn about simple interfacing techniques and the use of microcomputers as data-taking devices in physics experiments.

Finally, we must say that a similar experiment is being developed at Universidad Simón Bolívar (Venezuela) by Professor C L Ladera at the suggestion of one of us (ALSB).

Acknowledgments

We wish to thank C Carbajal and J Sandria for developing the machine-language subroutine used by the data-taking procedure and for their help in setting up the experiment. We also wish to thank F D Micha for her help in revising the manuscript.

References

Andrey L 1986 *Prog. Theor. Phys.* **75** 1258
Bai Lin H 1984 *Chaos* (Singapore: World Scientific)
Berry M V 1981 *Eur. J. Phys.* **2** 91
Briggs K 1987 *Am. J. Phys.* **55** 1083
Collet P and Eckmann J P 1980 *Iterated Maps of the Interval as Dynamical Systems* (Boston: Birkhäuser)
D'Humieres D, Beasley M R, Huberman B A and Libchaber A 1982 *Phys. Rev.* A **26** 3483
Feigenbaum M J 1978 *J. Stat. Phys.* **19** 25
——1979 *J. Stat. Phys.* **21** 669
——1980 *Los Alamos Sci.* **1** 4
——1983 *Physica* **7D** 16
Ford J 1983 *Phys. Today* **36** (4) 40
Gollub J P and Benson S V 1980 *J. Fluid Mech.* **100** 449
Hofstadter D R 1981 *Sci. Am.* **245** 22 (November)
Jensen R V 1987 *Am. Sci.* **75** 168
Kadanoff L P 1983 *Phys. Today* **36** (12) 46
Koch B P, Leven R W, Pompe B and Wilke G 1983 *Phys. Lett* **96A** 219
Linsay R S 1981 *Phys. Rev. Lett.* **47** 1349
Martien P, Pope S C, Scott P L and Shaw R S 1985 *Phys. Lett.* **110A** 399
May R M 1976 *Nature* **261** 459
Ott E 1981 *Rev. Mod. Phys.* **53** 635
Rössler O 1977 in *Synergetics: a workshop* ed. H Haken (Berlin: Springer) p 174
Ruelle D 1980 *La Recherche* **11** 133
Salas Brito A L and Vargas C 1986 *Rev. Mex. Fis.* **32** 357
Schuster H 1984 *Deterministic Chaos: An Introduction* (Weinhein: Physik Verlag)
Testa J, Pérez J and Jeffries C 1982 *Phys. Rev. Lett.* **48** 714
Viet O, Wesfreid J E and Guyon E 1983 *Eur. J. Phys.* **4** 72

Chaos, Strange Attractors, and Fractal Basin Boundaries in Nonlinear Dynamics

CELSO GREBOGI, EDWARD OTT, JAMES A. YORKE

Recently research has shown that many simple nonlinear deterministic systems can behave in an apparently unpredictable and chaotic manner. This realization has broad implications for many fields of science. Basic developments in the field of chaotic dynamics of dissipative systems are reviewed in this article. Topics covered include strange attractors, how chaos comes about with variation of a system parameter, universality, fractal basin boundaries and their effect on predictability, and applications to physical systems.

I N THIS ARTICLE WE PRESENT A REVIEW OF THE FIELD OF chaotic dynamics of dissipative systems including recent developments. The existence of chaotic dynamics has been discussed in the mathematical literature for many decades with important contributions by Poincaré, Birkhoff, Cartwright and Littlewood, Levinson, Smale, and Kolmogorov and his students, among others. Nevertheless, it is only recently that the wide-ranging impact of chaos has been recognized. Consequently, the field is now undergoing explosive growth, and many applications have been made across a broad spectrum of scientific disciplines—ecology, economics, physics, chemistry, engineering, fluid mechanics, to name several. Specific examples of chaotic time dependence include convection of a fluid heated from below, simple models for the yearly variation of insect populations, stirred chemical reactor systems, and the determination of limits on the length of reliable weather forecasting. It is our belief that the number of these applications will continue to grow.

We start with some basic definitions of terms used in the rest of the article.

Dissipative system. In Hamiltonian (conservative) systems such as arise in Newtonian mechanics of particles (without friction), phase space volumes are preserved by the time evolution. (The phase space is the space of variables that specify the state of the system.) Consider, for example, a two-dimensional phase space (q, p), where q denotes a position variable and p a momentum variable. Hamilton's equations of motion take the set of initial conditions at time $t = t_0$ and evolve them in time to the set at time $t = t_1$. Although the shapes of the sets are different, their areas are the same. By a dissipative system we mean one that does not have this property (and cannot be made to have this property by a change of variables). Areas should typically decrease (dissipate) in time so that the area of the final set would be less than the area of the initial set. As a consequence of this, dissipative systems typically are characterized by the presence of attractors.

Attractor. If one considers a system and its phase space, then the initial conditions may be attracted to some subset of the phase space (the attractor) as time $t \to \infty$. For example, for a damped harmonic oscillator (Fig. 1a) the attractor is the point at rest (in this case the origin). For a periodically driven oscillator in its limit cycle the limit set is a closed curve in the phase space (Fig. 1b).

Strange attractor. In the above two examples, the attractors were a point (Fig. 1a), which is a set of dimension zero, and a closed curve (Fig. 1b), which is a set of dimension one. For many other attractors the attracting set can be much more irregular (some would say pathological) and, in fact, can have a dimension that is not an integer. Such sets have been called "fractal" and, when they are attractors, they are called strange attractors. [For a more precise definition see (*1*).] The existence of a strange attractor in a physically interesting model was first demonstrated by Lorenz (*2*).

Dimension. There are many definitions of the dimension d (*3*). The simplest is called the box-counting or capacity dimension and is defined as follows:

$$d = \lim_{\epsilon \to 0} \frac{\ln N(\epsilon)}{\ln(1/\epsilon)} \quad (1)$$

where we imagine the attracting set in the phase space to be covered by small D-dimensional cubes of edge length ϵ, with D the dimension of the phase space. $N(\epsilon)$ is the minimum number of such cubes needed to cover the set. For example, for a point attractor (Fig. 1a), $N(\epsilon) = 1$ independent of ϵ, and Eq. 1 yields $d = 0$ (as it should). For a limit cycle attractor, as in Fig. 1b, we have that $N(\epsilon) \sim \ell/\epsilon$, where ℓ is the length of the closed curve in the figure (dotted line); hence, for this case, $d = 1$, by Eq. 1. A less trivial example is illustrated in Fig. 2, in the form of a Cantor set. This set is

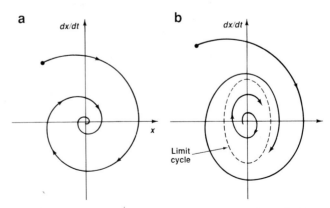

Fig. 1. (**a**) Phase-space diagram for a damped harmonic oscillator. (**b**) Phase-space diagram for a system that is approaching a limit cycle.

C. Grebogi is a research scientist at the Laboratory for Plasma and Fusion Energy Studies, E. Ott is a professor in the departments of electrical engineering and physics, and J. A. Yorke is a professor of mathematics and is the director of the Institute for Physical Science and Technology, University of Maryland, College Park, MD 20742.

Fig. 2. Construction of a Cantor set.

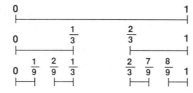

Fig. 3. Poincaré surface of section.

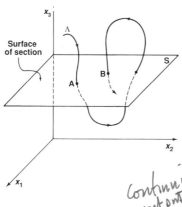

Fig. 4. The Hénon chaotic attractor. (**a**) Full set. (**b**) Enlargement of region defined by the rectangle in (a). (**c**) Enlargement of region defined by the rectangle in (b).

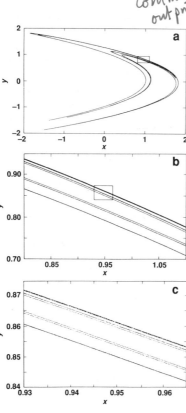

formed by taking the line interval from 0 to 1, dividing it in thirds, then discarding the middle third, then dividing the two remaining thirds into thirds and discarding their middle thirds, and so on ad infinitum. The Cantor set is the closed set of points that are left in the limit of this repeated process. If we take $\epsilon = 3^{-n}$ with n an integer, then we see that $N(\epsilon) = 2^n$ and Eq. 1 (in which $\epsilon \to 0$ corresponds to $n \to \infty$) yields $d = (\ln 2)/(\ln 3)$, a number between 0 and 1, hence, a fractal. The topic of the dimension of strange attractors is a large subject on which much research has been done. One of the most interesting aspects concerning dimension arises from the fact that the distribution of points on a chaotic attractor can be nonuniform in a very singular way. In particular, there can be an arbitrarily fine-scaled interwoven structure of regions where orbit trajectories are dense and sparse. Such attractors have been called multifractals and can be characterized by subsidiary quantities that essentially give the dimensions of the dense and sparse regions of the attractor. In this review we shall not attempt to survey this work. Several papers provide an introduction to recent work on the dimension of chaotic attractors (3–5).

Chaotic attractor. By this term we mean that if we take two typical points on the attractor that are separated from each other by a small distance $\Delta(0)$ at $t = 0$, then for increasing t they move apart exponentially fast. That is, in some average sense $\Delta(t) \sim \Delta(0)\exp(ht)$ with $h > 0$ (where h is called the Lyapunov exponent). Thus a small uncertainty in the initial state of the system rapidly leads to inability to forecast its future. [It is not surprising, therefore, that the pioneering work of Lorenz (2) was in the context of meteorology.] It is typically the case that strange attractors are also chaotic [although this is not always so; see (1, 6)].

Dynamical system. This is a system of equations that allows one, in principle, to predict the future given the past. One example is a system of first-order ordinary differential equations in time, $d\mathbf{x}(t)/dt = \mathbf{G}(\mathbf{x},t)$, where $\mathbf{x}(t)$ is a D-dimensional vector and $\mathbf{G}$ is a D-dimensional vector function of $\mathbf{x}$ and t. Another example is a map.

Map. A map is an equation of the form $\mathbf{x}_{t+1} = \mathbf{F}(\mathbf{x}_t)$, where the "time" t is discrete and integer valued. Thus, given $\mathbf{x}_0$, the map gives $\mathbf{x}_1$. Given $\mathbf{x}_1$, the map gives $\mathbf{x}_2$, and so on. Maps can arise in continuous time physical systems in the form of a Poincaré surface of section. Figure 3 illustrates this. The plane $x_3 = $ constant is the surface of section (S in the figure), and Λ denotes a trajectory of the system. Every time Λ pierces S going downward (as at points A and B in the figure), we record the coordinates (x_1, x_2). Clearly the coordinates of A uniquely determine those of B. Thus there exists a map, $B = F(A)$, and this map (if we knew it) could be iterated to find all subsequent piercings of S.

Chaotic Attractors

As an example of a strange attractor consider the map first studied by Hénon (7):

$$x_{n+1} = \alpha - x_n^2 + \beta y_n \quad (2)$$

$$y_{n+1} = x_n \quad (3)$$

Figure 4a shows the result of plotting 10^4 successive points obtained by iterating Eqs. 2 and 3 with parameters $\alpha = 1.4$ and $\beta = 0.3$ (and the initial transient is deleted). The result is essentially a picture of the chaotic attractor. Figure 4, b and c, shows successive enlargements of the small square in the preceding figure. Scale invariant, Cantor set–like structure transverse to the linear structure is evident. This suggests that we may regard the attractor in Fig. 4c, for example, as being essentially a Cantor set of approximately straight parallel lines. In fact, the dimension d in Eq. 1 can be estimated numerically (8) to be $d \cong 1.26$ so that the attractor is strange.

As another example consider a forced damped pendulum described by the equation

$$d^2\theta/dt^2 + \nu d\theta/dt + \omega_0^2 \sin\theta = f\cos(\omega t) \quad (4)$$

where θ is the angle between the pendulum arm and the rest position, ν is the coefficient of friction, ω_0 is the frequency of natural oscillation, and f is the strength of the forcing. In Eq. 4, the first term represents the inertia of the pendulum, the second term represents friction at the pivot, the third represents the gravitational force, and the right side represents an external sinusoidally varying torque of strength f and frequency ω applied to the pendulum at the pivot. In Fig. 5a, we plot the Poincaré surface of section of a strange

10.3 Chaos, Strange Attractors, and Fractal Basin Boundaries

Fig. 5. (a) Poincaré surface of section of a pendulum strange attractor. (b) Enlargement of region defined by rectangle in (a).

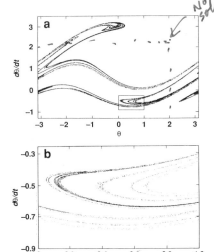

Fig. 6. Chaotic time series for pendulum shown as a plot of angular velocity versus time.

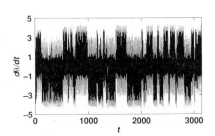

attractor for the pendulum, where we choose $\nu = 0.22$, $\omega_0 = 1.0$, $\omega = 1.0$, and $f = 2.7$ in Eq. 4. This surface of section is obtained by plotting 50,000 dots, one dot for every cycle of the forcing term, that is, one dot at every time $t = t_n = 2\pi n$ (where n is an integer). The strange attractor shown in Fig. 5a exhibits a Cantor set–like structure transverse to the linear structure. This is evident in Fig. 5b, which shows an enlargement of the square region in Fig. 5a. The dimension of this strange attractor in the surface of section is $d \cong 1.38$. Figure 6 shows the angular velocity $d\theta/dt$ as a function of t for the parameters of Fig. 5. Note the apparently erratic nature of this plot.

In general, the form of chaotic attractors varies greatly from system to system and even within the same system. This is indicated by the sequence of chaotic attractors shown in Fig. 7. All of these attractors were generated from the same map (9),

$$\psi_{n+1} = [\psi_n + \omega_1 + \epsilon P_1(\psi_n, \theta_n)] \bmod 1 \quad (5)$$

$$\theta_{n+1} = [\theta_n + \omega_2 + \epsilon P_2(\psi_n, \theta_n)] \bmod 1 \quad (6)$$

where P_1 and P_2 are periodic with period one in both their arguments. The P_1 and P_2 are the same in all of the cases shown in Fig. 7; only the parameters ω_1, ω_2, and ϵ have been varied. The results show the great variety of form and structure possible in chaotic attractors as well as their aesthetic appeal. Since ψ and θ may be regarded as angles, Eqs. 5 and 6 are a map on a two-dimensional toroidal surface. [This map is used in (9) to study the transition from quasiperiodicity to chaos.]

Because of the exponential divergence of nearby orbits on chaotic attractors, there is a question as to how much of the structure in these pictures of chaotic attractors (Figs. 4, 5, and 7) is an artifact due to chaos-amplified roundoff error. Although a numerical trajectory will diverge rapidly from the true trajectory with the same initial point, it has been demonstrated rigorously (10) in important cases [including the Hénon map (11)] that there exists a true trajectory with a slightly different initial point that stays near the noisy trajectory for a long time. [For example, for the Hénon map for a typical numerical trajectory computed with 14-digit precision there exists a true trajectory that stays within 10^{-7} of the numerical trajectory for 10^7 iterates (11).] Thus we believe that the apparently fractal structure seen in pictures such as Figs. 4, 5, and 7 is real.

The Evolution of Chaotic Attractors

In dissipative dynamics it is common to find that for some value of a system parameter only a nonchaotic attracting orbit (a limit cycle, for example) occurs, whereas at some other value of the parameter a chaotic attractor occurs. It is therefore natural to ask how the one comes about from the other as the system parameter is varied continuously. This is a fundamental question that has elicited a great deal of attention (9, 12–19).

To understand the nature of this question and some of the possible answers to it, we consider Fig. 8a, the so-called bifurcation diagram for the map.

$$x_{n+1} = C - x_n^2 \quad (7)$$

where C is a constant. Figure 8a can be constructed as follows: take $C = -0.4$, set $x_0 = -0.5$, iterate the map 100 times (to eliminate transients), then plot the next 1000 values of x; increase C by a small amount, say 0.001, and repeat what was done for $C = -0.4$; increase again, and repeat; and so on, until $C = 2.1$ is reached. We see from Fig. 8a that below a certain value, $C = C_0 = -0.25$, there is no attractor in $-2 < x < 2$. In fact, in this case all orbits go to $x \to -\infty$, hence the absence of points on the plot. This is also true for C above the "crisis value" $C_c = 2.0$. Between these two values there is an attractor. As C is increased we have an attracting orbit of "period one," which, at $C = 0.75$, bifurcates to a period-two attracting orbit ($x_\alpha \to x_\beta \to x_\alpha \to x_\beta \to \cdots$), which then bifurcates (at $C = 1.25$) to a period-four orbit ($x_a \to x_b \to x_c \to x_d \to x_a \to x_b \to x_c \to x_d \to x_a \to \cdots$). In fact, there are an infinite number of such bifurcations of period 2^n to period 2^{n+1} orbits, and these accumulate as $n \to \infty$ at a finite value of C, which we denote C_∞ (from Fig. 8a, $C_\infty \cong 1.4$). [The practical importance of this phenomenology was emphasized early on by May (12).]

What is the situation for $C_\infty < C < C_c$? Numerically what one sees is that for many C values in this range the orbits appear to be chaotic, whereas for others there are periodic orbits. For example, Fig. 8b shows an enlargement of Fig. 8a for C in the range $1.72 < C < 1.82$. We see what appear to be chaotic orbits below $C = C_0^{(3)} = 1.75$. However, just above this value, a period-three orbit appears, supplanting the chaos. The period-three orbit then goes through a period-doubling cascade, becomes chaotic, widens into a three-piece chaotic attractor, and then finally at $C = C_c^{(3)} \cong 1.79$ widens back into a single chaotic band. We call the region $C_0^{(3)} < C < C_c^{(3)}$ a period-three window. (Such windows, but of higher period, appear throughout the region $C_\infty < C < C_c$, but are not as discernible in Fig. 8a because they are much narrower than the period-three window.)

An infinite period-doubling cascade is one way that a chaotic attractor can come about from a nonchaotic one (13). There are also two other possible routes to chaos exemplified in Fig. 8, a and b. These are the intermittency route (14) and the crisis route (15).

Intermittency. Consider Fig. 8b. For C just above $C_0^{(3)}$ there is a period-three orbit. For C just below $C_0^{(3)}$ there appears to be a chaotic orbit. To understand the character of this transition it is useful to examine the chaotic orbit for C just below $C_0^{(3)}$. The character of this orbit is as follows: The orbit appears to be a period-three orbit for long stretches of time after which there is a short

Fig. 7. Sequence of chaotic attractors for system represented by Eqs. 5 and 6. Plot shows iterated mapping on a torus for different values of ω_1, ω_2, and ϵ. (**Top**) $\omega_1 = 0.54657$, $\omega_2 = 0.36736$, and $\epsilon = 0.75$. (**Center**) $\omega_1 = 0.45922$, $\omega_2 = 0.53968$, and $\epsilon = 0.50$. (**Bottom**) $\omega_1 = 0.41500$, $\omega_2 = 0.73500$, and $\epsilon = 0.60$.

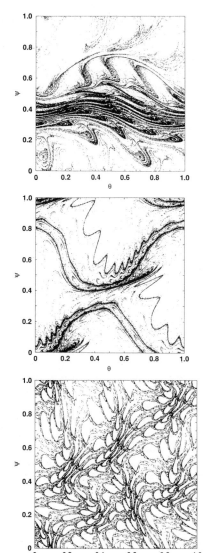

burst (the "intermittent burst") of chaotic-like behavior, followed by another long stretch of almost period-three behavior, followed by a chaotic burst, and so on. As C approaches $C_0^{(3)}$ from below, the average duration of the long stretches between the intermittent bursts becomes longer and longer (14), approaching infinity and proportional to $(C_0^{(3)} - C)^{-1/2}$ as $C \to C_0^{(3)}$. Thus the pure period-three orbit appears at $C = C_0^{(3)}$. Alternatively we may say that the attracting periodic attractor of period three is converted to a chaotic attractor as the parameter C decreases through the critical value $C_0^{(3)}$. It should be emphasized that, although our illustration of the transition to chaos by way of intermittency is within the context of the period-three window of the quadratic map given by Eq. 7, this phenomenon (as well as period-doubling cascades and crises) is very general; in other systems it occurs for other periods (period one, for example) in easily observable form.

Crises. From Fig. 8a we see that there is a chaotic attractor for $C < C_c = 2$, but no chaotic attractor for $C > C_c$. Thus, as C is lowered through C_c, a chaotic attractor is born. How does this occur? Note that at $C = C_c$ the chaotic orbit occupies the interval $-2 \leq x \leq 2$. If C is just slightly larger than C_c, an orbit with initial condition in the interval $-2 < x < 2$ will typically follow a chaotic-like path for a finite time, after which it finds its way out of the interval $-2 \leq x \leq 2$, and then rapidly begins to move to large negative x values (that is, it begins to approach $x = -\infty$). This is called a chaotic transient (15). The length of a chaotic transient will depend on the particular initial condition chosen. One can define a mean transient duration by averaging over, for example, a uniform distribution of initial conditions in the interval $-2 < x < 2$. For the quadratic map, this average duration is

$$\tau \sim 1/(C - C_c)^\gamma \qquad (8)$$

with the exponent γ given by $\gamma = 1/2$. Thus as C approaches C_c from above, the lifetime of a chaotic transient goes to infinity and the transient is converted to a chaotic attractor for $C < C_c$. Again, this type of phenomenon occurs widely in chaotic systems. For example, the model of Lorenz (2) for the nonlinear evolution of the Rayleigh-Bénard instability of a fluid subjected to gravity and heated from below has a chaotic onset of the crisis type and an accompanying chaotic transient. In that case, γ in Eq. 8 is $\gamma \sim 4$ (20). In addition, a theory for determining the exponent γ for two-dimensional maps and systems such as the forced damped pendulum has recently been published (21). Thus we have seen that the period doubling, intermittency, and crisis routes to chaos are illustrated by the simple quadratic map (Eq. 7).

We emphasize that, although a map was used for illustrating these routes, all of these phenomena are present in continuous-time systems and have been observed in experiments. As an example of chaotic transitions in a continuous time system, we consider the set of three autonomous ordinary differential equations studied by Lorenz (2) as a model of the Rayleigh-Bénard instability,

$$dx/dt = Py - Px \qquad (9)$$
$$dy/dt = -xz + rx - y \qquad (10)$$
$$dz/dt = xy - bz \qquad (11)$$

where P and b are adjustable parameters. Fixing $P = 10$ and $b = 8/3$ and varying the remaining parameter, r, we obtain numerical solutions that are clear examples of the intermittency and crisis types of chaotic transitions discussed above. We illustrate these in Fig. 9, a through d; the behavior of this system is as follows:

1) For r between 166.0 and 166.2 there is an intermittency transition from a periodic attractor ($r = 166.0$, Fig. 9a) to a chaotic attractor ($r = 166.2$, Fig. 9b) with intermittent turbulent bursts. Between the bursts there are long stretches of time for which the orbit oscillates in nearly the same way as for the periodic attractor (14) (Fig. 9a).

2) For a range of r values below $r = 24.06$ there are two periodic attractors, that represent clockwise and counterclockwise convections. For r slightly above 24.06, however, there are three attractors, one that is chaotic (shown in the phase space trajectory in Fig. 9c), whereas the other two attractors are the previously mentioned periodic attractors. The chaotic attractor comes into existence as r increases through $r = 24.06$ by conversion of a chaotic transient. Figure 9d shows an orbit in phase space executing a chaotic transient before settling down to its final resting place at one of the periodic attractors. Note the similarity of the chaotic transient trajectory in Fig. 9d with the chaotic trajectory in Fig. 9c.

The various routes to chaos have also received exhaustive experimental support. For instance, period-doubling cascades have been observed in the Rayleigh-Bénard convection (22, 23), in nonlinear circuits (24), and in lasers (25); intermittency has been observed in the Rayleigh-Bénard convection (26) and in the Belousov-Zhabotinsky reaction (27); and crises have been observed in nonlinear circuits (28–30), in the Josephson junction (31), and in lasers (32).

Finally, we note that period doubling, intermittency, and crises do not exhaust the possible list of routes to chaos. (Indeed, the

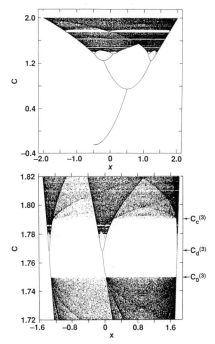

Fig. 8. (**Top**) Bifurcation diagram for the quadratic map. (**Bottom**) Period-three window for the quadratic map.

routes are not all known.) In particular, chaotic onsets involving quasiperiodicity have not been discussed here (*9, 16, 18*).

Universality

Universality refers to the fact that systems behave in certain quantitative ways that depend not on the detailed physics or model description but rather only on some general properties of the system. Universality has been examined by renormalization group (*33*) techniques developed for the study of critical phenomena in condensed matter physics. In the context of dynamics, Feigenbaum (*13*) was the first to apply these ideas, and he has extensively developed them, particularly for period doubling for dissipative systems. [See (*17*) for a collection of papers on universality in nonlinear dynamics.] For period doubling in dissipative systems, results have been obtained on the scaling behavior of power spectra for time series of the dynamical process (*34*), on the effect of noise on period doubling (*35*), and on the dependence of the Lyapunov exponent (*36*) on a system parameter. Applications of the renormalization group have also been made to intermittency (*19, 37*), and the breakdown of quasiperiodicity in dissipative (*18*) and conservative (*38*) systems.

As examples, two "universal" results can be stated within the context of the bifurcation diagrams (Fig. 8, a and b). Let C_n denote the value of C at which a period 2^n cycle period doubles to become a period 2^{n+1} cycle. Then, for the bifurcation diagram in Fig. 8a, one obtains

$$\lim_{n \to \infty} \frac{C_n - C_{n-1}}{C_{n+1} - C_n} = 4.669201\ldots \quad (12)$$

The result given in Eq. 12 is not restricted to the quadratic map. In fact, it applies to a broad class of systems that undergo period doubling cascades (*13, 39*). In practice such cascades are very common, and the associated universal numbers are observed to be well approximated by means of fairly low order bifurcations (for example, $n = 2,3,4$). This scaling behavior has been observed in many experiments, including ones on fluids, nonlinear circuits, laser systems, and so forth. Although universality arguments do not explain why cascades must exist, such explanations are available from bifurcation theory (*40*).

Figure 8b shows the period-three window within the chaotic range of the quadratic map. As already mentioned, there are an infinite number of such periodic windows. [In fact, they are generally believed to be dense in the chaotic range. For example, if k is prime, there are $(2^k - 2)/(2k)$ period-k windows.] Let $C_0^{(k)}$ and $C_c^{(k)}$ denote the upper and lower values of C bounding the period-k window and let $C_d^{(k)}$ denote the value of C at which the period-k attractor bifurcates to period $2k$. Then we have that, for typical k windows (*41*).

$$\lim_{k \to \infty} \frac{C_c^{(k)} - C_0^{(k)}}{C_d^{(k)} - C_0^{(k)}} \to 9/4 \quad (13)$$

In fact, even for the $k = 3$ window (Fig. 8b) the 9/4 value is closely approximated (it is $9/4 - 0.074\ldots$). This result is universal for one-dimensional maps (and possibly more generally for any chaotic dynamical process) with windows.

Fractal Basin Boundaries

In addition to chaotic attractors, there can be sets in phase space on which orbits are chaotic but for which points near the set move away from the set. That is, they are repelled. Nevertheless, such chaotic repellers can still have important macroscopically observable effects, and we consider one such effect (*42, 43*) in this section.

Typical nonlinear dynamical systems may have more than one time-asymptotic final state (attractor), and it is important to consider the extent to which uncertainty in initial conditions leads to uncertainty in the final state. Consider the simple two-dimensional phase space diagram schematically depicted in Fig. 10. There are two attractors denoted A and B. Initial conditions on one side of the boundary, Σ, eventually asymptotically approach B; those on the other side of Σ eventually go to A. The region to the left or right of Σ is the basin of attraction for attractor A or B, respectively, and Σ is the basin boundary. If the initial conditions are uncertain by an amount ϵ, then for those initial conditions within ϵ of the boundary we cannot say a priori to which attractor the orbit eventually tends.

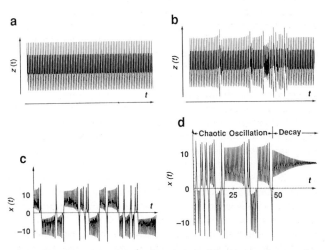

Fig. 9. Intermittency, crisis, and period doubling in continuous time systems. Intermittency in the Lorenz equations (**a**) $r = 166.0$; (**b**) $r = 166.2$. Crisis transition to a chaotic attractor in the Lorenz equations: (**c**) $r = 28$; (**d**) $r = 22$.

Fig. 10. A region of phase space divided by the basin boundary Σ into basins of attraction for the two attractors A and B. Points 1 and 2 are initial conditions with error ε.

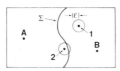

For example, in Fig. 10, points 1 and 2 are initial conditions with an uncertainty ε. The orbit generated by initial condition 1 is attracted to attractor B. Initial condition 2, however, is uncertain in the sense that the orbit generated by 2 may be attracted either to A or B. In particular, consider the fraction of the uncertain phase space volume within the rectangle shown and denote this fraction f. For the case shown in Fig. 10, we clearly have $f \sim \epsilon$. The main point we wish to make in what follows is that, from the point of view of prediction, much worse scalings of f with ϵ frequently occur in nonlinear dynamics. Namely, the fraction can scale as

$$f \sim \epsilon^\alpha \qquad (14)$$

with the "uncertainty exponent" α satisfying α < 1 (42, 43). In fact, α << 1 is fairly common. In such a case, a substantial reduction in the initial condition uncertainty, ε, yields only a relatively small decrease in the uncertainty of the final state as measured by f.

Although α is equal to unity for simple basin boundaries, such as that depicted in Fig. 10, boundaries with noninteger (fractal) dimension also occur. We use here the capacity definition of dimension, Eq. 1. In general, since the basin boundary divides the phase space, its dimension d must satisfy $d \geq D - 1$, where D is the dimension of the phase space. It can be proven that the following relation between the index α and the basin boundary dimension holds (42, 43)

$$\alpha = D - d \qquad (15)$$

For a simple boundary, such as that depicted in Fig. 10, we have $d = D - 1$, and Eq. 15 then gives α = 1, as expected. For a fractal basin boundary, $d > D - 1$, and Eq. 15 gives α < 1.

We now illustrate the above with a concrete example. Consider the forced damped pendulum as given by Eq. 4. For parameter values $\nu = 0.2$, $\omega_0 = 1.0$, $\omega = 1.0$, and $f = 2.0$, we find numerically that the only attractors in the surface of section (θ, dθ/dt) are the fixed points (−0.477, −0.609) and (−0.471, 2.037). They represent solutions with average counterclockwise and clockwise rotation at the period of the forcing. The cover shows a computer-generated picture of the basins of attraction for the two fixed point attractors. Each initial condition in a 1024 by 1024 point grid is integrated until it is close to one of the two attractors (typically 100 cycles). If an orbit goes to the attractor at θ = −0.477, a blue dot is plotted at the corresponding initial condition. If the orbit goes to the other attractor, a red dot is plotted. Thus the blue and red regions are essentially pictures of the basins of attraction for the two attractors to the accuracy of the grid of the computer plotter. Fine-scale structure in the basins of attraction is evident. This is a consequence of the Cantor-set nature of the basin boundary. In fact, magnifications of the basin boundary show that, as we examine it on a smaller and smaller scale, it continues to have structure.

We now wish to explore the consequences for prediction of this infinitely fine-scaled structure. To do this, consider an initial condition (θ, dθ/dt). What is the effect of a small change ε in the θ-coordinate? Thus we integrate the forced pendulum equation with the initial conditions (θ, dθ/dt), (θ, dθ/dt + ε), and (θ, dθ/dt − ε) until they approach one of the attractors. If either or both of the perturbed initial conditions yield orbits that do not approach the same attractor as the unperturbed initial condition, we say that (θ, dθ/dt) is uncertain. Now we randomly choose a large number of initial conditions and let $\hat{f}$ denote the fraction of these that we find to be uncertain. As a result of these calculations, we find that $\hat{f} \sim \epsilon^{\hat{\alpha}}$ where $\hat{\alpha} \cong 0.275 \pm 0.005$. If we assume that $\hat{f}$, determined in the way stated above, is approximately proportional to f [there is some support for this conjecture from theoretical work (44)], then α = $\hat{\alpha}$. Thus, from Eq. 15, the dimension of the basin boundary is $d \cong 1.725 \pm 0.005$. We conclude, from Eq. 14, that in this case if we are to gain a factor of 2 in the ability to predict the asymptotic final state of the system, it is necessary to increase the accuracy in the measurement of the initial conditions by a factor substantially greater than 2 (namely by $2^{1/0.275} \cong 10$). Hence, fractal basin boundaries (α < 1) represent an obstruction to predictability in nonlinear dynamics.

Some representative works on fractal basin boundaries, including applications, are listed in (42–47). Notable basic questions that have recently been answered are the following:

1) How does a nonfractal basin boundary become a fractal basin boundary as a parameter of the system is varied (45)? This question is similar, in spirit, to the question of how chaotic attractors come about.

2) Can fractal basin boundaries have different dimension values in different regions of the boundary, and what boundary structures lead to this situation? This question is addressed in (46) where it is shown that regions of different dimension can be intertwined on an arbitrarily fine scale.

3) What are the effects of a fractal basin boundary when the system is subject to noise? This has been addressed in the Josephson junction experiments of (31).

Conclusion

Chaotic nonlinear dynamics is a vigorous, rapidly expanding field. Many important future applications are to be expected in a variety of areas. In addition to its practical aspects, the field also has fundamental implications. According to Laplace, determination of the future depends only on the present state. Chaos adds a basic new aspect to this rule: small errors in our knowledge can grow exponentially with time, thus making the long-term prediction of the future impossible.

Although the field has advanced at a great rate in recent years, there is still a wealth of challenging fundamental questions that have yet to be adequately dealt with. For example, most concepts developed so far have been discovered in what are effectively low-dimensional systems; what undiscovered important phenomena will appear only in higher dimensions? Why are transiently chaotic motions so prevalent in higher dimensions? In what ways is it possible to use the dimension of a chaotic attractor to determine the dimension of the phase space necessary to describe the dynamics? Can renormalization group techniques be extended past the borderline of chaos into the strongly chaotic regime? These are only a few questions. There are many more, and probably the most important questions are those that have not yet been asked.

REFERENCES AND NOTES

1. C. Grebogi, E. Ott, S. Pelikan, J. A. Yorke, *Physica* **13D**, 261 (1984).
2. E. N. Lorenz, *J. Atmos. Sci.* **20**, 130 (1963).
3. J. D. Farmer, E. Ott, J. A. Yorke, *Physica* **7D**, 153 (1983).
4. J. Kaplan and J. A. Yorke, *Lecture Notes in Mathematics No. 730* (Springer-Verlag, Berlin, 1978), p. 228; L. S. Young, *Ergodic Theory Dyn. Syst.* **1**, 381 (1981).
5. P. Grassberger and I. Procaccia, *Phys. Rev. Lett.* **50**, 346 (1983); H. G. E. Hentschel and I. Procaccia, *Physica* **8D**, 435 (1983); P. Grassberger, *Phys. Lett.* **A97**, 227 (1983); T. C. Halsey et al., B. I. Shraiman, *Phys. Rev. A* **33**, 1141 (1986); C. Grebogi, E. Ott, J. A. Yorke, *ibid.* **36**, 3522 (1987).
6. A. Bondeson et al., *Phys. Rev. Lett.* **55**, 2103 (1985); F. J. Romeiras, A. Bondeson, E. Ott, T. M. Antonsen, C. Grebogi, *Physica* **26D**, 277 (1987).
7. M. Hénon, *Commun. Math. Phys.* **50**, 69 (1976).
8. D. A. Russell, J. D. Hanson, E. Ott, *Phys. Rev. Lett.* **45**, 1175 (1980).
9. C. Grebogi, E. Ott, J. A. Yorke, *Physica* **15D**, 354 (1985).

10. D. V. Anosov, *Proc. Steklov Ins. Math.* **90** (1967); R. Bowen, *J. Differ. Equations* **18**, 333 (1975).
11. S. M. Hammel, J. A. Yorke, C. Grebogi, *J. Complexity*, **3**, 136 (1987).
12. R. M. May, *Nature (London)* **261**, 459 (1976).
13. M. J. Feigenbaum, *J. Stat. Phys.* **19**, 25 (1978).
14. Y. Pomeau and P. Manneville, *Commun. Math. Phys.* **74**, 189 (1980).
15. C. Grebogi, E. Ott, J. A. Yorke, *Physica* **7D**, 181 (1983).
16. D. Ruelle and F. Takens, *Commun. Math. Phys.* **20**, 167 (1971).
17. P. Cvitanovic, Ed. *Universality in Chaos* (Hilger, Bristol, 1984).
18. For example, S. J. Shenker, *Physica* **5D**, 405 (1982); K. Kaneko, *Progr. Theor. Phys.* **71**, 282 (1984); M. J. Feigenbaum, L. P. Kadanoff, S. L. Shenker, *Physica* **5D**, 370 (1982); D. Rand, S. Ostlund, J. Sethna, E. Siggia, *Physica* **8D**, 303 (1983); S. Kim and S. Ostlund, *Phys. Rev. Lett.* **55**, 1165 (1985); D. K. Umberger, J. D. Farmer, I. I. Satija, *Phys. Lett. A* **114**, 341 (1986); P. Bak, T. Bohr, M. H. Jensen, *Phys. Scr.* **T9**, 50 (1985); P. Bak, *Phys. Today* **39** (No. 12), 38 (1987).
19. J. E. Hirsch, M. Nauenberg, D. J. Scalapino, *Phys. Lett. A* **87**, 391 (1982).
20. J. A. Yorke and E. D. Yorke, *J. Stat. Phys.* **21**, 263 (1979); in *Topics in Applied Physics* (Springer-Verlag, New York, 1981), vol. 45, p. 77.
21. C. Grebogi, E. Ott, J. A. Yorke, *Phys. Rev. Lett.* **57**, 1284 (1986).
22. A. Libchaber and J. Maurer, *J. Phys. (Paris)* **41**, C3-51 (1980); A. Libchaber, C. Laroche, S. Fauve, *J. Phys. (Paris) Lett.* **43**, L211 (1982).
23. J. P. Gollub, S. V. Benson, J. F. Steinman, *Ann. N.Y. Acad. Sci.* **357**, 22 (1980); M. Giglio, S. Musazzi, U. Perini, *Phys. Rev. Lett.* **47**, 243 (1981).
24. P. S. Linsay, *Phys. Rev. Lett.* **47**, 1349 (1981).
25. F. T. Arecchi, R. Meucci, G. Puccioni, J. Tredicce, *ibid.* **49**, 1217 (1982).
26. M. Dubois, M. A. Rubio, P. Bergé, *ibid.* **51**, 1446 (1983).
27. J. C. Roux, P. DeKepper, H. L. Swinney, *Physica* **7D**, 57 (1983).
28. C. Jeffries and J. Perez, *Phys. Rev. A* **27**, 601 (1983); S. K. Brorson, D. Dewey, P. S. Linsay, *ibid.* **28**, 1201 (1983).
29. H. Ikezi, J. S. deGrasse, T. H. Jensen, *ibid.* **28**, 1207 (1983).
30. R. W. Rollins and E. R. Hunt, *ibid.* **29**, 3327 (1984).
31. M. Iansiti et al., *Phys. Rev. Lett.* **55**, 746 (1985).
32. D. Dangoisse, P. Glorieux, D. Hannequin, *ibid.* **57**, 2657 (1986).
33. K. G. Wilson and J. Kogut, *Phys. Rep. C* **12**, 75 (1974); B. Hu, *ibid.* **91**, 233 (1982).
34. M. J. Feigenbaum, *Phys. Lett. A* **74**, 375 (1979); R. Brown, C. Grebogi, E. Ott, *Phys. Rev. A* **34**, 2248 (1986); M. Nauenberg and J. Rudnick, *Phys. Rev. B* **24**, 493 (1981); B. A. Huberman and A. B. Zisook, *Phys. Rev. Lett.* **46**, 626 (1981); J. D. Farmer, *ibid.* **47**, 179 (1981).
35. J. Crutchfield, M. Nauenberg, J. Rudnick, *Phys. Rev. Lett.* **46**, 933 (1981); B. Shraiman, C. E. Wayne, P. C. Martin, *ibid.*, p. 935.
36. B. A. Huberman and J. Rudnick, *ibid.* **45**, 154 (1980).
37. B. Hu and J. Rudnick, *ibid.* **48**, 1645 (1982).
38. L. P. Kadanoff, *ibid.* **47**, 1641 (1981); D. F. Escande and F. Doveil, *J. Stat. Phys.* **26**, 257 (1981); R. S. MacKay, *Physica* **7D**, 283 (1983).
39. P. Collet, J. P. Eckmann, O. E. Lanford III, *Commun. Math. Phys.* **76**, 211 (1980).
40. J. A. Yorke and K. A. Alligood, *ibid.* **100**, 1 (1985).
41. J. A. Yorke, C. Grebogi, E. Ott, L. Tedeschini-Lalli, *Phys. Rev. Lett.* **54**, 1095 (1985).
42. C. Grebogi, S. W. McDonald, E. Ott, J. A. Yorke, *Phys. Lett. A* **99**, 415 (1983).
43. S. W. McDonald, C. Grebogi, E. Ott, J. A. Yorke, *Physica* **17D**, 125 (1985).
44. S. Pelikan, *Trans. Am. Math. Soc.* **292**, 695 (1985).
45. C. Grebogi, E. Ott, J. A. Yorke, *Phys. Rev. Lett.* **56**, 1011 (1986); *Physica* **24D**, 243 (1987); F. C. Moon and G.-X. Li, *Phys. Rev. Lett.* **55**, 1439 (1985).
46. C. Grebogi, E. Kostelich, E. Ott, J. A. Yorke, *Phys. Lett.* **A118**, 448 (1986); *Physica* **25D**, 347 (1987); C. Grebogi, E. Ott, J. A. Yorke, H. E. Nusse, *Ann. N.Y. Acad. Sci.* **497**, 117 (1987).
47. C. Mira, *C. R. Acad. Sci.* **288A**, 591 (1979); C. Grebogi, E. Ott, J. A. Yorke, *Phys. Rev. Lett.* **50**, 935 (1983); R. G. Holt and I. B. Schwartz, *Phys. Lett.* **A105**, 327 (1984); I. B. Schwartz, *ibid.* **106**, 339 (1984); I. B. Schwartz, *J. Math. Biol.* **21**, 347 (1985); S. Takesue and K. Kaneko, *Progr. Theor. Phys.* **71**, 35 (1984); O. Decroly and A. Goldbeter, *Phys. Lett.* **A105**, 259 (1984); E. G. Gwinn and R. M. Westervelt, *Phys. Rev. Lett.* **54**, 1613 (1985); *Phys. Rev. A* **33**, 4143 (1986); Y. Yamaguchi and N. Mishima, *Phys. Lett.* **A109**, 196 (1985); M. Napiorkowski, *ibid.* **113**, 111 (1985); F. T. Arecchi, R. Badii, A. Politi, *Phys. Rev. A* **32**, 402 (1985); S. W. McDonald, C. Grebogi, E. Ott, J. A. Yorke, *Phys. Lett.* **A107**, 51 (1985); J. S. Nicolis and I. Tsuda, in *Simulation, Communication, and Control*, S. G. Tzafestas, Ed. (North-Holland, Amsterdam, 1985); J. S. Nicolis, *Rep. Prog. Phys.* **49**, 1109 (1986); J. S. Nicolis, *Kybernetes* **14**, 167 (1985).
48. This work was supported by the Air Force Office of Scientific Research, the U.S. Department of Energy, the Defense Advanced Research Projects Agency, and the Office of Naval Research.

Nonlinear forecasting as a way of distinguishing chaos from measurement error in time series

George Sugihara* & Robert M. May[†]

* Scripps Institution of Oceanography, University of California, San Diego, La Jolla, California 92093, USA
† Department of Zoology, Oxford University, Oxford, OX1 3PS, UK

An approach is presented for making short-term predictions about the trajectories of chaotic dynamical systems. The method is applied to data on measles, chickenpox, and marine phytoplankton populations, to show how apparent noise associated with deterministic chaos can be distinguished from sampling error and other sources of externally induced environmental noise.

TWO sources of uncertainty in forecasting the motion of natural dynamical systems, such as the annual densities of plant or animal populations, are the errors and fluctuations associated with making measurements (for example, sampling errors in estimating sizes, or fluctuations associated with unpredictable environmental changes from year to year), and the complexity of the dynamics themselves (where deterministic dynamics can easily lead to chaotic trajectories).

Here we combine some new ideas with previously developed techniques[1-7,16,24-26], to make short-term predictions that are

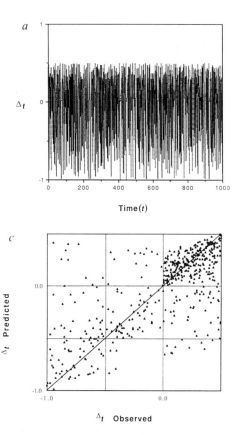

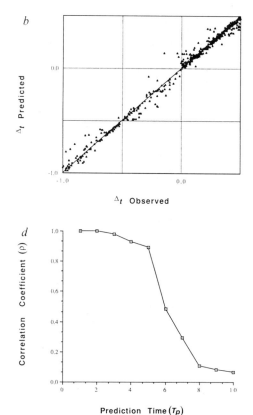

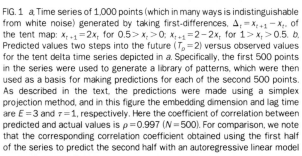

FIG. 1 a, Time series of 1,000 points (which in many ways is indistinguishable from white noise) generated by taking first-differences, $\Delta_t = x_{t+1} - x_t$, of the tent map: $x_{t+1} = 2x_t$ for $0.5 > x_t > 0$; $x_{t+1} = 2 - 2x_t$ for $1 > x_t > 0.5$. b, Predicted values two steps into the future ($T_p = 2$) versus observed values for the tent delta time series depicted in a. Specifically, the first 500 points in the series were used to generate a library of patterns, which were then used as a basis for making predictions for each of the second 500 points. As described in the text, the predictions were made using a simplex projection method, and in this figure the embedding dimension and lag time are $E = 3$ and $\tau = 1$, respectively. Here the coefficient of correlation between predicted and actual values is $\rho = 0.997$ ($N = 500$). For comparison, we note that the corresponding correlation coefficient obtained using the first half of the series to predict the second half with an autoregressive linear model (where the predictions are based on the weighted average of three linear maps, one for each of the three different τ-values that give the best results in such a linear scheme) is $\rho = 0.04$. c, Exactly as for Fig. b, except here the predictions are five time steps into the future ($T_p = 5$). The correlation coefficient between predicted and actual values is now $\rho = 0.89$ ($N = 500$). d, Summary of the trend between b and c, by showing ρ between predicted and observed values in the second half (second 500 points) of the time series of a, as a function of T_p. As in b and c, the simplex projection method here uses $E = 3$ and $\tau = 1$. That prediction accuracy (as measured by the coefficient of correlation between predicted and observed values) falls as predictions extend further into the future is a characteristic signature of a chaotic attractor.

10.4 Nonlinear Forecasting as a Way of Distinguishing Chaos...

based on a library of past patterns in a time series[1]. By comparing the predicted and actual trajectories, we can make tentative distinctions between dynamical chaos and measurement error: for a chaotic time series the accuracy of the nonlinear forecast falls off with increasing prediction-time interval (at a rate which gives an estimate of the Lyapunov exponent[3]), whereas for uncorrelated noise, the forecasting accuracy is roughly independent of prediction interval. For a relatively short time series, distinguishing between autocorrelated noise and chaos is more difficult; we suggest a way of distinguishing such 'coloured' noise from chaos in our scheme, but questions remain, at least for time series of finite length.

The method also provides an estimate of the number of dimensions, or 'active variables', of the attractor underlying a time series that is identified as chaotic. Unlike many current approaches to this problem (for example, that of Grassberger and Procaccia[8]), our method does not require a large number of data points, but seems to be useful when the observed time series has relatively few points (as is the case in essentially all ecological and epidemiological data sets).

Forecasting for a chaotic time series

Below, we outline the method and show how it works by applying it to a chaotic time series generated artificially from the deterministic 'tent map'. We then apply it to actual data on measles and chickenpox in human populations (which have been previously analysed using different techniques[9-13]) and on diatom populations. We conclude that the method may be capable of distinguishing chaos from measurement error even in such relatively short runs of real data.

As an example of the difficulties in short-range forecasting, we consider the chaotic time series shown in Fig. 1a. This time series was generated from the first-difference transformation $(x_{t+1} - x_t)$ on the deterministic tent map or triangular 'return map' (described in detail in the legend to Fig. 1a). Here and elsewhere is this report, we first-difference the data partly to give greater density in phase space to such chaotic attractors as may exist, and partly to clarify nonlinearities by reducing the effects of any short-term linear autocorrelations. It should be noted, however, that both in our artificial examples and in our later analysis of real data, we obtain essentially the same results if we work with the raw time series (without first-differencing). With the exception of a slight negative correlation between immediately adjacent values, the sequence in Fig. 1a is uncorrelated, and is in many ways indistinguishable from white noise: the null hypothesis of a flat Fourier spectrum cannot be rejected using Bartlett's Kolmogorov-Smirnov test, with $P = 0.85$. Because nonadjacent values in the time series are completely uncorrelated, standard statistical methods (that is, linear autoregression) cannot be used to generate predictions two or more steps into the future that are significantly better than the mean value (that is, zero) for the series.

Figure 1b and c show the results of local forecasting with the above data. The basic idea here, as outlined below, is that if deterministic laws govern the system, then, even if the dynamical behaviour is chaotic, the future may to some extent be predicted from the behaviour of past values that are similar to those of the present.

Specifically, we first choose an 'embedding dimension', E, and then use lagged coordinates to represent each lagged sequence of data points $\{x_t, x_{t-\tau}, x_{t-2\tau}, \ldots, x_{t-(E-1)\tau}\}$ as a point in this E-dimensional space; for this example we choose $\tau = 1$, but the results do not seem to be very sensitive to the value of τ, provided that it is not large[14,15]. For our original time series, shown in Fig. 1a, each sequence for which we wish to make a prediction—each 'predictee'—is now to be regarded as an E-dimensional point, comprising the present value and the $E-1$ previous values each separated by one lag time τ. We now locate all nearby E-dimensional points in the state space, and choose a minimal neighbourhood defined to be such that the predictee is contained within the smallest simplex (the simplex with minimum diameter) formed from its $E+1$ closest neighbours; a simplex containing $E+1$ vertices (neighbours) is the smallest simplex that can contain an E-dimensional point as an interior point (for points on the boundary, we use a lower-dimensional simplex of nearest neighbours). The prediction is now obtained by projecting the domain of the simplex into its range, that is by keeping track of where the points in the simplex end up after p time steps. To obtain the predicted value, we compute where the original predictee has moved within the range of this simplex,

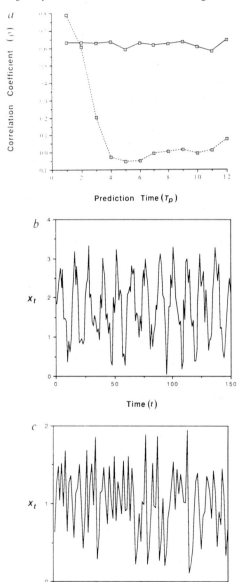

FIG. 2 a, Solid line shows ρ between predicted and observed values for the second half of the time series defined in b (which is, in fact, a sine wave with additive noise) as a function of T_p. As discussed in the text, the accuracy of the prediction, as measured by ρ, shows no systematic dependence on T_p. By contrast, the time series shown in c (which is the sum of two separate tent map series) does show the decrease in ρ with increasing T_p, as illustrated by the dashed line, that is characteristic of a chaotic sequence. Both curves are based on the simplex methods described in the text, with $E=3$ and $\tau=1$. b, First 150 points in the time series generated by taking discrete points on a sine wave with unit amplitude ($x_t = \sin(0.5t)$), and adding a random variable chosen (independently at each step) uniformly from the interval [−0.5, 0.5]. That is, the series is generated as a 'sine wave + 50% noise'. c, Time series illustrated here is generated by adding together two independent tent map sequences.

giving exponential weight to its original distances from the relevant neighbours. This is a nonparametric method, which uses no prior information about the model used to generate the time series, only the information in the output itself. It should apply to any stationary or quasi-ergodic dynamic process, including chaos. This method is a simpler variant of several more complicated techniques explored recently by Farmer and Sidorowich[3] and by Casdagli[7].

Figure 1b compares predicted with actual results, two time steps into the future. Figure 1c makes the same comparison, but at five time steps into the future. There is obviously more scatter in Fig. 1c than in Fig. 1b. Figure 1d quantifies how error increases as we predict further into the future in this example, by plotting the conventional statistical coefficient of correlation, ρ, between predicted and observed values as a function of the prediction-time interval, T_p (or the number of time steps into the future, p). Such decrease in the correlation coefficient with increasing prediction time is a characteristic feature of chaos (or equivalently, of the presence of a positive Lyapunov exponent, with the magnitude of the exponent related to the rate of decrease of ρ with T_p). This property is noteworthy, because it indicates a simple way to differentiate additive noise from deterministic chaos: predictions with additive noise that is uncorrelated (in the first-differences) will seem to have a fixed amount of error, regardless of how far, or close, into the future one tries to project, whereas predictions with deterministic chaos will tend to deteriorate as one tries to forecast further into the future. Farmer and Sidorowich[3,16] have derived asymptotic results (for very long time series, $N \gg 1$) that describe how this error typically propagates, over time, in simple chaotic systems. The standard correlation coefficient is one of several alternative measures of the agreement between predicted and observed values; results essentially identical to those recorded in Figs 1–6 can be obtained with other measures (such as the mean squared difference between predicted and observed values as a ratio to the mean squared error).

Forecasting with uncorrelated noise

Figure 2a (solid line) shows that, indeed, this signature of ρ decreasing with T_p does not arise when the erratic time series is in fact a noisy limit cycle. Here we have uncorrelated additive noise superimposed on a sine wave (Fig. 2b). Such uncorrelated noise is reckoned to be characteristic of sampling variation. Here the error remains constant as the simplex is projected further into the future; past sequences of roulette-wheel numbers that are similar to present ones tell as much or little about the next spin as the next hundredth spin. By contrast, the dashed line in Fig. 2a represents ρ as a function of T_p, for a chaotic sequence generated as the sum of two independent runs of tent map; that is, for the time series illustrated in Fig. 2c. Although the two time series in Fig. 2b and Fig. 2c can both look like the sample functions of some random process, the characteristic signatures in Fig. 2a distinguish the deterministic chaos in Fig. 2c from the additive noise in Fig. 2b.

The embedding dimension

The predictions in Figs 1 and 2 are based on an embedding dimension of $E = 3$. The results are, however, sensitive to the choice of E. Figure 3a compares predicted and actual results for the tent map two time steps ahead ($T_p = 2$), as in Fig. 1b, except that now $E = 10$ (versus $E = 3$ in Fig. 1b). Clearly the predictions are less accurate with this higher embedding dimension. More generally, Fig. 3b shows ρ between predicted and actual results one time step into the future ($T_p = 1$) as a function of E, for two different choices of the lag time ($\tau = 1$ and $\tau = 2$). It may seem surprising that having potentially more information—more data summarized in each E-dimensional point, and a higher-dimensional simplex of neighbours of the predictee—reduces the accuracy of the predictions; in this respect, these results differ from results reported by Farmer and Sidorowich for parametric forecasting involving linear interpo-

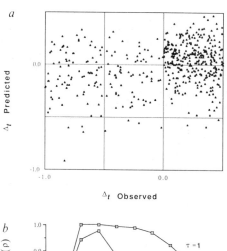

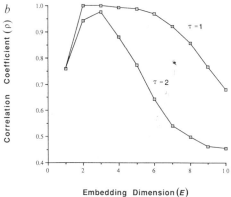

FIG. 3 a, Similar to Fig. 1b, this figure shows predictions one time step into the future ($T_p = 1$) versus observed values, for the second 500 points in the tent map time series of Fig 1a, with the difference that here we used an embedding dimension $E = 10$ (in contrast to $E = 3$ in Fig. 1b; the lag time remains unchanged at $\tau = 1$). As discussed in the text, the accuracy of the prediction deteriorates as E gets too large $\rho = 0.25$, $N = 500$). b, Correlation coefficient between predicted and observed results, ρ, is shown as a function of E for predictions one time step into the future ($T_p = 1$). The relationship is shown for $\tau = 1$ and $\tau = 2$. The figure indicates how such empirical studies of the relation between ρ and E may be used to assess the optimal E.

lation to construct local polynomial maps[3]. We think this effect is caused by contamination of nearby points in the higher-dimensional embeddings with points whose earlier coordinates are close, but whose recent (and more relevant) coordinates are distant. If this is so, our method may have additional applications as a trial-and-error method of computing an upper-bound on the embedding dimension, and thence on the dimensionality of the attractor (see also refs 2, 6, 7).

Problems and other approaches

We have applied these ideas to a variety of other 'toy models', including the quadratic map along with other first-order difference equations and time series obtained by taking points at discrete time intervals from continuous chaotic systems such as those of the Lorenz and Rossler models (in which the chaotic orbits are generated by three coupled, nonlinear differential equations). The results for ρ as a function of T_p are in all cases very similar to those shown in Fig. 1d. Even in more complicated cases, such as those involving the superposition of different chaotic maps, we observe a decline in ρ versus T_p; here, however, the signature can show a step pattern, with each step corresponding to the dominant Lyapunov exponent for each map.

So far, we have compared relationships between ρ and T_p for chaotic time series with the corresponding relations for white noise. More problematic, however, is the comparison with ρ-T_p relationships generated by coloured noise spectra, in which there are significant short-term autocorrelations, although not long-term ones. Such autocorrelated noise can clearly lead to correlations, ρ, between predicted and observed values that decrease as T_p lengthens. Indeed, it seems likely that a specific pattern of autocorrelations could be hand-tailored, to mimic any given relationship between ρ and T_p (such as that shown in Fig. 1d) obtained from a finite time series. We conjecture, however, that such an artificially designed pattern of autocorrelation would in general give a flatter ρ-versus-E relationship than those of simple chaotic time series corresponding to low-dimensional attractors (for example, see Fig. 3b). Working from the scaling relations for error versus T_p in chaotic systems[3,16], Farmer (personal communication) has indeed suggested that asymptotically (for very large N), the ρ-T_p relationships generated by autocorrelated noise may characteristically scale differently from those generated by deterministic chaos. Although we have no solution to this central problem—which ultimately may not have any general solution, at least for time series of the sizes found in population biology—we suggest that an observed time series may tentatively be regarded as deterministically chaotic if, in addition to a decaying ρ-T_p signature, the correlation, ρ, between predicted and observed values obtained by our methods is significantly better than the corresponding correlation coefficient obtained by the best-fitting autoregressive linear predictor (see also, ref. 16). For the tent map, as detailed in the legend to Fig. 1, b and c, our nonlinear method gives ρ values significantly better than those from autoregressive linear models (composed of the weighted average of the three best linear maps).

Most previous work applying nonlinear theory to experimental data begins with some estimate of the dimension of the underlying attractor[2-7]. The usual procedure (for exceptions, see refs 2, 6, 7, 25) is to construct a state-space embedding for the time series, and then to calculate the dimension of the putative attractor using some variant of the Grassberger-Procaccia algorithm[8]. A correlation integral is calculated that is essentially the number of points in E space separated by a distance less than l, and the power-law behaviour of this correlation integral (l^ν) is then used to estimate the dimension, D, of the attractor ($D \geq \nu$). This dimension is presumed to give a measure of the effective number of degrees of freedom or 'active modes' of the system. An upper bound on a minimal embedding dimension (which can be exceeded when the axes of the embed-

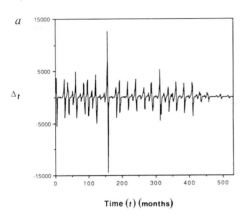

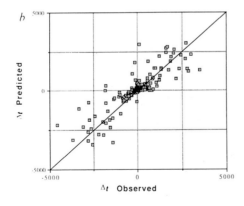

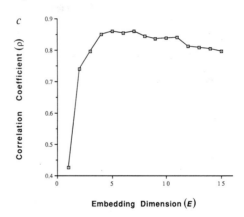

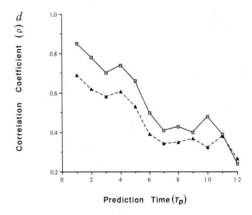

FIG. 4 a, Time series generated by taking first differences, $x_{t+1} - x_t$, of the monthly number of cases of measles reported in New York City between 1928 and 1972 (the first 532 points in the sequence shown here). After 1963, the introduction of immunization against measles had a qualitative effect on the dynamics of infection; this can be seen in the later part of the sequence illustrated here. b, Using the methods described earlier, the first part of the series in Fig. 4a (216 points, 1928 to 1946)) was used to construct a library of past patterns, which were then used as a basis for predicting forward from each point in the second part of the series, from 1946 to 1963. Predicted and observed values are shown here for predictions one time step into the future ($T_p = 1$), using $E = 6$ and $\tau = 1$. The correlation coefficient between predicted and observed values is $\rho = 0.85$ ($P < 10^{-5}$ for $N = 216$). For comparison, the corresponding prediction based on an autoregressive linear model (composed of five optimal linear maps, compare Fig. 1b) gives $\rho = 0.72$ (which is significantly different from $\rho = 0.85$ at the $P < 0.0005$ level). c, As in Fig. 3b, ρ between predicted and observed results, is shown as a function of E (for $T_p = 1$ and $\tau = 1$). This figure indicates an optimal embedding dimension of $E \sim 5$-7, corresponding to a chaotic attractor with dimension 2-3. d, Here ρ, between predicted and observed results for measles, is shown as a function of T_p (for $E = 6$ and $\tau = 1$). For the points connected by the solid lines, the predictions are for the second half of the time series (based on a library of patterns compiled from the first half). For the points connected by the dashed lines, the forecasts and the library of patterns span the same time period (the first half of the data). The similarity between solid and dashed curves indicates that secular trends in underlying parameters do not introduce significant complications here. The overall decline in prediction accuracy with increasing time into the future is, as discussed in the text, a signature of chaotic dynamics as distinct from uncorrelated additive noise.

ding are not truly orthogonal) is $E_{min} < 2D + 1$, where D is the attractor dimension[8,15]. The scaling regions used to estimate power laws by these methods are typically small and, as a consequence, such calculations of dimension involve only a small fraction of the points in the series (that is, they involve only a small subset of pairs of points in the state space). In other words, the standard methods discard much of the information in a time series, which, because many natural time series are of limited size, can be a serious problem. Furthermore, the Grassberger–Procaccia and related methods are somewhat more qualitative, requiring subjective judgement about whether there is an attractor of given dimensions. Prediction methods, by contrast, have the advantage that standard statistical criteria can be used to evaluate the significance of the correlation between predicted and observed values. As Farmer and Sidorowich[3,16], and Casdagli[7], have also emphasized, prediction methods should provide a more stringent test of underlying determinism in situations of given complexity. Prediction is, after all, the *sine qua non* of determinism.

Time series from the natural world

Measles. For reported cases of measles in New York City, there is a monthly time series extending from 1928 (ref. 17). After 1963, immunization began to alter the intrinsic dynamics of this system, and so we use only the data from 1928 to 1963 ($N = 432$). These particular data have received a lot of attention recently, and they are the focus of a controversy about whether the dynamics reflect a noisy limit cycle[9] or low-dimensional chaos superimposed on a seasonal cycle[10-13]. In particular, the data have been carefully studied by Schaffer and others[13-16,27], who have tested for low-dimensional chaos using a variety of methods, including the Grassberger–Procaccia algorithm[13], estimation of Lyapunov exponents[13], reconstruction of Poincare return maps[10,11], and model simulations[12,13]. Although it is not claimed that any of these tests are individually conclusive, together they support the hypothesis that the measles data are described by a two- to three-dimensional chaotic attractor.

Figure 4a shows the time series obtained by taking first differences, $X_{t+1} - X_t$, of these data. As discussed above, the first difference was taken to 'whiten' the series (that is, reduce autocorrelation) and to diminish any signals associated with simple cycles (a possibility raised by proponents of the additive noise hypothesis[9]). We then generated our predictions by using the first half of the series (216 points) to construct an ensemble of points in an E-dimensional state space, that is, to construct a library of past patterns. The resulting information was then used to predict the remaining 216 values in the series, along the lines described above, for each chosen value of E. Figure 4b, for example, compares predicted and observed results, one time step into the future ($T_p = 1$ month), with $E = 6$. Figure 4c shows ρ between predicted and observed results as a function of E for $T_p = 1$. Taking the optimal embedding dimension to be that yielding the highest correlation coefficient (or least error) between prediction and observation in one time step, it is seen from Fig. 4c that $E \approx 5\text{--}7$. This accords with previous estimates[10-13] made using various other methods, and is consistent with the finding of an attractor with dimension $D = 2\text{--}3$.

The points joined by the solid lines in Fig. 4d show ρ as a function of T_p (for $E = 6$). Prediction error seems to propagate in a manner consistent with chaotic dynamics. This result, in

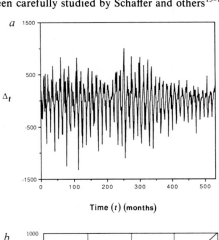

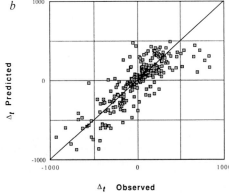

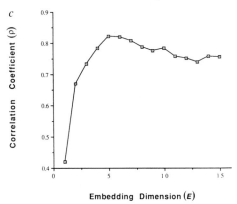

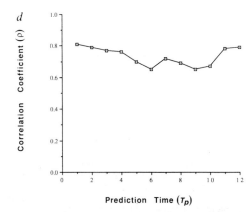

FIG. 5 a, As for Fig. 4a, except the time series comes from taking first-differences of the monthly numbers of reported cases of chickenpox in New York City from 1928 to 1972. b, As in Fig. 4b, predicted and observed numbers of cases of chickenpox are compared, the predictions being one time step into the future, $T_p = 1$ (here, $E = 5$ and $\tau = 1$). The correlation coefficient between predicted and observed values is $\rho = 0.82$; an autoregressive linear model alternatively gives predictions which have $\rho = 0.84$. In contrast to Fig. 4b for measles, here there is no significant difference between our prediction technique and standard linear autoregressive methods. c, Correlation coefficient between predicted and observed results for chickenpox, ρ, shown as a function of E for predictions one time step into the future ($T_p = 1$ and $\tau = 1$). d, Compare with Fig. 4d; ρ, between predicted and observed values, as a function of T_p (with $E = 5$ and $\tau = 1$) is shown. Here the lack of dependence of ρ on T_p, which is in marked contrast with the pattern for measles in Fig. 4d, indicates pure additive noise (superimposed on a basic seasonal cycle).

10.4 Nonlinear Forecasting as a Way of Distinguishing Chaos...

combination with the significantly better performance ($P < 0.0005$) of our nonlinear predictor as compared with an optimal linear autoregressive model (see legend to Fig. 4b) agrees with the conclusion that the noisy dynamics shown in Fig. 4a are, in fact, deterministic chaos[10-13].

For data from the natural world, as distinct from artificial models, physical or biological parameters, or both, can undergo systematic changes over time. In this event, libraries of past patterns can be of dubious relevance to an altered present and even-more-different future. In a different context, there is the example of how secular trends in environmental variables can complicate an analysis of patterns of fluctuation in the abundance of bird species[18,19]. We can gauge the extent to which secular trends might confound our forecasting methods in the following way. Rather than using the first half of the time series to compile the library of patterns, and the second half to compute correlations between predictions and observations, we instead investigate the case in which the library and forecasts span the same time period. Therefore we focus our predictions in the first half of the series, from which the library was drawn. To avoid redundancy, however, between our forecasts and the model, we sequentially exclude points from the library that are in the neighbourhood of each predictee (specifically, the $E\tau$ points preceding and following each forecast). The points connected by the dashed lines in Fig. 4d show the ρ versus T_p relationship that results from treating the measles data in this way (again with $E=6$). The fairly close agreement between these results (for which the library of patterns and the forecasts span the same time period) and those of the simpler previous analysis (the solid line in Fig. 4d) indicates that within these time frames, secular trends in underlying parameters are not qualitatively important.

Chickenpox. Figure 5a-d repeat the process just described for measles, but now for monthly records of cases of chickenpox in New York City from 1949 to 1972 (ref. 20). Figure 5a shows the time series of differences, $X_{t+1} - X_t$. The 532 points in Fig. 5a are divided into two halves, with the first half used to construct the library, on which predictions are made for the second 266 points. These predictions are compared with the actual data points, as shown for predictions one time step ahead ($T_p = 1$ month) in Fig. 5b. In Fig. 5b, $E = 5$; Fig. 5c shows that an optimum value of E, in the sense just defined, is about 5 to 6. By contrast with Fig. 4d for measles, ρ between predicted and observed results for chickenpox shows no dependence on T_p: one does as well at predicting the incidence next year as next month. Moreover, the optimal linear autoregressive model performs as well as our nonlinear predictor. We take this to indicate that chickenpox has a strong annual cycle (as does

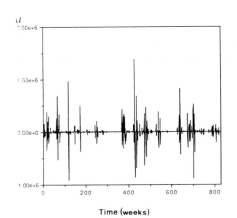

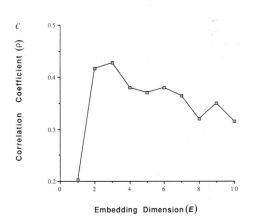

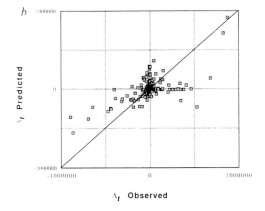

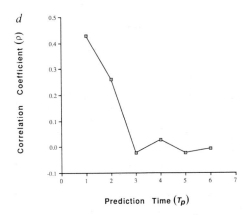

FIG. 6 *a*, Time series of first differences, $x_{t+1} - x_t$, of the weekly numbers of diatoms in seawater samples taken at Scripps Pier, San Diego, from 1929 to 1939 ($N = 830$). *b*, Using the first half of the time series in *a* to construct a library of patterns, we use the simplex projection methods described in the text to predict one week ($T_p = 1$) into the future from each point in the second half of the series ($N = 415$); here $E = 3$, and $\tau = 1$. The correlation coefficient between predicted and observed values is $\rho = 0.42$ ($P < 10^{-4}$ for $N = 415$); the best autoregressive linear predictions (composed of three optimal linear maps) give $\rho = 0.13$, which is significantly less than the nonlinear result ($P < 0.0005$). *c*, As in Figs 4c and 5c, ρ between predicted and observed values is shown as a function of the choice of E for predictions two time steps into the future ($T_p = 2$ and $\tau = 1$). This figure indicates an optimal E of about 3, consistent with an attractor of dimension about 2. *d*, As in Figs 4d and 5d, ρ is shown as a function of the T_p (for $E = 3$ and $\tau = 1$). Here the correlation coefficient decreases with increasing prediction interval, in the manner characteristic of chaotic dynamics generated by a some low-dimensional attractor. That ρ is about 50% at best, however, indicates that roughly half the variance in the time series comes from additive noise. The dynamics of this system therefore seem to be intermediate between those of measles (for which Fig. 4d indicates deterministically chaotic dynamics) and chickenpox (for which Fig. 5d indicates purely additive noise superimposed on a seasonal cycle).

measles), with the fluctuations being additive noise (in contrast to measles, for which the fluctuations derive mainly from the dynamics).

The contrast between measles and chickenpox can be explained on biological grounds[21]. Measles has a fairly high 'basic reproductive rate' ($R_0 = 10$–20), and, after a brief interval of infectiousness, recovered individuals are immune and uninfectious for life; these conditions tend to produce long-lasting 'interepidemic' oscillations, with a period of about 2 years, even in the simplest models[22]. This, in combination with seasonal patterns, makes it plausible that measles has complex dynamics. Chickenpox is less 'highly reproductive' (with R_0 values of about 8 to 10), and may recrudesce as shingles in later life; this makes for an infection less prone to show periodicities other than basic seasonal ones associated with schools opening and closing, and therefore indicates seasonal cycles with additive noise. Furthermore, reporting was compulsory for measles but not for chickenpox over the time period in question, which itself would be likely to make sampling error greater for chickenpox. Whatever the underlying biological explanation, the patterns in Figs $4d$ and $5d$ differ in much the same way as those illustrated in Fig. $2a$ for the artificially generated time series of Fig. 2, a and b.

Marine plankton. A time series is provided by Allen's weekly record of marine planktonic diatoms gathered at Scripps Pier, San Diego, between 1920 and 1939 ($N = 830$). With the exception of the work of Tont[23], this collection of information has been little analysed, and not at all in the light of contemporary notions about nonlinear dynamics. The data comprise weekly totals of the numbers of individuals of all diatom species, tallied in daily seawater samples collected over ~20 years. As for our analysis of the measles and chickenpox data above, we do not 'smooth' the diatom data in any of the usual ways, although we take first-differences for reasons stated earlier. The resulting time series is shown in Fig. $6a$.

The results of using the first half of the diatom series to predict the second half are shown in the usual way in Fig. $6b$. Figure $6c$ shows ρ between predicted and observed results looking one time step ahead ($T_p = 1$), as a function of E. The optimum embedding dimension seems to be about 3. This value for E is consistent with our independent analysis of the data using the Grassberger–Procaccia algorithm, which indicates that $D \simeq 2$. Figure $6d$ shows ρ as a function of T_p (for $E = 3$). The consistent decay in predictive power as one extrapolates further into the future is consistent with the dynamics of the diatom population being partly governed by a chaotic attractor. This view is supported by the significantly better fit of the nonlinear predictor as compared with the optimal linear autoregressive model ($P < 0.0005$). We note, however, that deterministic chaos at best accounts for about 50% of the variance, with the rest presumably deriving from additive noise; the relatively low dimension of the attractor for diatoms compared with measles makes it plausible that the noisier fit of predicted weekly fluctuations in diatoms, versus the predicted monthly fluctuations in measles, reflects a much higher sampling variance for diatoms than for reported measles cases.

Conclusion

The forecasting technique discussed here is phenomenological in that it attempts to assess the qualitative character of a system's dynamics—and to make short-range predictions based on that understanding—without attempting to provide an understanding of the biological or physical mechanisms that ultimately govern the behaviour of the system. This often contrasts strongly with the laboratory and field-experiment approaches that are used to elucidate detailed mechanisms by, for example, many population biologists. The approach outlined here splits the time series into two parts, and makes inferences about the dynamical nature of the system by examining the way in which ρ (the correlation coefficient between predicted and observed results for the second part of the series) varies with prediction interval, T_p, and embedding dimension, E; given the low densities of most time series in population biology, we share Ruelle's[28] lack of confidence in a direct assessment of the dimension of any putative attractor by Grassberger–Procaccia or other algorithms. Our approach works with artificially generated time series (for which we know the actual dynamics, and the underlying mechanisms, by definition), and it seems to give sensible answers with the observed time series for measles, chickenpox and diatoms (deterministic chaos in one case, seasonal cycles with additive noise in another, and a mixture of chaos and additive noise in the third). We hope to see the approach applied to other examples of noisy time series in population biology, and in other disciplines in which time series are typically sparse. □

Received 12 July 1989; accepted 19 February 1990.

1. Lorenz, E. N. *J. atmos. Sci.* **26**, 636–646 (1969).
2. Tong, H. & Lim, K. S. *Jl R. statist. Soc.* **42**, 245–292 (1980).
3. Farmer, J. D. & Sidorowich, J. J. *Phys. Rev. Lett.* **62**, 845–848 (1987).
4. Priestly, M. B. *J. Time Series Analysis* **1**, 47–71 (1980).
5. Eckman, J. P. & Ruell, D. *Rev. mod. Phys.* **57**, 617–619 (1985).
6. Crutchfield, J. P. & MacNamara, B. S. *Complex Systems* **1**, 417–452 (1987).
7. Casdagli, M. *Physica D* **35**, 335–356 (1989).
8. Grassberger, P. & Procaccia, I. *Phys. Rev. Lett.* **50**, 346–369 (1983).
9. Schwartz, I. *J. math. Biol.* **21**, 347–361 (1985).
10. Schaffer, W. M. & Kot, M. *J. theor. Biol.* **112**, 403–407 (1985).
11. Schaffer, W. M. & Kot, M. in *Chaos: An Introduction* (ed. Holden, A. V.) (Princeton Univ. Press, 1986).
12. Schaffer, W. M., Ellner, S. & Kot, M. *J. math. Biol.* **24**, 479–523 (1986).
13. Schaffer, W. M., Olsen, L. F., Truty, G. L., Fulmer, S. L. & Graser, D. J. in *From Chemical to Biological Organization* (eds Markus, M., Muller, S. C. & Nicolis, G.) (Springer-Verlag, New York, 1988).
14. Yule, G. U. *Phil. Trans. R. Soc.* **A226**, 267–278 (1927).
15. Takens, F. in *Dynamical Systems and Turbulence* (Springer-Verlag, Berlin, 1981).
16. Farmer, J. D. & Sidorowich, J. J. in *Evolution, Learning and Cognition* (ed. Lee, Y. C.) (World Scientific, New York, 1989).
17. London, W. P. & Yorke, J. A. *Am. J. Epidem.* **98**, 453 (1973).
18. Pimm, S. L. & Redfearn, A. *Nature* **334**, 613–614 (1988).
19. Lawton, J. H. *Nature* **334**, 563 (1988).
20. Helsenstein, U. *Statist. Med.* **5**, 37–47 (1986).
21. Anderson, R. M. & May, R. M. *Nature* **318**, 323–329 (1985).
22. Anderson, R. M., Grenfell, B. T. & May, R. M. *J. Hyg.* **93**, 587–608 (1984).
23. Tont, S. A. *J. mar. Res.* **39**, 191–201 (1981).
24. Varosi, F., Grebogi, C. & Yorke, J. A. *Phys. Lett.* **A124**, 59–64 (1987).
25. Abarbanel, H. D., Kadtke, J. B. & Brown, R. *Phys. Rev.* **B41**, 1782–1807 (1990).
26. Mees, A. I. Research Report No. 8 (Dept Mathematics, University of Western Australia, 1989).
27. Drepper, F. R. in *Erodynamics* (eds Wolff, W., Soeder, C. J. & Drepper, F. R.) (Springer-Verlag, New York, 1988).
28. Ruelle, D. *Proc. R. Soc. A* (in the press).

ACKNOWLEDGEMENTS. We thank Henry Abarbanel, Martin Casdagli, Sir David Cox, Doyne Farmer, Arnold Mandell, John Sidorowich and Bill Schaffer for helpful comments, John McGowan and Sargun Tont for directing us to Allen's phytoplankton records, Alan Trombla for computing assistance, and the United States NSF and the Royal Society for supporting this research.

CONTROLLING CHAOS

The extreme sensitivity and complex behavior that characterize chaotic systems prohibit long-range prediction of their behavior but paradoxically allow one to control them with tiny perturbations.

Edward Ott and Mark Spano

A violent order is disorder; and
A great disorder is an order. These
Two things are one.
—Wallace Stevens, *Connoisseur of Chaos* (1942)

Scientists in many fields are recognizing that the systems they study often exhibit a type of time evolution known as chaos. Its hallmark is wild, unpredictable behavior, a state often perplexing and unwelcome to those who encounter it. Indeed this highly structured and deterministic phenomenon was in the past frequently mistaken for noise and viewed as something to be avoided in most applications. Recently researchers have realized that chaos can actually be advantageous in many situations and that when it is unavoidably present, it can often be controlled to obtain desired results. In this article, we present some of the basic ideas behind the feedback control of chaos, review a few illustrative experimental results and assess the status and future promise of the field.[1]

Dynamical systems

A dynamical system is one whose evolution is deterministic in the sense that its future motion is determined by its current state and past history. The system may be as simple as a swinging pendulum or as complicated as a turbulent fluid.

Figure 1 shows a simple mechanical system set up at the Naval Surface Warfare Center by William Ditto, now at the Georgia Institute of Technology, and his coworkers. The system consists of a magnetoelastic metal ribbon clamped at its lower end. The Young's modulus of the ribbon can be decreased by more than an order of magnitude by applying a magnetic field parallel to the ribbon, thereby causing it to buckle.[2] This highly nonlinear system is placed in a vertical magnetic field of the form $H(t) = H_{dc} + H_{ac}\cos(2\pi ft)$, where f is on the order of 1 hertz. The position of one point on the ribbon is monitored by an optical sensor located at a spot near the ribbon's base. With appropriate choices of the dc and ac field amplitudes,

the temporal motion of this simple system is observed to be chaotic. Here time is a continuous variable t. In other dynamical systems time can be a discrete integer-valued variable n. A continuous-time dynamical system like the ribbon can be represented by ordinary differential equations.

An example of a discrete-time dynamical system is a d-dimensional map,

$$\mathbf{y}_{n+1} = \mathbf{G}(\mathbf{y}_n) \qquad (1)$$

where $\mathbf{y}$ is d-dimensional. Given an initial condition $\mathbf{y}_0$ at time $n = 0$, the system state at time $n = 1$ is $\mathbf{y}_1 = \mathbf{G}(\mathbf{y}_0)$; at time $n = 2$, it is $\mathbf{y}_2 = \mathbf{G}(\mathbf{y}_1)$; and so on.

Physicists are used to dealing with continuous-time systems. Newton's equations of motion, Maxwell's equations and Schrödinger's equation are all formulated in continuous time. However, continuous-time dynamical systems can often profitably be reduced to discrete-time systems.

For example, assume that we sample the state of the periodically driven magnetoelastic ribbon (figure 1) once every drive period—at the times $t = T, 2T, 3T, \ldots, nT$, where $T = 1/f$. Then $\mathbf{y}_n$ denotes the system state at time nT. Because the system is deterministic, the state at time $(n + 1)T$ is uniquely determined by the state at time nT. That is, an equation of the form of equation 1 holds. It turns out to be quite easy to do this sampling for driven experimental systems by strobing the experimental data acquisition at the drive frequency.

In experiments it is sometimes difficult to measure the full state vector of the system. As an extreme example, assume that one can measure only a single scalar function of the system state. In this case, it has been shown that a delay coordinate embedding provides a useful representation of the system state.[3] For example, for a discrete-time system, if the observed scalar is w_n, then the delay coordinate vector $\mathbf{W}_n$ replaces the vector $\mathbf{y}_n$ from equation 1, where

$$\mathbf{W}_n = [w_n, w_{n-1}, w_{n-2}, \ldots, w_{n-(q-1)}] \qquad (2)$$

The number of delays q should be large enough to reproduce the dynamics of the system. Essentially the information contained in a measurement of all of a system's variables at a single time is reconstructed from measurements of a single variable at q different times.

A common feature of (non-Hamiltonian) dynamical systems is the presence of "attractors." If typical initial conditions located in some region of phase space (or a

---

EDWARD OTT *is a professor of physics and of electrical engineering and a member of the Institute for Plasma Research and the Institute for Systems Research at the University of Maryland, College Park.* MARK SPANO *is a senior research physicist at the Naval Surface Warfare Center, in Silver Spring, Maryland, and a visiting scholar at the Applied Chaos Laboratory of the Georgia Institute of Technology, in Atlanta.*

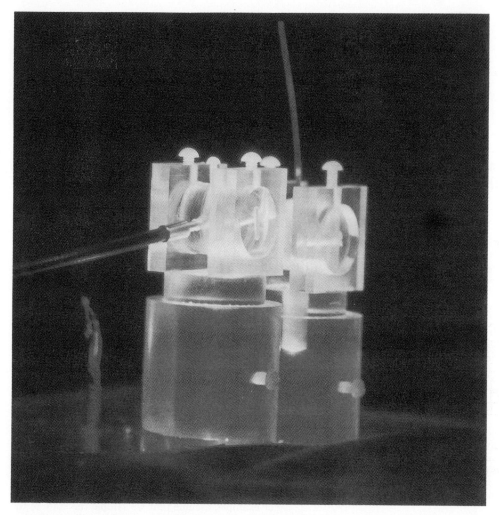

MAGNETOELASTIC RIBBON that undergoes chaotic motion. The 10-cm-long metallic glass ribbon changes its Young's modulus by more than an order of magnitude in response to an applied magnetic field. The related change in its stiffness causes it to buckle under the force of gravity. An optical sensor measures its position at a single point near its base once each drive period. **FIGURE 1**

delay coordinate embedding) approach a set of values asymptotically with time, we call that set the attractor. A chaotic attractor is a geometric object that is neither pointlike nor space filling. Chaotic attractors typically have fractal (noninteger) dimensions and so are often called strange attractors. Figure 2b shows a chaotic attractor (gray) from the magnetoelastic ribbon experiment pictured in figure 1. All initial conditions eventually evolve into motion confined to this set of points.

Once on the attractor, the system state bounces around it ergodically. That is, the state eventually comes arbitrarily close to any point on the attracting set. Even when the actual state space dimension, (the dimensionality of **y** in equation 1) is large (or, as in the case of a spatially continuous system like the ribbon, infinite), the fractal dimension of the strange attractor is often fairly low (between 1 and 2 for the ribbon attractor in figure 2b). For the most part, chaos control techniques have been formulated and implemented for low-dimensional strange attractors (attractor dimensions less than four or five), and we will limit our discussion to this situation.

Sensitivity and orbit complexity

The two most common ways of characterizing chaos are exponential sensitivity and orbit complexity. These are not independent properties, but rather two sides of the same chaos coin. Either may be viewed as a way of defining chaos.

Exponential sensitivity refers to the fact that if we consider two chaotic orbits initially displaced only slightly from each other, then the displacement between the two orbits grows exponentially with time. What might have been a very tiny separation between the orbits eventually becomes a large displacement (on the order of the attractor size). Thus small errors eventually defeat any attempt to predict the exact longtime evolution of a chaotic system.

Orbit complexity means that many different kinds of motion are possible on a chaotic attractor. One manifestation of this is that chaotic attractors typically have embedded within them an infinity of unstable periodic orbits. By a periodic orbit we mean an orbit that repeats itself after some characteristic time. The colored regions in figure 2b mark the location of periodic orbits on the attractor. In particular, an initial condition $\mathbf{y}_0$ in the center of the green dots maps to a $\mathbf{y}_1$ located at the same point, $\mathbf{y}_1 = \mathbf{y}_0$, which maps to a $\mathbf{y}_2$ also at the same point, $\mathbf{y}_2 = \mathbf{y}_0$; this point is a period-one orbit. An initial condition in the center of one of the red regions maps to the center of the other red region and then back to the original red region; the orbit cycles repeatedly between the two red regions, which makes it a period-two orbit. The four blue regions in figure 2b show the location of a period-four orbit.

The periodic orbits embedded in a chaotic attractor are all unstable in that if one displaces the system state slightly from a periodic orbit, this displacement grows exponentially in time. Thus periodic orbits are typically

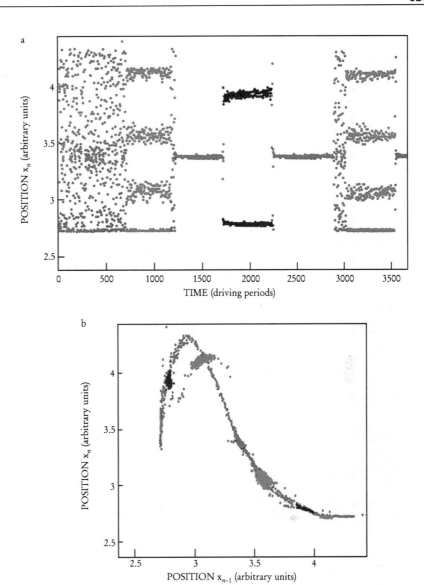

CHAOTIC AND CONTROLLED MOTION of a magnetoelastic ribbon prone to chaotic motion. **a:** Position of a point on the ribbon, plotted as a time series. **b:** The same data plotted as a delay coordinate embedding using a delay of one drive period. The data in gray indicate chaotic motion; green, period-one control; red, period-two control; blue, period-four control. Note that because the period-four control is achieved by altering the control signal once every four driving periods, the sensitivity of the chaos to small perturbations (including noise) causes the data to become progressively more noisy, until it approaches the original unstable fixed point (lowest trace), where the correcting signal is applied anew. Effective higher-period control requires that the control signal be adjusted more often—every few drive periods. This is the case for the diode resonator (in figure 4). **FIGURE 2**

not observed in a free-running chaotic system. Very infrequently a chaotic orbit may, in its ergodic wandering, approach close to a given periodic orbit. If this happens, the orbit may approximately follow the periodic orbit for a few cycles, but it subsequently moves away, resuming its wandering over the chaotic attractor.

Control based on unstable periodic orbits

Both exponential sensitivity and orbit complexity provide points of view that are relevant to controlling the motion on a chaotic attractor. In most of what follows we will concentrate on orbit complexity—in particular, the presence of embedded unstable periodic orbits. Later, when we briefly discuss the idea of targeting, exponential sensitivity will assume center stage.

Because we wish to control the motion using only small perturbations, we do not expect to be able to create orbits very different from those already allowed by the dynamics on the strange attractor. Thus we will seek to exploit the already existing unstable periodic orbits embedded in the attractor.

The approach is as follows:[4] We first determine some of the low-period unstable orbits that are embedded in the attractor. For each such orbit we determine the system performance that would result if that periodic orbit were actually followed by the system. (If we use a laser as an example, the relevant measure of performance might be its output power at a given wavelength.) Typically, some of the periodic orbits will yield improved performance compared to the free-running chaotic motion, and some will not. We tailor our small time-dependent controls in such a way as to stabilize one of the unstable periodic orbits that yields improved performance.

Loosely speaking, we can think of the controls as small kicks that place the actual orbit back onto the desired unstable periodic orbit. We apply these kicks whenever we sense that the actual orbit has wandered slightly away from the desired orbit. This wandering might occur, for example, because the orbit is unstable and noise or some other perturbation is present. Because chaotic orbits are ergodic on the attractor, they eventually wander close to the desired periodic orbit and then because of this proximity, can be captured by a small control. Once captured, the required controls remain small—on the order of the inherent system noise.

If chaos control is practical in a system, then the

presence of chaos can be an advantage. Any one of a number of different orbits can be stabilized, and one can select the orbit that gives the best system performance. Thus we have the flexibility of easily switching the system behavior by controlling a different periodic orbit. On the other hand, if the attractor is not chaotic but is, say, periodic, then small perturbations typically can change the orbit only slightly. We are stuck with whatever system performance the stable periodic orbit gives, and we have no option for substantial improvement, short of making large alterations in the system.

To be concrete, one might wish to build chaos into a system for the same reason that the newer "fly-by-wire" fighter planes are designed to be unstable: The built-in instability gives the planes more maneuverability and lets them respond quickly to the pilot's commands. In a similar fashion, building chaos into a system might provide it with more flexibility than would otherwise be possible.

Methods for controlling chaos

There are various methods by which one may control a chaotic system. One such method is based on the fact that the points on the attractor do not bounce around randomly, but instead approach and depart from the vicinity of unstable periodic orbits in a highly structured fashion. In a region near each periodic orbit, the system state point tends to move toward the periodic orbit from certain positions (stable manifolds) and to move away from the periodic orbit from other positions (unstable manifolds). The exponentially increasing speed with which the state point moves in each direction is governed by the stable and unstable eigenvalues, respectively.

For simplicity, consider a two-dimensional map that has one stable direction (implying an eigenvalue with magnitude less than unity) and one unstable direction (with eigenvalue greater than unity). This situation is analogous to a ball rolling under the influence of gravity on a saddle-shaped surface. Chaos control reduces to finding a way to move this saddle so that the ball remains balanced on the saddle's (unstable) equilibrium point.

The procedure goes as follows (see figure 3): When a point begins to move away from the desired periodic orbit (the saddle's equilibrium point) along the unstable manifold, we shift the saddle slightly (by perturbing some system parameter) so that the point now lies on the stable manifold. The natural motion of the system will now tend to move it toward the unstable periodic point rather than away from it. In a perfect world we could turn off the perturbation when the system arrives at the desired periodic orbit. But small errors in our control calculations and the noise inherent in any experimental system will tend to knock the system off this desired orbit again, so we will repeat the previous step as needed to keep the system under control.

An important point to remember is that this procedure does not require a model of the system. All that is needed is to determine experimentally the local geometry around the chosen unstable periodic point: the position of the saddle, the stable and unstable directions, the steepness of these directions (that is, the stable and unstable eigenvalues) and, finally, the shift in the position of the saddle with a small change in some system parameter. As has

---

CHAOS CONTROL VS. CHAOS MAINTENANCE

William Ditto of the Georgia Institute of Technology, Steven Schiff of Children's Hospital in Washington, DC, and I recently conducted experiments on the hippocampus of the temporal lobe of the rat brain.[16] When bathed in artificial cerebrospinal fluid containing high levels of potassium, the brain exhibits spontaneous bursts of synchronized neuronal activity in a portion of the hippocampus designated CA3. These bursts can trigger seizure-like discharges in a nearby region, CA1. As in the heart experiments, we measured and plotted the interburst intervals as an embedding. Employing the same techniques as in the heart experiment, control was achieved. (See figure below.)

The relevance of these experiments to such brain-seizure disorders as epilepsy is not clear. It may be that controlling these bursts to make them more periodic may actually increase the seizure activity. As an alternative, we implemented a new technique for the maintenance of chaos in this system. Sometimes dubbed "anti-control," this technique is actually a form of control that keeps the system *away* from low-period orbits. The figure shows a crude application for the neural system. Much improved versions have been detailed computationally by Weiming Yang and his coworkers[17] at Florida Atlantic University and the University of Maryland and implemented experimentally by Visarath In and his coworkers at Georgia Tech.[18]

— MARK SPANO

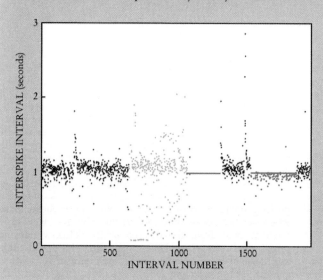

DATA FROM A RAT BRAIN hippocampus during exposure to high levels of potassium. The data in red indicate chaotic interludes; green and blue indicate double-pulse control and single-pulse control, respectively. Also shown (in pink) is a related method known variously as "anti-control" or the "maintenance of chaos."

been demonstrated in many cases, these can be determined in real time during an experiment.

Real experiments are always subject to noise, both in the system itself and in the measurement process. As might be expected, the basic effect is that noise increases the minimum size of the control perturbations required for effective operation.

Controlling chaos experimentally

Chaos control techniques of the type just described have been implemented in a large number of experiments across a wide variety of fields. Here are a few representative examples.

A simple mechanical system. As mentioned above, using discrete time measurements, a simple delay coordinate embedding was constructed for the experiment in figures 1 and 2 by plotting the current position of the ribbon against its position one drive cycle previously ($q = 2$ in equation 2). The time evolution of the system reflected in this embedding was used to identify the stable and unstable manifolds. The application of small perturbations to the dc magnetic field, according to the scheme described above, controlled the system onto an unstable period-one motion embedded in the attractor. By embedding the data with delays of two, three, four or more drive cycles, orbits of higher period were also located. The same control technique was used to select out these higher period motions.

Figure 2a shows the time series data moving from chaos to period-four control to period-one control to period-two control and so on, with short interludes of chaos between each instance of control. These interludes arise because of the necessity to wait for the ergodicity of the chaotic system to bring it from the vicinity of the previous control point to the neighborhood of the new control point. Such waiting times can be mitigated by employing a technique known as targeting, which we will describe later.

Electronics. Earle Hunt of Ohio University used modification of the method described above to control a driven, chaotic diode resonator circuit.[5] In this method, one periodically applies a control proportional to deviations of the chaotic variable from a set point. The coefficient of proportionality is determined empirically by varying the control parameter until control is achieved. Although the choice of periodic orbits is less easily made, the method is simple, entails little overhead and enables control at frequencies exceeding 50 kHz. As with the original control scheme, unstable periodic orbits may be stabilized with an extremely small feedback control. However, if a slightly larger feedback signal is allowed, one may also stabilize orbits of very high periods. Figure 4a shows a delay coordinate embedding for Hunt's system when it oscillates chaotically, while figure 4b shows the same embedding during period-62 control.

Lasers. Rajarshi Roy and his coworkers at Georgia performed an experiment of great practical interest.[6] They examined the output of a solid-state Nd-doped yttrium–aluminum–garnet laser with a KTP frequency-doubling crystal in the laser cavity. As the pumping power of the laser was increased, the output became chaotic. The system was observed to undergo chaotic relaxation oscillations with a characteristic time scale of roughly 10 microseconds. Using an embedding constructed with this natural time as the delay, the researchers were able to control the system. Figure 5a shows the laser intensity as a function of time during chaotic oscillation. The corresponding fast-Fourier transform is broad and of lower amplitude than that obtained during period-one control, as seen in figure 5b. We see that the output power at specific Fourier frequencies can be boosted significantly by these methods.

Chemical systems. The oscillatory Belousov–Zhabotinsky chemical reaction becomes chaotic for suitable values of the flow rates of reactants into the reaction tank. The system may be monitored electrically by measuring the voltage of a bromide electrode placed in the tank. Ken Showalter and his group at West Virginia University were able to stabilize both period-one and period-two oscillations in the reactor by adjusting one of the flow rates using the same method used on the magnetoelastic ribbon.[7]

Roger Rollins and his coworkers at Ohio University applied chaos control to an electrochemical cell.[7] Their

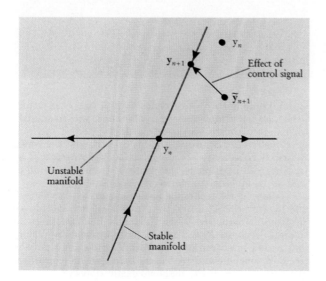

SYSTEM STATE POINT'S TENDENCIES near any periodic orbit. By moving the saddle, the control signal puts y_{n+1} on the stable manifold of y_*. **FIGURE 3**

DIODE RESONATOR DATA. The system consists of two coupled diode resonators, each a series combination of an inductor and a p–n junction diode. It is driven sinusoidally at 50–100 kHz. **a:** Poincaré section formed by sampling the currents through the two branches each drive cycle. **b:** Stabilized period-62 orbit. Only half of the attractor is shown. (Courtesy of Earle Hunt, Ohio University.) **FIGURE 4**

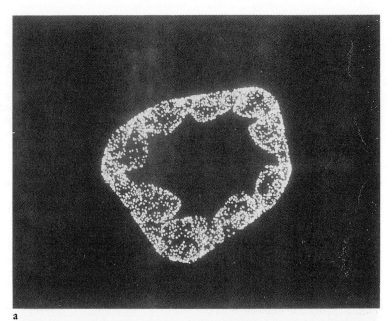

a

b

approach, based on the work of Ute Dressler and Gregor Nitsche of the Daimler-Benz Research Institute in Germany,[4] uses a recursive method to make available a wider choice of system parameters for control.

Heart tissue. An experiment on an *in vitro* rabbit heart septum used the drug ouabain to induce arrhythmias in the autonomous beating of the heart tissue.[8] A discrete-time embedding was constructed using the intervals between heartbeats as the system variable of interest. For this induced arrhythmia, the presence of deterministic chaos was confirmed by the observation of repeated approaches of the system state to a period-one orbit, with each approach along the same stable direction and with corresponding departures along the same unstable direction. In this case, however, it was not possible to find a system parameter that would move the system state point onto the stable manifold. But it *was* possible to intervene directly in the system by injecting a premature heartbeat at an interval timed to place the system state point onto the stable manifold. The dynamics of the system then naturally tended to carry it toward the (unstable) period-one motion. However, because it was possible only to shorten the interbeat interval and not to lengthen it, the control achieved was, at best, period three. Further control experiments are studying an artificially perfused canine heart undergoing ventricular fibrillation.[9]

New directions

The arena of chaos control is growing at an accelerating pace. Here we present some new directions; one of the more exciting and controversial results appears in the box on page 37.

Communications. In work by Scott Hayes and his coworkers at the Army Research Laboratory, a chaotically behaving oscillator is manipulated by a small control so that the oscillator output can carry information. As an example, they controlled a signal from an oscillator whose free-running state is a sequence of positive and negative peaks. Taking these peaks to represent the binary digits 1 and 0, respectively, they used the controlled signal to encode information.[10] The same idea should be applicable to other kinds of signal sources, such as a chaotic laser.

Tracking. One situation often encountered in experiments is that the system undergoes some slow change with time (sometimes called "drift"), either intentionally or unintentionally. Once chaos control has been established, an important concern is whether it will be possible to maintain control despite this drift. Recent methods developed by Ira Schwartz and his coworkers at the Naval Research Laboratory accommodate such slow changes by tracking the location of the controlled unstable orbit as well as its stability properties.[11]

This tracking technique has been used by Tom Carroll and his coworkers (also of the Naval Research Lab) in an electronic circuit, by Zelda Gills and her coworkers at Georgia Tech in a multimode laser system, by Valery Petrov and his coworkers at West Virginia University in the B–Z reaction and by Visarath In and his coworkers at Georgia Tech in the magnetoelastic ribbon.[12]

Targeting. As we saw in the ribbon experiment, it might take an unacceptably long time for the ergodically wandering, uncontrolled system to come close enough to the desired orbit to be captured by a small control. Given an initial condition A and a small target region B, how can we apply small controlling perturbations in such a

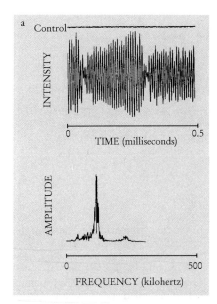
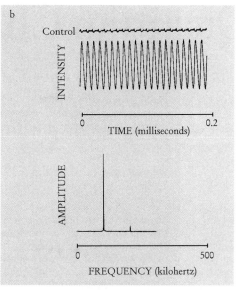

LASER DATA. a: Chaotic fluctuations in the intensity of a Nd-doped yttrium–aluminum–garnet laser and the corresponding fast-Fourier transform of that data. **b:** The same laser during period-one control. The Fourier transform has narrowed dramatically and increased significantly in amplitude. (Courtesy of Rajarshi Roy, Georgia Institute of Technology.) **FIGURE 5**

way as to bring the orbit quickly to B? This should be possible because of the exponential sensitivity of chaotic orbits to small perturbations. If the small perturbations are carefully chosen, then one might hope to effect a desired large change. Such "targeting" has been developed and experimentally implemented.[13]

An example predating these considerations, but nevertheless illustrating the general principle, was provided by NASA scientists, who were able to turn the International Sun–Earth Explorer-3 spacecraft into the International Cometary Explorer by redirecting it from its orbit near Earth to an encounter with a comet halfway across the solar system, using only small amounts of fuel.[14] This first-ever cometary encounter was made possible by utilizing the chaotic sensitivity of the three-body system in celestial mechanics, those bodies being the Earth, the Moon and the spacecraft.

The last few years have seen substantial theoretical, numerical and experimental research demonstrating the feasibility of controlling systems that behave chaotically.[15] The remaining question is whether it will prove possible to move from laboratory demonstrations on model systems to real-world situations of economic, engineering and societal importance. The ubiquity of chaotic dynamics leads us to suspect that this will indeed be the case.

References

1. For reviews of controlling chaos, see T. Shinbrot, C. Grebogi, E. Ott, J. A. Yorke, Nature **363**, 411 (1993); W. L. Ditto, L. Pecora, Sci. Am., August 1993, p. 78; G. Chen, X. Dong, Int. J. Bifurcations and Chaos **3**, 1363 (1993); E. Hunt, G. Johnson, IEEE Spectrum, November 1993, p. 32; R. Roy, Z. Gills, K. S. Thornburg, Opt. Photon. News, May 1994, p. 8.
2. W. L. Ditto, S. N. Rauseo, M. L. Spano, Phys. Rev. Lett. **65**, 3211 (1990).
3. F. Takens, in *Dynamical Systems and Turbulence*, D. A. Rand, L. S. Young, eds., Springer-Verlag, Berlin (1981), p. 366. For reviews of this topic, see H. D. I. Abarbanel, R. Brown, J. J. Sidorowich, L. S. Tsimiring, Rev. Mod. Phys. **65**, 1331 (1993); E. Ott, T. Sauer, J. A. Yorke, eds., *Coping with Chaos*, Wiley, New York (1994), E. Ott, *Chaos in Dynamical Systems*, Cambridge U. P., New York (1993).
4. E. Ott, C. Grebogi, J. A. Yorke, Phys. Rev. Lett. **64**, 1996 (1990). For a discussion of control using delay coordinate embeddings, see U. Dressler, G. Nitsche, Phys. Rev. Lett. **68**, 1 (1992).
5. E. R. Hunt, Phys. Rev. Lett. **67**, 1953 (1991). R. W. Rollins, P. Parmananda, P. Sherard, Phys. Rev. E **47**, R780 (1993).
6. R. Roy, T. W. Murphy Jr, T. D. Maier, Z. Gills, Phys. Rev. Lett. **68**, 1259 (1992).
7. V. Petrov, V. Gaspar, J. Masere, K. Showalter, Nature **361**, 240 (1993). P. Parmananda, P. Sherard, R. W. Rollins, H. D. Dewald, Phys. Rev. E **47**, R3003 (1993).
8. A. Garfinkel, M. L. Spano, W. L. Ditto, J. Weiss, Science **257**, 1230 (1992).
9. F. X. Witkowski, K. M. Kavanagh, P. A. Penkoske, R. Plonsey, M. L. Spano, W. L. Ditto, D. T. Kaplan, preprint, available from Spano.
10. S. Hayes, C. Grebogi, E. Ott, A. Mark, Phys. Rev. Lett. **73**, 1781 (1994). S. Hayes, C. Grebogi, E. Ott, Phys. Rev. Lett. **70**, 3031 (1993).
11. I. Schwartz, I. Triandaf, Phys. Rev. A **46**, 7439 (1992).
12. T. L. Carroll, I. Triandaf, I. Schwartz, L. Pecora, Phys. Rev. A **46**, 6189 (1992). Z. Gills, C. Iwata, R. Roy, I. B. Schwartz, I. Triandaf, Phys. Rev. Lett. **69**, 3169 (1992). V. Petrov, M. J. Crowley, K. Showalter, Phys. Rev. Lett. **72**, 2955 (1994). V. In, W. L. Ditto, M. L. Spano, Phys. Rev. E **51**, 2689 (1995).
13. T. Shinbrot, E. Ott, C. Grebogi, J. A. Yorke, Phys. Rev. Lett. **65**, 3215 (1990). T. Shinbrot, W. Ditto, C. Grebogi, E. Ott, M. Spano, J. A. Yorke, Phys. Rev. Lett. **68**, 2863 (1992).
14. R. Farquhar, D. Muhonen, L. C. Church, J. Astronaut. Sci. **33**, 235 (1985).
15. For discussion of a nonfeedback technique for eliminating chaos, see A. Hubler, Helv. Phys. Acta **62**, 343 (1989); E. A. Jackson, Phys. Lett. A **151**, 478 (1990) and references therein. An example of another nonfeedback technique that attempts to entrain a chaotic system to an externally applied periodic perturbation is discussed in Z. Qu, G. Hu, G. Yang, G. Qin, Phys. Rev. Lett. **74**, 1736 (1995) and references therein.
16. S. J. Schiff, K. Jerger, D. H. Duong, T. Chang, M. L. Spano, W. L. Ditto, Nature **370**, 615 (1994).
17. W. Yang, M. Ding, A. J. Mandell, E. Ott, Phys. Rev. E **51**, 102 (1995).
18. V. In, S. Mahan, W. L. Ditto, M. L. Spano, Phys. Rev. Lett. (1995), in press. ■

Quantum Chaos

*Does chaos lurk in the smooth, wavelike quantum world?
Recent work shows that the answer is yes—symptoms of chaos enter
even into the wave patterns associated with atomic energy levels*

by Martin C. Gutzwiller

In 1917 Albert Einstein wrote a paper that was completely ignored for 40 years. In it he raised a question that physicists have only recently begun asking themselves: What would classical chaos, which lurks everywhere in our world, do to quantum mechanics, the theory describing the atomic and subatomic worlds? The effects of classical chaos, of course, have long been observed—Kepler knew about the irregular motion of the moon around the earth, and Newton complained bitterly about the phenomenon. At the end of the 19th century, the American astronomer George William Hill demonstrated that the irregularity is the result entirely of the gravitational pull of the sun. Shortly thereafter, the great French mathematician-astronomer-physicist Henri Poincaré surmised that the moon's motion is only a mild case of a congenital disease affecting nearly everything. In the long run, Poincaré realized, most dynamic systems show no discernible regularity or repetitive pattern. The behavior of even a simple system can depend so sensitively on its initial conditions that the final outcome is uncertain [see "The Amateur Scientist," page 144].

At about the time of Poincaré's seminal work on classical chaos, Max Planck started another revolution, which would lead to the modern theory of quantum mechanics. The simple systems that Newton had studied were investigated again, but this time on the atomic scale. The quantum analogue of the humble pendulum is the laser; the flying cannonballs of the atomic world consist of beams of protons or electrons, and the rotating wheel is the spinning electron (the basis of magnetic tapes). Even the solar system itself is mirrored in each of the atoms found in the periodic table of the elements.

Perhaps the single most outstanding feature of the quantum world is its smooth and wavelike nature. This feature leads to the question of how chaos makes itself felt when moving from the classical world to the quantum world. How can the extremely irregular character of classical chaos be reconciled with the smooth and wavelike nature of phenomena on the atomic scale? Does chaos exist in the quantum world?

Preliminary work seems to show that it does. Chaos is found in the distribution of energy levels of certain atomic systems; it even appears to sneak into the wave patterns associated with those levels. Chaos is also found when electrons scatter from small molecules. I must emphasize, however, that the term "quantum chaos" serves more to describe a conundrum than to define a well-posed problem.

Considering the following interpretation of the bigger picture may be helpful in coming to grips with quantum chaos. All our theoretical discussions of mechanics can be somewhat artificially divided into three compartments [*see illustration on page 80*]—although nature recognizes none of these divisions.

Elementary classical mechanics falls in the first compartment. This box contains all the nice, clean systems exhibiting simple and regular behavior, and so I shall call it R, for regular. Also contained in R is an elaborate mathematical tool called perturbation theory, which is used to calculate the effects of small interactions and extraneous disturbances, such as the influence of the sun on the moon's motion around the earth. With the help of perturbation theory, a large part of physics is understood nowadays as making relatively mild modifications of regular systems. Reality, though, is much more complicated; chaotic systems lie outside the range of perturbation theory, and they constitute the second compartment.

Since the first detailed analyses of the systems of the second compartment were done by Poincaré, I shall name this box P in his honor. It is stuffed with the chaotic dynamic systems that are the bread and butter of science [see "Chaos," by James P. Crutchfield, J. Doyne Farmer, Norman H. Packard and Robert S. Shaw; SCIENTIFIC AMERICAN, December 1986]. Among these systems are all the fundamental problems of mechanics, starting with three, rather than only two, bodies interacting with one another, such as the earth, moon and sun, or the three atoms in the water molecule, or the three quarks in the proton.

Quantum mechanics, as it has been practiced for about 90 years, belongs in the third compartment, called Q. After the pioneering work of Planck, Einstein and Niels Bohr, quantum mechanics was given its definitive form in four short years, starting in 1924. The seminal work of Louis de Broglie, Werner Heisenberg, Erwin Schrödinger, Max Born, Wolfgang Pauli and Paul Dirac has stood the test of the laboratory without the slightest lapse. Miraculously, it provides physics with a mathematical framework that, according to Dirac, has yielded a deep understanding of "most of physics and all of chemistry." Nevertheless, even though most

---

MARTIN C. GUTZWILLER is a member of the research staff at the IBM Thomas J. Watson Research Center in Yorktown Heights, N.Y. He also serves as adjunct professor of metallurgy at the Columbia University School of Engineering. Born in Switzerland, he received his education in Swiss public schools and at the Federal Institute of Technology in Zurich, where he obtained a degree in physics and mathematics in 1950. His Ph.D. in physics was awarded by the University of Kansas in 1953. Gutzwiller joined IBM in 1960 after working for seven years in the Exploration and Production Research Laboratory of Shell Oil Company. At IBM his focus was first on the interaction of electrons in metals and is now on the relations between classical and quantum mechanics. He is also interested in celestial mechanics and the history of astronomy and physics. In addition, he is an amateur musician, a collector of old science books and a hiker.

10.6 Quantum Chaos

physicists and chemists have learned how to solve special problems in quantum mechanics, they have yet to come to terms with the incredible subtleties of the field. These subtleties are quite separate from the difficult, conceptual issues having to do with the interpretation of quantum mechanics.

The three boxes R (classic, simple systems), P (classic chaotic systems) and Q (quantum systems) are linked by several connections. The connection between R and Q is known as Bohr's correspondence principle. The correspondence principle claims, quite reasonably, that classical mechanics must be contained in quantum mechanics in the limit where objects become much larger than the size of atoms. The main connection between R and P is the Kolmogorov-Arnold-Moser (KAM) theorem. The KAM theorem provides a powerful tool for calculating how much of the structure of a regular system survives when a small perturbation is introduced, and the theorem can thus identify perturbations that will cause a regular system to undergo chaotic behavior.

Quantum chaos is concerned with establishing the relation between boxes P (chaotic systems) and Q (quantum systems). In establishing this relation, it is useful to introduce a concept called phase space. Quite amazingly, this concept, which is now so widely exploited by experts in the field of dynamic systems, dates back to Newton.

The notion of phase space can be found in Newton's *Mathematical Princi-*

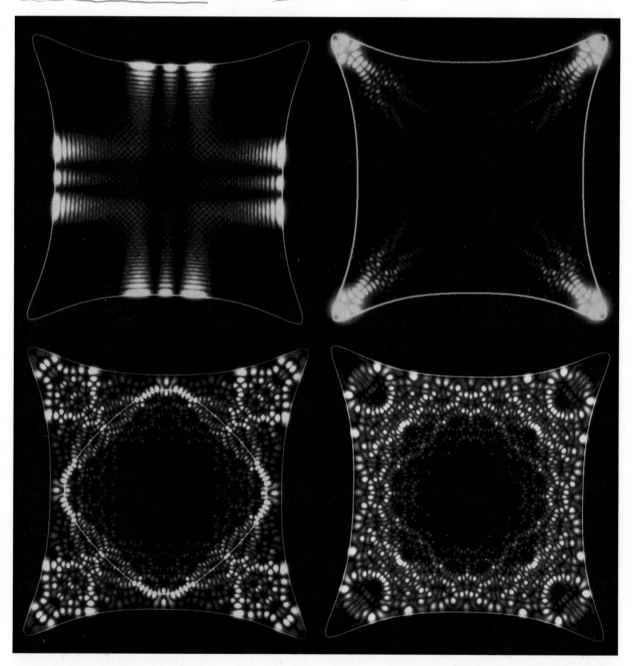

STATIONARY STATES, or wave patterns, associated with the energy levels of a Rydberg atom (a highly excited hydrogen atom) in a strong magnetic field can exhibit chaotic qualities. The states shown in the top two images seem regular; the bottom two are chaotic. At the bottom left, the state lies mostly along a periodic orbit; at the bottom right, it does not and is difficult to interpret, except for the four mirror symmetries with respect to the vertical, horizontal and two diagonal lines.

SCIENTIFIC AMERICAN *January 1992* 79

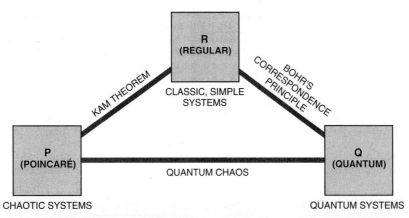

MECHANICS is traditionally (and artificially) divided into the three compartments depicted here, which are linked together by several connections. Quantum chaos is concerned with establishing the relation between boxes P and Q.

ples of *Natural Philosophy,* published in 1687. In the second definition of the first chapter, entitled "Definitions," Newton states (as translated from the original Latin in 1729): "The quantity of motion is the measure of the same, arising from the velocity and quantity of matter conjointly." In modern English, this means that for every object there is a quantity, called momentum, which is the product of the mass and velocity of the object.

Newton gives his laws of motion in the second chapter, entitled "Axioms, or Laws of Motion." The second law says that the change of motion is proportional to the motive force impressed. Newton relates the force to the change of momentum (not to the acceleration, as most textbooks do).

Momentum is actually one of two quantities that, taken together, yield the complete information about a dynamic system at any instant. The other quantity is simply position, which determines the strength and direction of the force. Newton's insight into the dual nature of momentum and position was put on firmer ground some 150 years later by two mathematicians, William Rowan Hamilton and Karl Gustav Jacob Jacobi. The pairing of momentum and position is no longer viewed in the good old Euclidean space of three dimensions; instead it is viewed in phase space, which has six dimensions, three dimensions for position and three for momentum.

The introduction of phase space was a wonderful step from a mathematical point of view, but it represents a serious setback from the standpoint of human intuition. Who can visualize six dimensions? In some cases, fortunately, the phase space can be reduced to three or, even better, two dimensions.

Such a reduction is possible in examining the behavior of a hydrogen atom in a strong magnetic field. The hydrogen atom has long been a highly desirable system because of its simplicity: a lone electron moves around a lone proton. And yet the classical motion of the electron becomes chaotic when the magnetic field is turned on. How can we claim to understand physics if we cannot explain this basic problem?

Under normal conditions, the electron of a hydrogen atom is tightly bound to the proton. The behavior of the atom is governed by quantum mechanics. The atom is not free to take on any arbitrary energy; it can take on only discrete, or quantized, energies. At low energies, the allowed values are spread relatively far apart. As the energy of the atom is increased, the atom grows bigger, because the electron moves farther from the proton, and the allowed energies get closer together. At high enough energies (but not too high, or the atom will be stripped of its electron!), the allowed energies get very close together into what is effectively a continuum, and it now becomes fair to apply the rules of classical mechanics.

Such a highly excited atom is called a Rydberg atom [see "Highly Excited Atoms," by Daniel Kleppner, Michael G. Littman and Myron L. Zimmerman; SCIENTIFIC AMERICAN, May 1981]. Rydberg atoms inhabit the middle ground between the quantum and the classical worlds, and they are therefore ideal candidates for exploring Bohr's correspondence principle, which connects boxes Q (quantum phenomena) and R (classic phenomena). If a Rydberg atom could be made to exhibit chaotic behavior in the classical sense, it might provide a clue as to the nature of quantum chaos and thereby shed light on the middle ground between boxes Q and P (chaotic phenomena).

A Rydberg atom exhibits chaotic behavior in a strong magnetic field, but to see this behavior we must reduce the dimension of the phase space. The first step is to note that the applied magnetic field defines an axis of symmetry through the atom. The motion of the electron takes place effectively in a two-dimensional plane, and the motion around the axis can be separated out; only the distances along the axis and from the axis matter. The symmetry of motion reduces the dimension of the phase space from six to four.

Additional help comes from the fact that no outside force does any work on the electron. As a consequence, the total energy does not change with time. By focusing attention on a particular value of the energy, one can take a three-dimensional slice—called an energy shell—out of the four-dimensional phase space. The energy shell allows one to watch the twists and turns of the electron, and one can actually see something resembling a tangled wire sculpture. The resulting picture can be simplified even further through a simple idea that occurred to Poincaré. He suggested taking a fixed two-dimensional plane (called a Poincaré section, or a surface of section) through the energy shell and watching the points at which the trajectory intersects the surface. The Poincaré section reduces the tangled wire sculpture to a sequence of points in an ordinary plane.

A Poincaré section for a highly excited hydrogen atom in a strong magnetic field is shown on the opposite page. The regions of the phase space where the points are badly scattered indicate chaotic behavior. Such scattering is a clear symptom of classical chaos, and it allows one to separate systems into either box P or box R.

What does the Rydberg atom reveal about the relation between boxes P and Q? I have mentioned that one of the trademarks of a quantum mechanical system is its quantized energy levels, and in fact the energy levels are the first place to look for quantum chaos. Chaos does not make itself felt at any particular energy level, however; rather its presence is seen in the spectrum, or distribution, of the levels. Perhaps somewhat paradoxically, in a nonchaotic quantum system the energy levels are distributed randomly and without correlation, whereas the energy levels of a chaotic quantum system exhibit strong correlations [*see top illustration on page 82*]. The levels of the regular system are often close to one another, because a regular system is composed of smaller subsystems that are completely decoupled. The energy levels of the chaotic system, however, almost seem to be aware of one another and try to keep a safe distance. A chaotic system cannot be decomposed; the motion along one coordinate axis is always coupled to what happens along the other axis.

The spectrum of a chaotic quantum system was first suggested by Eugene P. Wigner, another early master of quantum mechanics. Wigner observed, as had many others, that nuclear physics does not possess the safe underpinnings of atomic and molecular physics; the origin of the nuclear force is still not clearly understood. He therefore asked whether the statistical properties of nuclear spectra could be derived from the assumption that many parameters in the problem have definite, but unknown, values. This rather vague starting point allowed him to find the most probable formula for the distribution. Oriol Bohigas and Marie-Joya Giannoni of the Institute of Nuclear Physics in Orsay, France, first pointed out that Wigner's distribution happens to be exactly what is found for the spectrum of a chaotic dynamic system.

Chaos does not seem to limit itself to the distribution of quantum energy levels, however; it even appears to work its way into the wavelike nature of the quantum world. The position of the electron in the hy-

POINCARÉ SECTION OF A HYDROGEN ATOM in a strong magnetic field has regions (*orange*) where the points of the electron's trajectory scatter wildly, indicating chaotic behavior. The section is a slice out of phase space, an abstract six-dimensional space: the usual three for the position of a particle and an additional three for the particle's momentum.

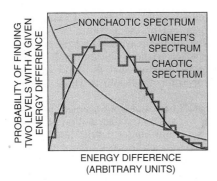

ENERGY SPECTRUM, or distribution of energy levels, differs markedly between chaotic and nonchaotic quantum systems. For a nonchaotic system, such as a molecular hydrogen ion (H_2^+), the probability of finding two energy levels close to each other is quite high. In the case of a chaotic system such as a Rydberg atom in a strong magnetic field, the probability is low. The chaotic spectrum closely matches the typical nuclear spectrum derived many years ago by Eugene P. Wigner.

drogen atom is described by a wave pattern. The electron cannot be pinpointed in space; it is a cloudlike smear hovering near the proton. Associated with each allowed energy level is a stationary state, which is a wave pattern that does not change with time. A stationary state corresponds quite closely to the vibrational pattern of a membrane that is stretched over a rigid frame, such as a drum.

The stationary states of a chaotic system have surprisingly interesting structure, as demonstrated in the early 1980s by Eric Heller of the University of Washington. He and his students calculated a series of stationary states for a two-dimensional cavity in the shape of a stadium. The corresponding problem in classical mechanics was known to be chaotic, for a typical trajectory quickly covers most of the available ground quite evenly. Such behavior suggests that the stationary states might also look random, as if they had been designed without rhyme or reason. In contrast, Heller discovered that most stationary states are concentrated around narrow channels that form simple shapes inside the stadium, and he called these channels "scars" [see illustration on opposite page]. Similar structure can also be found in the stationary states of a hydrogen atom in a strong magnetic field [see illustration on page 79]. The smoothness of the quantum wave forms is preserved from point to point, but when one steps back to view the whole picture, the fingerprint of chaos emerges.

It is possible to connect the chaotic signature of the energy spectrum to ordinary classical mechanics. A clue to the prescription is provided in Einstein's 1917 paper. He examined the phase space of a regular system from box R and described it geometrically as filled with surfaces in the shape of a donut; the motion of the system corresponds to the trajectory of a point over the surface of a particular donut. The trajectory winds its way around the surface of the donut in a regular manner, but it does not necessarily close on itself.

In Einstein's picture, the application of Bohr's correspondence principle to find the energy levels of the analogous quantum mechanical system is simple. The only trajectories that can occur in nature are those in which the cross section of the donut encloses an area equal to an integral multiple of Planck's constant, h (2π times the fundamental quantum of angular momentum, having the units of momentum multiplied by length). It turns out that the integral multiple is precisely the number that specifies the corresponding energy level in the quantum system.

Unfortunately, as Einstein clearly saw, his method cannot be applied if the system is chaotic, for the trajectory does not lie on a donut, and there is no natural area to enclose an integral multiple of Planck's constant. A new approach must be sought to explain the distribution of quantum mechanical energy levels in terms of the chaotic orbits of classical mechanics.

Which features of the trajectory of classical mechanics help us to understand quantum chaos? Hill's discussion of the moon's irregular orbit because of the presence of the sun provides a clue. His work represented the first instance where a particular periodic orbit is found to be at the bottom of a difficult mechanical problem. (A periodic orbit is like a closed track on which the system is made to run; there are many of them, although they are isolated and unstable.) Inspiration can also be drawn from Poincaré, who emphasized the

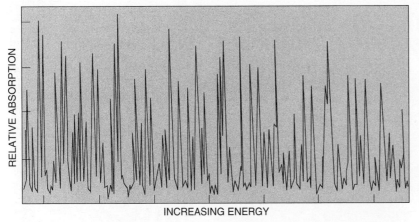

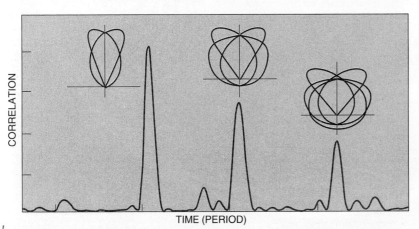

ABSORPTION OF LIGHT by a hydrogen atom in a strong magnetic field appears to vary randomly as a function of energy (top), but when the data are analyzed according to the mathematical procedure called Fourier analysis, a distinct pattern emerges (bottom). Each peak in the bottom panel has associated with it a specific classical periodic orbit (red figures next to peaks).

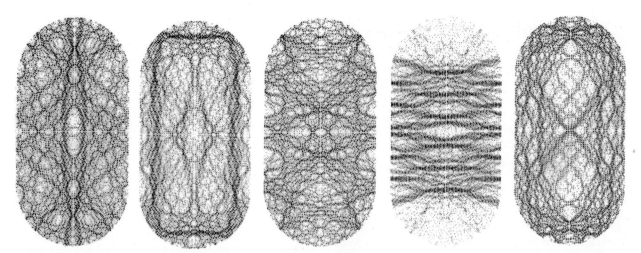

PARTICLE IN A STADIUM-SHAPED BOX has chaotic stationary states with associated wave patterns that look less random than one might expect. Most of the states are concentrated around narrow channels that form simple shapes, called scars.

general importance of periodic orbits. In the beginning of his three-volume work, *The New Methods of Celestial Mechanics,* which appeared in 1892, he expresses the belief that periodic orbits "offer the only opening through which we might penetrate into the fortress that has the reputation of being impregnable." Phase space for a chaotic system can be organized, at least partially, around periodic orbits, even though they are sometimes quite difficult to find.

In 1970 I discovered a very general way to extract information about the quantum mechanical spectrum from a complete enumeration of the classical periodic orbits. The mathematics of the approach is too difficult to delve into here, but the main result of the method is a relatively simple expression called a trace formula. The approach has now been used by a number of investigators, including Michael V. Berry of the University of Bristol, who has used the formula to derive the statistical properties of the spectrum.

I have applied the trace formula to compute the lowest two dozen energy levels for an electron in a semiconductor lattice, near one of the carefully controlled impurities. (The semiconductor, of course, is the basis of the marvelous devices on which modern life depends; because of its impurities, the electrical conductivity of the material is halfway between that of an insulator, such as plastic, and that of a conductor, such as copper.) The trajectory of the electron can be uniquely characterized by a string of symbols, which has a straightforward interpretation. The string is produced by defining an axis through the semiconductor and simply noting when the trajectory crosses the axis. A crossing to the "positive" side of the axis gets the symbol +, and a crossing to the "negative" side gets the symbol −.

A trajectory then looks exactly like the record of a coin toss. Even if the past is known in all detail—even if all the crossings have been recorded—the future is still wide open. The sequence of crossings can be chosen arbitrarily. Now, a periodic orbit consists of a binary sequence that repeats itself; the simplest such sequence is (+ −), the next is (+ + −), and so on. (Two crossings in a row having the same sign indicate that the electron has been trapped temporarily.) All periodic orbits are thereby enumerated, and it is possible to calculate an approximate spectrum with the help of the trace formula. In other words, the quantum mechanical energy levels are obtained in an approximation that relies on quantities from classical mechanics only.

The classical periodic orbits and the quantum mechanical spectrum are closely bound together through the mathematical process called Fourier analysis [see "The Fourier Transform," by Ronald N. Bracewell; SCIENTIFIC AMERICAN, June 1989]. The hidden regularities in one set, and the frequency with which they show up, are exactly given by the other set. This idea was used by John B. Delos of the College of William and Mary and Dieter Wintgen of the Max Planck Institute for Nuclear Physics in Heidelberg to interpret the spectrum of the hydrogen atom in a strong magnetic field.

Experimental work on such spectra has been done by Karl H. Welge and his colleagues at the University of Bielefeld, who have excited hydrogen atoms nearly to the point of ionization, where the electron tears itself free of the proton. The energies at which the atoms absorb radiation appear to be quite random [see upper part of bottom illustration on opposite page], but a Fourier analysis converts the jumble of peaks into a set of well-separated peaks [see lower part of bottom illustration on opposite page]. The important feature here is that each of the well-separated peaks corresponds precisely to one of several standard classical periodic orbits. Poincaré's insistence on the importance of periodic orbits now takes on a new meaning. Not only does the classical organization of phase space depend critically on the classical periodic orbits, but so too does the understanding of a chaotic quantum spectrum.

So far I have talked only about quantum systems in which an electron is trapped or spatially confined. Chaotic effects are also present in atomic systems where an electron can roam freely, as it does when it is scattered from the atoms in a molecule. Here energy is no longer quantized, and the electron can take on any value, but the effectiveness of the scattering depends on the energy.

Chaos shows up in quantum scattering as variations in the amount of time the electron is temporarily caught inside the molecule during the scattering process. For simplicity, the problem can be examined in two dimensions. To the electron, a molecule consisting of four atoms looks like a small maze. When the electron approaches one of the atoms, it has two choices: it can turn left or right. Each possible trajectory of the electron through the molecule can be recorded as a series of left and right turns around the atoms, until the particle finally emerges. All of the trajectories are unstable: even a

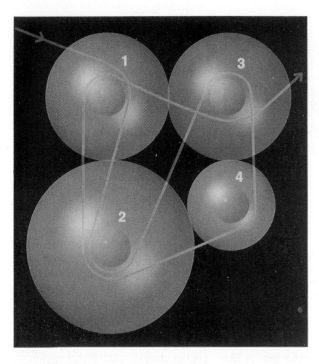
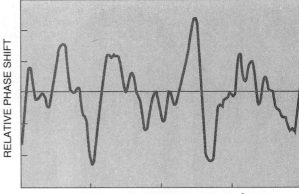

TRAJECTORY OF AN ELECTRON through a molecule during scattering can be recorded as a series of left and right turns around the atoms making up the molecule (*left*). Chaotic variation (*above*) characterizes the time it takes for a scattered electron of known momentum to reach a fixed monitoring station. Arrival time varies as a function of the electron's momentum. The variation is smooth when changes in the momentum are small but exhibits a complex chaotic pattern when the changes are large. The quantity shown on the vertical axis, the phase shift, is a measure of the time delay.

minute change in the energy or the initial direction of the approach will cause a large change in the direction in which the electron eventually leaves the molecule.

The chaos in the scattering process comes from the fact that the number of possible trajectories increases rapidly with path length. Only an interpretation from the quantum mechanical point of view gives reasonable results; a purely classical calculation yields nonsensical results. In quantum mechanics, each classical trajectory of the electron is used to define a little wavelet that winds its way through the molecule. The quantum mechanical result follows from simply adding up all such wavelets.

Recently I have done a calculation of the scattering process for a special case in which the sum of the wavelets is exact. An electron of known momentum hits a molecule and emerges with the same momentum. The arrival time for the electron to reach a fixed monitoring station varies as a function of the momentum, and the way in which it varies is what is so fascinating about this problem. The arrival time fluctuates smoothly over small changes in the momentum, but over large changes a chaotic imprint emerges, which never settles down to any simple pattern [*see right part of illustration above*].

A particularly tantalizing aspect of the chaotic scattering process is that it may connect the mysteries of quantum chaos with the mysteries of number theory. The calculation of the time delay leads straight into what is probably the most enigmatic object in mathematics, Riemann's zeta function. Actually, it was first employed by Leonhard Euler in the middle of the 18th century to show the existence of an infinite number of prime numbers (integers that cannot be divided by any smaller integer other than one). About a century later Bernhard Riemann, one of the founders of modern mathematics, employed the function to delve into the distribution of the primes. In his only paper on the subject, he called the function by the Greek letter zeta.

The zeta function is a function of two variables, x and y (which exist in the complex plane). To understand the distribution of prime numbers, Riemann needed to know when the zeta function has the value of zero. Without giving a valid argument, he stated that it is zero only when x is set equal to $1/2$. Vast calculations have shown that he was right without exception for the first billion zeros, but no mathematician has come even close to providing a proof. If Riemann's conjecture is correct, all kinds of interesting properties of prime numbers could be proved.

The values of y for which the zeta function is zero form a set of numbers that is much like the spectrum of energies of an atom. Just as one can study the distribution of energy levels in the spectrum, so can one study the distribution of zeros for the zeta function. Here the prime numbers play the same role as the classical closed orbits of the hydrogen atom in a magnetic field: the primes indicate some of the hidden correlations among the zeros of the zeta function.

In the scattering problem the zeros of the zeta function give the values of the momentum where the time delay changes strongly. The chaos of the Riemann zeta function is particularly apparent in a theorem that has only recently been proved: the zeta function fits locally any smooth function. The theorem suggests that the function may describe all the chaotic behavior a quantum system can exhibit. If the mathematics of quantum mechanics could be handled more skillfully, many examples of locally smooth, yet globally chaotic, phenomena might be found.

FURTHER READING

DYNAMICAL CHAOS. Edited by M. V. Berry, I. C. Percival and N. O. Weiss in *Proceedings of the Royal Society of London*, Vol. A413, No. 1844, pages 1–199; September 8, 1987.

THE HYDROGEN ATOM IN A UNIFORM MAGNETIC FIELD: AN EXAMPLE OF CHAOS. Harald Friedrich and Dieter Wintgen in *Physics Reports*, Vol. 183, No. 2, pages 37–79; November 1989.

CHAOS IN CLASSICAL AND QUANTUM MECHANICS. Martin C. Gutzwiller. Springer-Verlag, 1990.

CELESTIAL MECHANICS ON A MICROSCOPIC SCALE. T. Uzer, David Farrelly, John A. Milligen, Paul E. Raines and Joel P. Skelton in *Science*, Vol. 253, pages 42–48; July 5, 1991.

CHAOS AND QUANTUM PHYSICS. Edited by A. Voros, M.-J. Giannoni and J. Zinn-Justin. Elsevier Science Publishers (in press).

How random is a coin toss?

In examining the differences between orderly and chaotic behavior in the solutions of nonlinear dynamical problems, we are led to explore algorithmic complexity theory, the computability of numbers and the measurability of the continuum.

Joseph Ford

We ought then to regard the present state of the Universe as the effect of its preceding state and as the cause of its succeeding state.
— Laplace

The true logic of this world is the calculus of probabilities.
— Maxwell

Probabilistic and deterministic descriptions of macroscopic phenomena have coexisted for centuries. During the period 1650–1750, for example, Newton developed his calculus of determinism for dynamics while the Bernoullis simultaneously constructed their calculus of probability for games of chance and various other many-body problems. In retrospect, it would appear strange indeed that no major confrontation ever arose between these seemingly contradictory world views were it not for the remarkable success of Laplace in elevating Newtonian determinism to the level of dogma in the scientific faith. Thereafter, probabilistic descriptions of classical systems were regarded as no more than useful conveniences to be invoked when, for one reason or another, the deterministic equations of motion were difficult or impossible to solve exactly. Moreover, these probabilistic descriptions were presumed derivable from the underlying determinism, although no one ever indicated exactly how this feat was to be accomplished.

Despite this clearly stated orthodoxy in classical physics, science has nonetheless exhibited a great deal of unease and ambivalence regarding the coexistence of apparently random and apparently determinate behavior. Telltale evidence of this ambivalence appears in the almost capricious decisions which catalog systems as predominantly random or determinate.

Roulette wheel spins, dice throws and idealized coin tosses are universally presumed completely random despite their obvious underlying deter-

Joseph Ford is Regents' Professor of Physics at the Georgia Institute of Technology and is an editor of *Physica D: Nonlinear Phenomena.*

minism. Weather, human behavior and the stock market are, on the other hand, commonly regarded as strictly deterministic, notwithstanding their seemingly frivolous unpredictability. But perhaps nowhere in science does there exist greater confusion over the random–determinate question than that which arises for analytic Hamiltonian systems

$$H = H_0(q_k, p_k) + \lambda H_1(q_k, p_k) \qquad (1)$$

where H_0 describes an analytically exactly solvable system with N degrees of freedom, the small parameter λ determines the strength of the perturbation H_1, and the argument (q_k, p_k) is shorthand for the full argument $(q_1, \ldots q_N, p_1, \ldots p_N)$. The traditional folklore of this topic asserts that Hamiltonians of this form are analytically solvable and determinate when the number of degrees of freedom N is small; when N is large, statistical mechanics and the law of large numbers are presumed valid. Doubts regarding this folklore immediately arise, however, when one recalls the notorious insolubility of the three-body problem or even the nonseparable two-body problem, when one considers Poincaré's warning that, independent of N, Hamiltonian systems quite generally have no well-behaved constants of the motion other than H itself, when one notes that no analytic solution for dice throwing has ever been used to derive the laws of probability, and finally when one notices the infinite class of analytically solvable many-body problems for arbitrarily large N which can easily be obtained using classical perturbation theory.[1] Quite obviously, there is confusion here over the random–determinate conundrum, but much worse, there is even greater confusion over what kind of behavior is to be expected of the solutions of any specific Hamiltonian. Yet incredibly enough, the classrooms, the textbooks, the teachers, and the researchers of science have maintained a remarkable, almost total silence[2] on these matters

for many decades, as if the Hamiltonian of equation 1 had an incurable disease unmentionable in polite society. But then around 1950, some three hundred years after the birth of Newton, a new multidisciplinary area, now called nonlinear dynamics, began a concerted effort to solve some of the deeper puzzles presented by these Hamiltonians. The following few paragraphs briefly discuss one new result of especial relevance to this paper. More comprehensive presentations appear in tutorial review papers by Joel Lebowitz and Oliver Penrose[3] and by Michael Berry.[4]

Contemporary results

The success of astronomical perturbation theory for the solar system and other few-body problems and the equal success of statistical mechanics for many-body problems is *prima facie* evidence supporting the existence of a transition from orderly to highly erratic orbital motion in Hamiltonian systems as particle number is increased. However, this evidence provides little insight into the root cause of the transition or into the detailed structure of the resulting erratic orbits. Although nonlinear dynamics has much to say concerning these matters,[3,4] let me confine myself here to establishing as understandably as possible that deterministic Hamiltonian systems can exhibit orbits whose phase-space wanderings are so erratic that the words "unpredictable," "chaotic," and even "random" leap to mind. For this purpose, the startingly simple Hamiltonian with two degrees of freedom

$$H = \tfrac{1}{2}(p_1^2 + q_1^2 + p_2^2 + q_2^2) + q_1^2 q_2 - q_2^3/3 \qquad (2)$$

admirably suits our purpose. When the system energy E is sufficiently small, equation 2 very nearly describes two uncoupled, harmonic oscillators; as the system energy is increased, the nonlinear cubic coupling begins to exert a noticeable effect on orbital motion. This Hamiltonian was brought to public attention in the much quoted, now classic paper[5] of Michel Henon and Carl Heiles. A reader who finds the following brief resume confusing could do no better than seek clarification in the original source.

The astronomers Henon and Heiles recognized that if the Hamiltonian in equation 2 were analytically solvable in the same sense as all the problems in advanced classical mechanics textbooks, then this system would possess two functionally independent, well-behaved constants of the motion. In consequence, system orbits would be constrained to lie on two-dimensional integral surfaces in the four-dimensional system phase space. On the other hand, if the Hamiltonian ever gives rise to statistical or chaotic behavior, then H itself must be the only well-behaved constant of the motion, and the system orbits may wander freely over part or all of the three-dimensional energy surface. In short, one regards the orbits of this Hamiltonian as orderly if they lie on two-dimensional surfaces and erratic or chaotic if they wander freely over a three-dimensional surface. At this point, a wary reader may be tempted to consider this emphasis on two- as opposed to three-dimensional surfaces as artificial and irrelevant to the issue at hand; such is in fact not the case, as we now show.

Henon and Heiles numerically integrated the equations of motion for their Hamiltonian, obtaining many orbits at various energies. To make it visually obvious whether a given orbit moves on a two-dimensional or a three-dimensional surface in the four-dimensional phase space, they plotted the points at which the orbits intersect the (q_2, p_2) plane. If a given orbit lies on a two-dimensional surface, then the points at which it intersects the (q_2, p_2) plane would lie on a curve. Alternatively, if the orbit roams freely over part or all of a three-dimensional surface, then the points at which the orbit crosses the (q_2, p_2) plane would fill some area in the plane. The figure on page 42 shows the points at which orbits of various energies cross the $(q_2 p_2)$ plane. At an energy $E = 1/12$, we observe that each orbit generates a curve, indicating that all motion at this energy is orderly and that the Hamiltonian is analytically solvable or integrable. At an energy $E = 1/6$, Henon and Heiles found the then startling result that even simple systems can exhibit apparently chaotic behavior. The energy $E = 1/8$ shows the transition from order to chaos. For both these energies the splatter of dots was generated by a single orbit. It is

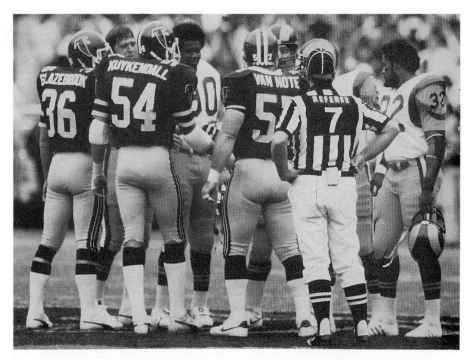

Random behavior. The handshake and coin toss, opening ceremonies at a football game. National Football League regulations, like the rules for many other games, provide for a coin toss to determine a random and unpredictable starting order of play. (Photo by Scott Cunningham.)

Orderly behavior. The Federal Reserve Board and others claim that the motion of money can be controlled. Despite numerous unanticipated financial crises, analysts perceive no similarity between the behavior of money on Wall Street and in Las Vegas; this view is shared by casino owners, who run an avowedly random business but never lose. (Photo by E. C. Topple, NYSE.)

10.7 How Random is a Coin Toss?

the complete lack of apparent order in the patterns which causes one to think of "chaos" or "random." Indeed, if one watches a video terminal sequentially plot the orbital dots at $E = 1/6$, one is unable to discern any spatial or temporal order. In short, the transition of orbital motion from a two-dimensional to a three-dimensional surface carries with it the introduction of a chaotic element in the motion. We can perhaps begin to gain an intuitive understanding of these matters by noticing that points initially close in the (q_2, p_2) plane separate linearly with time when $E = 1/12$, whereas at $E = 1/6$, initially close points separate at an exponential rate. Similar behavior is seen at a glance in the photo on page 44, which shows the transition to turbulence in the flow of cigarette smoke. By analogy, if a set of chaotic Henon–Heiles orbits could be similarly visualized, they would look for all the world like an incredibly mixed up plate of spaghetti. Finally, it must be mentioned that chaos of the Henon–Heiles type is now known to be a commonplace in Hamiltonian dynamics and elsewhere.

Chaotic orbits

By chaotic orbit, we have thus far meant little more than that the orbit does not lie on a smooth invariant integral surface of dimension less than the energy surface itself. But in what precise sense is such an orbit more random or less deterministic than one lying on a smooth, lower-dimensional invariant surface? Nonlinear dynamics provides a whole hierarchy of answers;[3,4] here we discuss only one relevant to our central theme. Specifically, let us imagine that a given energy surface for a dynamical system has been completely partitioned into a finite set of non-overlapping numbered cells, as if we were trying to make this energy surface resemble a wheel of fortune. Suppose now that someone having precise knowledge of a system orbit begins sequentially revealing to us at one-second intervals the number of the cell in which the system state then resides, somewhat analogous to being told at one-second intervals the number under the pointer of a wheel of fortune. If we consider the orbits of a nonchaotic system—a simple harmonic oscillator, say, or a low-energy orbit of the Henon–Heiles system—we will find that the numbers fall into a fairly regular pattern. Even if the period of the oscillator is not an integral number of seconds, we will see a regular progression in the cell numbers once we have accumulated data over several periods. For chaotic orbits—such as those of the Henon–Heiles system for larger energies, or for a particle of cigarette smoke above the transition to turbulence—no such regularity ever becomes apparent. We might as well be looking at the numbers from a wheel of fortune. That is, for a chaotic orbit we can at best give only probabilities for transitions from one cell number to the next, we cannot predict the transitions.

We can thus define an orbit to be chaotic if complete knowledge of which cells the system occupied in the past, that is, all the cell numbers up to the present t, does not allow us to determine the cell number at $t+1$ or any other future time. (This definition may appear unorthodox, but we will see its usefulness.) For chaotic orbits, the coarse-grained past does not uniquely determine the coarse-grained future; nothing less than a full sequence of finite-precision measurements made at finite intervals from $t = -\infty$ to $+\infty$ can ever specify a precise orbit, and the specification may not be unique even then. For nonchaotic orbits, the coarse-grained past does uniquely and completely determine the future; non-chaotic systems retain a coarse-grained determinism even in the presence of finite observational precision.

It may seem that the chaotic unpredictability described here arises solely from the self-inflicted ignorance of the precise state of the system imposed by

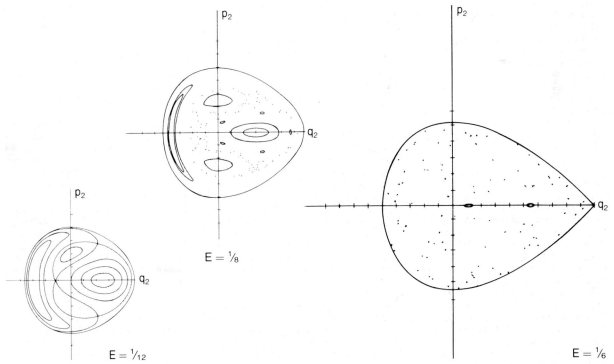

Computed orbits for the nonlinear Hamiltonian of equation 2. The graphs show the intersections of the orbits with the (q_2, p_2) plane. At left the system energy is $1/12$. Each of the curves in the plot consists of intersection points for a single orbit; the fact that the curves are clearly continuous and closed indicates that the orbits lie on two-dimensional surfaces in the four-dimensional (q_1, q_2, p_1, p_2) phase space. In the central figure the system energy is $1/8$. Some orbits still lie on two-dimensional surfaces; the erratically splattered dots, however, were generated by a single orbit that fills a three-dimensional volume, its energy being the only constant of the motion. In the graph at right the energy is $1/6$. The dots in the figure were all generated by a single orbit, which clearly wanders freely over a region of the three-dimensional energy surface. At this energy, almost all pairs of points that start close together in the (q_2, p_2) plane separate exponentially with time.

our division of the phase space into finite cells—that is, that the unpredictability is due to the coarse-graining. Let me hasten to reemphasize that this same ignorance induces no similar unpredictability on nonchaotic orbits and to point out that no restrictions on the fineness of the partition have been made. That is, nothing less than the full cell-number sequence can uniquely determine a chaotic orbit no matter how small the cells are chosen in the finite partition. If a physical system is proceeding along a precise chaotic orbit and if a finite number of observations having finite precision are made, then, no matter how many the observations nor how great their accuracy, these measurements will appear random and will show no evidence of the presumed underlying determinism. This then provides the first hint that, for chaotic systems, Newtonian determinism may be only a theorist's unattainable dream. Almost all dynamical systems are now known to exhibit chaotic orbits.

A simple model

To one untutored in the algorithmic art of generating pseudorandom numbers, forward iteration of the first-order difference equation

$$X_{n+1} = 2X_n \quad (\text{Mod } 1) \qquad (3)$$

appears guilelessly determinate. In this equation (Mod 1) means drop the integer part; equation 3 is thus a mapping of the unit interval upon itself. This simple difference equation has the equally simple analytic solution

$$X_n = 2^n X_0 \quad (\text{Mod } 1) \qquad (4)$$

An even more imaginative form of the solution may be obtained by writing the initial value X_0 as a base-two, binary numeral in the form

$$X_0 = 0.11000011101111\ldots \qquad (5)$$

Forward iterates of equation are now generated merely by moving the decimal point sequentially to the right, each time dropping the integer part to the left of the decimal point. It is truly difficult to imagine a more transparently deterministic system than the one specified by equation 3, and yet almost all its orbits are chaotic!

To see this, first note that all iterates of equation 3 lie on the unit interval (0,1) and hence that the "energy surface" here is precisely this unit interval. We partition this "energy surface" into two cells, a left cell ($0 \leqslant X < \frac{1}{2}$) and a right cell ($\frac{1}{2} \leqslant X < 1$). One can now easily recognize from the form of the binary representation in equation 5 that a given iterate X_n is in the left or the right cell of the unit interval as its first digit to the right of the moving decimal point is a zero or a one. If now someone knowing a precise orbit of the system tells us only whether the sequential iterates X_n are in the right or left cell, we can write down a cell number sequence composed of zeros and ones, zero meaning the left cell and one the right cell. When completed, this cell number sequence will be identical to the binary decimal expansion of X_0 for the given orbit. Simply put, the person knowing the orbit is in essence merely reading out the binary string for X_0 and the listener (or observer) is merely copying it down. Since we cannot in general determine future digits of X_0 from any past finite part of the digit string for X_0, the true orbit is chaotic by our previous definition.

The chaotic orbits generated by equation 3 permit us to expose a quite general unpredictability in chaotic orbit behavior that comes perilously close to true randomness.

Consider again the procedure in which someone with precise knowledge of a given orbit in effect reads us the sequential digits in X_0. Is there any way we can definitively ascertain whether this person is sequentially telling us the first binary digit in each X_n computed from equation 3 or whether he is obtaining this digital string merely by flipping an honest coin having no underlying determinism? In seeking to answer this, let us first consider the binary digit strings for each X_0 in the set of all possible X_0 on the unit interval. Then, if we let unity represent heads and zero tails, we perceive that the set of all X_0 provides a set of semi-infinite digit strings which is one to one with the set of all possible semi-infinite coin-toss sequences. But this clearly means that the digit string in the binary representation of X_0 is as random as a coin-toss sequence. In turn, despite the underlying determinism assigned to the system described by equation 3, if we make only a coarse-grained left–right determination on sequential iterates X_n, then these deterministically computed iterates hop between left and right according to a rule that cannot be distinguished from a truly random coin-toss sequence. Here sequential left–right measurements are totally uncorrelated and yield a so-called Bernoulli process; if we divide the unit interval into many equal cells, each orbit would generate cell-number sequences representing Markov processes.

No hint of determinism would appear as long as the partition is finite. Now let us throw caution to the winds and consider infinite partitions having zero cell size. Here a cell number sequence is just the precise X_n sequence itself. Nonetheless, this deterministic X_n sequence is in fact also a set of random numbers as one can see from the presumed randomness of the digit string of the initial X_0. Because of this property, equation 3 is sometimes used as a computer algorithm for obtaining pseudorandom numbers—pseudorandom because a computer performs only finite arithmetic.

These considerations point toward a body of theory[6] in nonlinear dynamics that rigorously proves that strictly deterministic systems can in varying degrees mimic true randomness. But I have not yet outlined the rigorous theory that can confirm the possibility that a deterministic system can actually also be truly random. To move toward that goal, let us recall that chaotic orbits can, under a partitioning of phase space, always be associated with sequences of integer cell numbers. Moreover, if the cell-number digit string for an orbit can be shown to be random, then the orbit itself is also random.[7] But this immediately raises the question of how any specified digit string such as the decimal representation for π or the digit string in a table of random numbers can ever be termed truly random. Here we are at last being led to one of the deepest questions in all of probability theory: the definition of randomness in given digit strings.

Algorithmic complexity theory

Without loss of generality, we may confine our attention to binary digit sequences and, initially, we shall consider only finite sequences. Each single binary digit carries one bit of information by definition. Therefore, an n-digit binary sequence could provide n bits of information; however, frequently the digits in the sequence are obviously correlated, and the information contained in the n-digit sequence can be expressed by a much shorter sequence. The shorter digit sequence could, for example, be a relatively brief computer code that can generate the longer original n-digit string on some machine **M**. With these thoughts in mind, Andrey N. Kolmogorov, Gregory J. Chaitin, and Ray J. Solomonov independently defined[8] the complexity $K_\mathbf{M}^{(n)}$ of an n-digit sequence as the integer bit length of the shortest computer program that can print the given sequence using machine **M**. Later, Kolmogorov showed, in essence, that there is a universal computer providing the minimum $K_\mathbf{M}^{(n)}$; thus, there is no generality lost by dropping the subscript **M**. Consider now the complexity $K^{(n)}$ of the simple sequence consisting of all ones. A minimal program might read "PRINT 1, n times." The bit length of this program is very nearly equal to $\log_2(n)$ for sufficiently large n. Indeed, for any sequence calculable via repetition of some finite computer algorithm, its complexity $K^{(n)}$ also very nearly equals $\log_2(n)$ as n become large. On the other hand, the complexity of an n-digit sequence $\{G_k\}$ can never appreciably exceed n since the sequence can always be produced by the computer program "PRINT $G_1, G_2 \ldots G_n$" which contains very nearly n bits for large n. By the definition of complexity, n-digit sequences which actually have this maximum complexity cannot be calculated by any algorithm whose bit length is appreciably less than the bit length of the sequence itself. The information contained in a sequence of maximum complexity has thus been irreducibly encoded, and the simplest way to specify the sequence is to provide a copy of it. As a consequence, the sequential digits in these maximum-complexity sequences are so incalculable and hence so unpredictable that the word "random" seems inescapable. Thus, let us follow Kolmogorov and others and define a finite, specified digit string to be random provided it has maximum complexity. Using this definition, it may be shown that most finite digit strings are random.

As the digitial length of the binary sequence tends to infinity, one might be tempted, as Kolmogorov originally was, to define a random infinite sequence as one for which $K^{(n)}$ goes as n for large n. Unfortunately, Per Martin-Löf proved this to be an empty definition because $K^{(n)}$ for chaotic sequences can oscillate appreciably below the expected value on the order of n even as n tends to infinity. As a reasonable way to eliminate or "damp out" these oscillations, let us define[8] the complexity of an infinite sequence by

$$K = \lim_{n \to \infty} [K^{(n)}/n] \quad (6)$$

Although there are minor problems of convergence here, the limit may be shown to exist in general. Using a slight abuse of language, we shall sometimes continue to speak of maximum complexity for an infinite sequence when its K computed from equation 6 is non-zero. As before, a sequence having maximum complexity is defined to be random. The virtue of this definition rests with its intuitive appeal. Infinite sequences having maximum complexity are so unpredictable they are incalculable by any finite algorithm; moreover, the simplest way to specify such sequences is to provide copies of them. On the other hand, this definition suffers the defect of not clearly revealing its complete equivalence to earlier definitions of randomness. We now move to close this loophole. In the following discussion, I shall always mean infinite sequences when I refer to sequences.

A "calculable test" is one expressible by a finite algorithm. Because randomness implies a certain lack of order, and because disorder can occur in infinite variety, there is no single calculable test that can rigorously prove a sequence to be random. Individual calculable tests form necessary but not sufficient tests for randomness. However, by combining all possible such calculable tests into a single composite test, we can determine when a sequence passes every humanly computable test for randomness. We shall call this the "Universal Test" for randomness. We now know that almost all sequences having maximum complexity pass the Universal Test. This theorem, which Martin-Löf proved in 1966, justifies our defining a sequence to be random when it has maximum complexity since no human will ever be able to distinguish this definition from earlier ones. Finally, Martin-Löf has proved that almost all sequences actually have maximum complexity and are therefore random. This proof immediately implies that almost all the individual orbits defined by equation 3 are truly random in addition to being strictly deterministic, a point which we discuss more fully below.

Returning now to Hamiltonian dynamics, recall that a chaotic orbit is defined as one yielding a cell number sequence whose future is not determined by its past. Quite clearly, a chaotic orbit generates a random cell-number sequence having maximal complexity.[9] The chaotic orbit itself is therefore truly random in the sense that such an orbit cannot be computed by any finite algorithm and its information content is both infinite and incompressible. Finally, for chaotic orbits, Newtonian dynamics must deterministically compute random cell number sequences. In general then, Newton's laws are merely formal, humanly incalculable algorithms. For centuries, randomness has been deemed a useful, but subservient citizen in a deterministic universe. Algorithmic complexity theory and nonlinear dynamics together establish the fact that determinism actually reigns only over a quite finite domain; outside this small haven of order lies a largely uncharted, vast wasteland of chaos where determinism has faded into an ephemeral memory of existence theorems and only randomness survives.

As additional consequences of algorithmic complexity theory, we note the mildly amusing fact that π and e do not yield random digit strings since both can be calculated via indefinite repetition of short algorithms, implying $K = 0$ in equation 6. Moreover, if we define an incalculable number[10] as one whose digit string has maximum complexity, then almost all real numbers are incalculable and cannot be computed by any finite algorithm. As Mark Kac phrases it,[11] most numbers in the continuum are not definable using a finite number of words. The continuum therefore has the distinction of being a well-defined collection of mostly undefinable objects. Thus, despite its illusions, science can actually compute at most only with the dense set of calculable numbers having zero complexity (as given by equation 6). Paradoxically, while this dense set of calculable numbers has measure zero, it is not a countable set as may be easily proved.

How random is a coin toss?

We now apply the notions developed thus far to a specific example. An idealized, deterministic "coin toss" is quite well described by the difference in equation 3. Given an X_0, we may deterministically compute sequential X_n and thence determine whether we obtain 1 or 0, heads or tails, at each iteration according to the first binary digit of each X_n. The process is strictly deterministic; existence and uniqueness theorems are easily derived for it. But equally, this process is completely random because the digit strings for almost all X_0 are random. Yet how can this be? Why is there not a contradiction in terms here? Doesn't full determinism preclude randomness? Doesn't complete randomness preclude any underlying determinism?

Of itself, equation 3 is merely a finite algorithm which upon iteration computes all X_n given X_0; no lack of determinism exists in this iteration process. Equally, no lack of determinism is to be found in the existence and uniqueness theorem where, of course, X_0 is again presumed given. Determinism or its lack in the coin-toss sequence thus rests soley upon the minor details of specifying or determining X_0, a task of such apparent triviality that no mention of it appears in the literature. But is this task in fact a trivial one? In response, algorithmic complexity theory points out that almost all numbers X_0 are not only incalculable but also undefinable. Despite three centuries of prejudice to the contrary, specifying or determining X_0 is now seen to require superhuman skills. One can maintain determinism for equation 3 or for the idealized coin it describes only if one presumes an ability to compute infinite algorithms of maximum complexity and to understand definitions having infinite word lengths. One must also presume an ability to make infinitesimal distinctions—as fine as any we expect from gods—for equation 3, like all its fully chaotic brothers, is harshly intolerant of even the slightest error: Any initial uncertainty in the specification of X_0 will grow exponentially with increasing iteration number n. Because of this exponential error growth, determinism is, from a practical point of view, at best only a temporally local property which rapidly vanishes without a trace under an avalanche of overwhelming error.

Popular belief to the contrary notwithstanding, the "coin" of equation 3 now emerges as completely random, but has determinism thereby been completely eliminated? Certainly not, for the traditional notions of probability theory contain just as large an unexamined tacit assumption as do the traditional notions of determinism. Specifically, the notion that randomness precludes determinism is based on the assumption that infinite computational or observational precision is impossible. If infinite observational and computational skills be assumed, then equation 3 provides a deterministic scheme for computing a random coin-toss process. Historically, theories of determinism have failed to recognize that their tacit assumption of infinite precision was sufficient to compute random orbits just as probability theories failed to recognize that infinite precision could provide a link to determinism. We shall return shortly to the question of whether or not infinite precision is a physically meaningful concept.

As a final note here, let us briefly seek to allay a confusion that might arise from our earlier discussion of the difference equation 3. Specifically, I have emphasized that the randomness of iterates arises strictly out of the randomness in the digit string for each initial condition X_0. Randomness thus appears to depend solely upon the random and incalculable character of the initial state X_0. Yet a thoughtful reader will immediately note that most solutions to difference equations evolve from initial data X_0 having random digit strings. Why then, according to our arguments, aren't all difference equations random? To answer this question without a lengthy foray into technical details, let us note that although iterates of a difference equation are at liberty to depend sensitively upon the random character of the initial state X_0, they are not required to. For example, consider the simple difference equation or mapping

$$X_{n+1} = X_n + b \quad (\text{Mod } 1) \qquad (7)$$

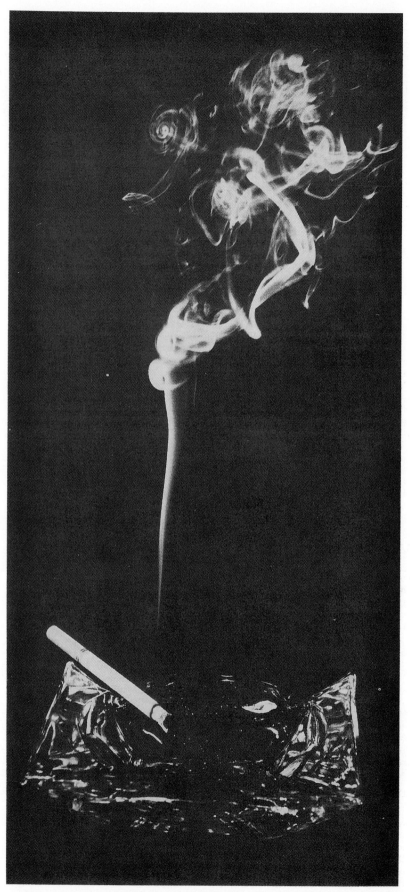

Transition from laminar to turbulent flow in a rising column of cigarette smoke. The laminar flow here is analogous to the smooth orbit structure of the Henon–Heiles system shown in the left graph on page 42. The turbulent flow at higher Reynolds number gives a hint of the orbit structure underlying the high–energy behavior of the Henon-Heiles system shown in the right-hand graph. The experiment shown in this photo may also illustrate one of the few medically acceptable uses of a cigarette. (Photo by Vincent Malette, Georgia Tech.)

10.7 How Random is a Coin Toss?

where b is an irrational number. Here the solution reads

$$X_n = nb + X_0 \quad (\text{Mod } 1) \qquad (8)$$

Because the number b is irrational, the iterates X_n of Eq. (7) are rigorously known to be dense and uniformly distributed on the unit interval. Yet this weakest and non-random version of "chaos," called ergodicity,[3,4] in no way depends on any randomness in the digit string for X_0. For, suppose there is a small error ΔX_0, in the initial state X_0, then from equation 8 we immediately find that $\Delta X_n = \Delta X_0$ and hence that there is no growth of error upon iteration. Thus equation 7 maps entire finite intervals as rigid units; in consequence, digital randomness (or its lack) of an interior point X_0 is clearly irrelevant to the subsequent orbital history. The essential point to notice here is that fully chaotic, random systems propagate error exponentially. The orbits of such systems thus are forced to depend with exquisite sensitivity upon the precise initial state and its random character. For nonchaotic systems, error propagates less rapidly and, as mentioned earlier, even a coarse-grained past suffices to determine precisely a coarse-grained future. To summarize, a chaotic orbit is random and incalculable; its information content is both infinite and incompressible. In seeking to specify a chaotic orbit, one is at liberty to place the requisite infinite amount of information in the initial state X_0, in the governing difference-equation algorithm, or in some mixture of the two; it is purely a matter of taste.

Consequences

Almost all physical theories—including quantum mechanics—are characterized by deterministic rate equations for continuum variables. Advocates of these theories defend use of the continuum chiefly on the grounds that nothing in principle presently limits measurement accuracy for any single variable. Nonetheless, not even the staunchest of these advocates suggests that our observational precision is or ever will be infinite. But without this infinite precision, the continuum becomes, physically speaking, meaningless, as complexity theory so clearly shows. Much more than a minor pedagogical issue or a small mathematical footnote is involved here. For speaking strictly in laboratory terms, an initial observational error in a nonchaotic system increases slowly as some power of the time t for subsequent observa-

tions; these systems therefore generally allow us to maintain the fiction of determinism and the continuum, at least over laboratory time scales. Unfortunately, nonchaotic systems are very nearly as scarce as hen's teeth, despite the fact that our physical understanding of nature is largely based upon their study. On the other hand, for the much more common, overwhelmingly dominant class of chaotic systems, initial observational error grows exponentially, in general, and determinism and the continuum become meaningless on an impressively short human time scale. For example, even if we assume an error in the X_0 of equation 3 as small as 10^{-31}, determinism is nonetheless completely lost at about the hundredth iteration.

What consequences flow from these statements? Can algorithmic complexity theory provide road guides for our ongoing trip to the future?

Newtonian dynamics has, over the centuries, twice foundered on assumptions that something was infinite when in fact it was not: the speed of light, c, and the reciprocal of Planck's constant $1/h$. Reformulations omitting these infinities led first to special relativity and then to quantum mechanics. Complexity theory now reveals a third tacitly assumed infinity in classical dynamics, namely the assumption of infinite computational and observational precision. In consequence, Newtonian dynamics now faces a third reformulation whose impact on science may be as significant as the first two. Moreover, quantum mechanics, itself the second revolution, will not long remain exempt from the upcoming third revolution, for quantum mechanics also assumes infinite computational and observational precision. But in addition, quantum theory is now seen to be incomplete, just as Einstein suggested. For, if quantum investigators have the power of infinite precision required to compute the precise time evolution of continuum variables, then clearly they can also deterministically compute all the random variables that appear in the presently incomplete theory. Alternatively, if infinite precision be abandoned, then reformulation of quantum mechanics will be just as extensive as that of classical dynamics.

Over the centuries, we have, with great reluctance, recognized discreteness and finiteness in this universe. Centuries ago, the Greeks suggested that the matter continuum should be replaced by discrete atoms, and, eventually, Avagadro counted the atoms in a box, finding their number to be finite. In this century, Einstein denied the Newtonian notion of an infinite speed, and Planck deprived us of our energy continuum. More recently, Heisenberg reminded us of limits on observational

Newtonian but random processes. The photos on these pages illustrate process that may be described by deterministic equations of motion, but whose later state depends so exquisitely on the initial conditions that it is unpredictable. These processes thus serve as models for randomness, in spite of their underlying Newtonian character.

precision for conjugate variables. And finally, algorithmic complexity theory is gently asserting that no variable can ever be measured precisely. The number continuum is, physically speaking, a fiction.

Complexity theory can guide us toward a humanly meaningful number set that does not involve the assumption of infinite precision, and which can replace the number continuum—although complexity theory can not, of course, provide us with a natural bound on observational precision.

The first numbers to be deleted from the continuum are the incalculable irrational numbers having positive Kolmogorov complexity. These totally inhuman numbers require infinite information to compute, to store, and to define. Next to go are the seemingly innocent numbers whose computation requires infinite repetition of a finite algorithm. The problem here is that the digit strings for these numbers also contain an infinite amount of information ($\log_2 n$ as n tends to infinity), require infinite storage capacity, and take an infinite time to compute. Following this second elimination, each number in the remaining countably dense set has only a finite number of digits in its decimal representation. The individual remaining numbers are now quite acceptable, but the set as a whole is not, because it still contains infinite information. We thus eliminate the last infinities, the infinitely large

$$\infty = (1 + 1 + 1 + \ldots)$$

and the infinitely small

$$0 = (1 + 1 + 1 + \ldots)^{-1}$$

In addition, we insist that the algebraic difference between any two set members be bounded away from these two "infinities." The number continuum is now reduced to a bounded, finite points set which, without loss of generality, may be taken to be a finite set of integers. It is this number set which suggests itself as the base for reformulating physical theories. Perhaps the most striking feature of this reformulation that we can anticipate is that all physical variables will be quantized. In closing, much of what I have said in this entire article[12] can be summarized in the following fable.

Long ago, before the beginning of time, the gods gave man the integer 1 to provide amusement for idle moments. Thereupon, man became so delighted with the number 1 that he sought and found a duplicate, calling it 2, and another duplicate, calling it 3. On a subsequent day, as man was contemplating the truth and beauty of the integer N, his mate arrived with a tempting fruit which the serpent who supplied it had called $\aleph_0$. Immediately upon the first bite, man became intoxicated and his mind reached and fleetingly grasped the meaning of $(1 + 1 + 1 + \ldots)$, but by morning he retained only the empty symbols ....

Moral: Do not let your reach exceed your grasp.

* * *

To Russia with Love: This article is the end product of many tutorial hours and numerous encouragements generously bestowed upon the author by his colleague and friend, Professor Boris V. Chirikov of the Soviet Union. Praise should be shipped FOB, Novosibirsk; blame should be forwarded directly to Atlanta.

References

1. A clear and highly readable discussion is J. Moser, Memoirs Am. Math. Soc., No. 81 (1968).
2. A. S. Wightman has also lamented this seeming conspiracy of silence in *Perspectives in Statistical Physics*, H. J. Raveche, ed., North-Holland, Amsterdam (1981).
3. J. L. Lebowitz, O. Penrose, PHYSICS TODAY, February 1973, page 23.
4. M. Berry in *Topics in Nonlinear Dynamics*, S. Jorna, ed., A.I.P. Conf. Proc. **46**, (1978), page 16.
5. M. Henon, C. Heiles, Astron. J. **69**, 73 (1964).
6. In addition to references 1–4, see the reference list of R.G.H. Helleman in *Fundamental Problems in Statistical Physics*, Vol V, E.G.D. Cohen, ed., North-Holland, Amsterdam (1980).
7. A. A. Brudno, Usp. Mat. Nauk **33**, 207 (1978).
8. G. J. Chaitin, Sci. Am., May 1975 page 47; A. K. Zvonkin, L. A. Levin, Usp. Mat. Nauk. **25**, 85 (1970); P. Martin-Lof, J. Information and Control **9**, 602 (1966); V. M. Alekseev, M. V. Yakobson, Phys. Repts. **75**, 287 (1981).
9. It perhaps should be mentioned here that entries in a cell-number sequence need not be statistically independent despite any impression to the contrary created by the text itself. See Martin-Lof, reference 8.
10. Our definition of incalculable number is not the same as the computer theorist's definition of uncomputable number, although the terms are related.
11. Semi-private barroom conversations held at various conference watering holes around the world.
12. Readers with a historical bent may have already recognized many parallels of fact or spirit between the present paper and numerous earlier articles dating back at least to Maxwell. But perhaps the most complete and striking parallel of all is to the paper "Is Classical Mechanics In Fact Deterministic?" by Max Born, *Physics In My Generation*, Springer-Verlag, New York (1969), page 78. □

R K BULLOUGH

SOLITONS

Solitons are mathematical objects which have excited theoreticians because of their wide ranging applications in physics. They appear as solutions of particular nonlinear wave equations which often have a certain universal significance. In this article Professor Bullough indicates the importance of solitons to modern physics

The first recorded observation of a soliton was that made by John Scott Russell in August 1834 when he saw 'a rounded smooth well defined heap of water' detach itself from the prow of a stopped barge and proceed without change of shape or diminution of speed for over two miles along a channel, supposedly the Edinburgh–Glasgow canal. In the following ten years Russell dubbed this object 'the wave of translation' or 'great solitary wave', created similar objects in smaller water channels only six inches deep and divined the empirical formula $c^2 = g(h+k)$ connecting the speed c of the disturbance with its height k above the undisturbed water surface and with h the undisturbed depth (notice that bigger disturbances travel faster than smaller ones). He disagreed strongly with Airy who proposed a different formula for c, and observed the break up of a larger disturbance into two of his solitary waves travelling at different speeds, the taller preceding the smaller as his formula demands. Waves of permanent shape in one dimension are now called solitary waves – most especially and appropriately when they are roughly bell shaped and isolated from other waves. The solitary wave Russell saw and named in 1834 is, however, a soliton as it has the collision properties we associate with solitons. These collision properties are also the origin of the break up into solitary waves Russell observed.

Russell believed that in his solitary wave he had found a universal phenomenon. In a posthumous book published in 1885 entitled *The Wave of Translation in the Oceans of Water, Air and Ether* he used his formula $c^2 = g(h+k)$ to compute from the velocity of sound the depth of the atmosphere (5 mile) and from the velocity of light the depth of the universe (5×10^{17} mile). The book does not have the permanent value of his early work published in the British Association 'Report on waves' (1844), but Russell may have divined more than he knew. In the 12 years since the word soliton was coined in 1965 we have seen its appearance in every branch of physics – not least in sound waves in crystals and plasmas and as a natural model of a particle in particle physics.

There are now two sorts of soliton, the classical c-number soliton, and the quantal, that is quantised, soliton. I shall say a little about each type. A rough description of a classical soliton is that of a solitary wave which shows great stability in collision with other solitary waves. A solitary wave, as we have seen, does not change its shape: it is a disturbance $u(x-ct)$ which travels by translating along the x axis with speed c. This disturbance is in one space dimension x only: I come to the more general case later. This solution is no surprise if the wave equation governing its motion is simple enough: $u(x-ct)$ solves the linear dispersionless wave equation $u_x + cu_t = 0$ (subscripts denote partial differentiation with respect to that variable) for *any* shape of the function u. But all physicists know that most systems are dispersive and in the case of the wave equation $u_t - u_{xxx} = 0$, for example, only the solutions $\cos k(x-Vt)$ and $\sin k(x-Vt)$ have permanent profile. These waves have $V = k^2$ and arbitrary disturbances disperse as the large k modes run away from the small k ones.

Nonlinearity and solitons

Nonlinearity introduces a new feature. The simplest nonlinear wave is perhaps $u_t + uu_x = 0$. This is a wave in which the speed is the disturbance u itself. Thus points of large u overtake points of small u, the wave shocks and ultimately breaks. The solitary wave solution of a nonlinear dispersive wave equation cleverly balances the nonlinearity against the dispersion so that the wave retains its shape. A remarkable example is the solitary wave solution $12\xi^2 \text{sech}^2 \xi(x - 4\xi^2 t)$ of the Korteweg–de Vries (KdV) equation $u_t + uu_x + u_{xxx} = 0$. The KdV describes the propagation of gravity waves in shallow water and its solitary wave is precisely that observed by Scott Russell in 1834. Notice again how the speed $4\xi^2$ is proportional to the amplitude $12\xi^2$ and increases with it[1].

There is no superposition principle for nonlinear wave equations: two solutions do not add to form another solution. Thus if two solitary waves of the KdV collide we would expect each to scatter off the other and some new disturbance emerge. In practice this does not happen: the solitary waves simply pass through each other and emerge essentially unchanged. It is this collisional stability which characterises the soliton. Figure 1 shows results obtained by Ikezi, Taylor and Baker in 1970 for ion acoustic waves in a plasma. The waves are governed by the KdV or an equation closely related to it. The figure well illustrates the behaviour of ion acoustic solitons travelling in the same or opposite directions, though the former collision is rather heavily damped. Notice that break up is also an aspect of collision: this is why Scott Russell could observe a break up into two of his solitary water waves. The number and speeds of the emergent solitons depend on the initial disturbance.

This behaviour is striking but it might simply be an oddity of one small corner of physics. This could not be wholly true because the KdV is a good approximate equation governing any weakly nonlinear, weakly dispersive, system. The KdV, or a modified form of it, governs shallow water waves, ion acoustic waves, Alfven waves in a cold collisionless plasma and the propagation of sound waves in anharmonic crystals, for example. Its importance to crystal lattice theory is that it explains the recurrence phenomena observed in numerical work by Fermi, Pasta and Ulam on one-dimensional lattices (the FPU problem). An initial excitation in a single linear mode spikes and breaks up into solitons: these pass through each other and can ultimately reassemble to simulate the initial excitation. Energy swings between only a few crystal modes and then is largely restored to the initially excited mode. The relevance of this nonergodic behaviour to thermal transport in real crystals is not yet fully understood. But Debye's idea, that energy excited in

---

1 The speed does not match Russell's empirical formula because the KdV equation has been written in a moving coordinate system.

any crystal mode diffuses to all other modes in a time characterising the thermal conductivity of the crystal, is not applicable to one-dimensional anharmonic crystals. Figure 2 shows the spiking of a harmonic crystal wave into at least eight solitons and is taken from the article in which the word soliton was coined (Zabusky N J and Kruskal M D 1965 *Phys. Rev. Lett.* **15** 240).

Nonlinear Schrödinger equation

If solitons were simply solutions of the KdV equation there would be little more to say. But in 1971 Zakharov and Shabat published an exact analytical solution of the nonlinear Schrödinger (NLS) equation $i u_t = -u_{xx} - 2u|u|^2$ (in which u now takes on complex values). The NLS looks like the time-dependent equation of wave mechanics $i\psi_t = -\psi_{xx} + V(x)\psi$ in which the potential $V(x) = -2|\psi|^2$ and is determined by the disturbance itself (the usual factor $h^2/8\pi^2 m$ on $-\psi_{xx}$ is scaled away; all equations I discuss are scaled in some way). The NLS has soliton, that is *multi-soliton*, solutions, just as the KdV has.

The connection between Schrödinger's famous equation and the NLS raises interesting points concerning the quantisation of the NLS; but as the NLS is not relativistically invariant its quantised form has so far not proved important. The classical form has a large number of physical applications, however, which have nothing to do with quantum mechanics. Gravity waves in deep water are governed by the NLS. The interesting solutions are *envelope* solitons. Here a bell shaped hyperbolic secant envelope modulates a harmonic (cosine) wave. Figure 3 shows observations reported by Yuen and Lake (*Phys. Fluids* 1975 **18** 956). On the left is shown a single envelope soliton; in the centre a single envelope breaks up into two envelope solitons; on the right the single envelope soliton collides with the second envelope which continues to break up undisturbed. In the case of deep water a dozen harmonic oscillations lie within the sech envelope. This is the origin of the fisherman's warning that the seventh wave to come will be the largest!

The NLS also governs the self-focusing of intense laser light in a dielectric. Intuitively one sees that if the refractive index depends on the light intensity the NLS could result. The simplest behaviour arises in a steady, time-independent situation when the electric field envelope ε satisfies $i\varepsilon_z = \varepsilon_{xx} + \varepsilon_{yy} + 2\varepsilon|\varepsilon|^2$: x, y and z are space coordinates but the direction of propagation z is 'time like'. If we consider x and z only, for the moment, laser light at $z = 0$ breaks up into hyperbolic secant solitons following straight line tracks at different angles to the z axis. The tangents of these angles are the 'speeds' of the solitons. The intensity across a single track is a sech^2 and the laser light has broken up into filaments with these intensity profiles. Figure 4 shows a side view of damage tracks created by filamentation, now in two space dimensions, induced in Perspex by a 20 TW pulse from a Nd:YAG/glass laser/amplifier prior to the formation of plasma.

In the interaction of a laser with plasma itself radiation pressure can blow plasma out of regions occupied by intense light. Regions emptied of plasma have been called cavitons, and these have been observed whilst 'photon bubbles' have also been talked about. Filamentation may influence the transfer of energy from a laser to imploding plasma and extensive numerical studies of it form a part of the current laser–fusion programme at Los Alamos. Studies of optical filamentation are also in hand in the United Kingdom.

Two other applications of the NLS of particular interest to theoreticians are a recent demonstration that a model one-dimensional Heisenberg ferromagnet[2] can be mapped on to the NLS and an earlier one that a hydrodynamical vortex problem also maps onto the NLS. These results illustrate that the exact solution of one nonlinear problem can also solve a number of other apparently unrelated model physical problems. These particular examples do not end the list of applications of the NLS to physics. Typically it governs any weakly nonlinear strongly dispersive phenomenon.

Sine–Gordon equation

Perhaps the most striking of the one-dimensional wave equations with soliton solutions is the sine–Gordon equation (SG) $u_{xx} - u_{tt} = \sin u$. Particle physicists, and plasma physicists for that matter, will recognise in the linearised form of the SG the Klein–Gordon equation $u_{xx} - u_{tt} = u$. The sobriquet for the nonlinear equation may be due to Martin Kruskal who with colleagues at Princeton solved the KdV in 1967, but it may not (on this and other matters compare Sydney Coleman's (1975) 'Classical lumps and their quantum descendents' Lectures at the 1975 International School of Subnuclear Physics 'Ettore Majorana').

The SG first appeared in physics in the work of Frenkel on crystal dislocations in 1939. It appeared in differential geometry before 1882 when Bäcklund found an important transformation for it. The Bäcklund transformation adds one more soliton to any multisoliton solution of the SG including the zero or 'vacuum' solution – it creates a particle. As the equation is relativistically invariant (it does not change its form under Lorentz transformations) it is an obvious choice as a model field theory in particle physics. It turns out that any relativistically invariant generalised Klein–Gordon equation $u_{xx} - u_{tt} = F(u)$ has soliton solutions if and essentially only if $F(u)$ is $\sin u$. This puts the SG in a very special

[2] The one-dimensional Heisenberg model of a ferromagnet consists of spins σ_i occupying lattice sites $i=1,\ldots,N$ and interacting, with first neighbours only, with interaction $-\lambda\sigma_i\cdot\sigma_{i+1}$ (λ is a coupling constant, $i=1,\ldots,N$ labels the spin). The related Ising model was solved in two dimensions by Lars Onsager in 1944. The one-dimensional model which maps onto the NLS is a continuum approximation to the Heisenberg model in which the lattice spacing is reduced to zero. In a different continuum limit the problem maps onto the quantised sine–Gordon equation discussed below.

place in particle physics. I return to this application of the SG shortly.

The SG can be modelled by a line of coupled pendulums undergoing large oscillations including complete rotations. Recall that a single pendulum is governed by $u_{tt} = -\sin u$: it is only the more usual small oscillations case which is governed by $u_{tt} = -u$ and exemplifies simple harmonic motion. If the pendulum is inverted ($u = -\pi$) and then nudged from this unstable equilibrium point it turns through $u = 0$ to return eventually (very eventually for it takes an infinite time) to $u = +\pi$.

Figure 1 Ion acoustic solitons: **a** collision of solitons in the same direction; **b** in opposite directions. The smaller soliton is at rest in case **a** and the larger approaches it with speed v (*from Ikezi, Taylor and Baker* 1970 *Phys. Rev. Lett.* **25** 11)

Figure 2 Spiking of a harmonic wave in an anharmonic crystal at $t = 0$ (A) into eight solitons at later times $t = t_B$ (B) and $t = 3\cdot 6 t_B$ (C)

Figure 3 Envelope solitons in deep water (*from Yuen H C and Lake B M* 1975 *Phys. Fluids* **18** 956)

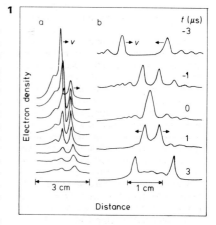

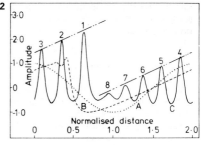

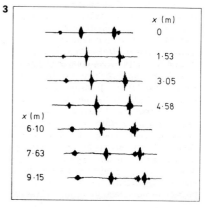

11.1 Solitons

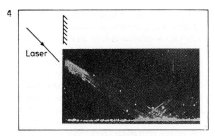

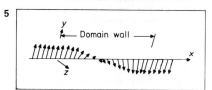

Figure 4 Self-focusing damage tracks induced in Perspex by a 2×10^{13} W cm^{-2} laser pulse. Note the totally internally reflected filaments

Figure 5 Reversal of magnetisation in a domain wall. The magnetic crystal is uniaxial with the direction of easy magnetisation along y

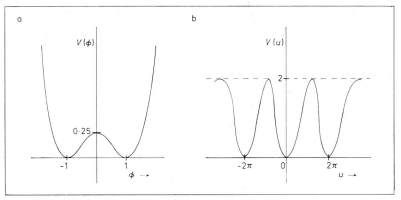

Figure 6 Potential energy functions associated with: **a** the φ-four equation; **b** the sine–Gordon equation

This unstable to unstable motion is $u = -\pi + 4\tan^{-1} e^t$. There is a corresponding static solution in x, a solution of $u_{xx} = +\sin u$, namely $u = 4\tan^{-1} e^x$. This is called a 'kink'. In a ferromagnet, spins interact through a combination of an exchange interaction and a magnetic anisotropy field. The combination yields $u_{xx} = \sin u$ and the kink is a description of the Bloch wall which lies between two domains of oppositely directed magnetisation. The directions of the spins are given by $\tfrac{1}{2}u$ and within the wall spins twist smoothly from 0 to π as figure 5 shows.

The Bloch wall is stable because it connects energetically *stable* equilibrium points at $u = 0$ and $u = 2\pi$. It can move and in one perhaps imperfect approximation the sG results. The moving solution is easily obtained by a Lorentz transformation to velocity V

$$u = 4\tan^{-1}[\pm(x-Vt)/(1-V^2)^{1/2}]. \quad (1)$$

Two signs are included to describe kinks ($+$) and antikinks ($-$): walls can twist in either a right-handed or left-handed way.

Other equations have wall solutions like this: the 'ϕ-four equation' $\phi_{xx} - \phi_{tt} = -\phi + \phi^3$ has the kink (antikink) solutions $\phi = \pm \tanh[(x-Vt)(1-V^2)^{-1/2}/2^{1/2}]$. Comparison with the sG shows that it has potential energy $\tfrac{1}{4} + \tfrac{1}{4}\phi^4 - \tfrac{1}{2}\phi^2$ (recall the pendulum has potential energy $(1-\cos u)$ and this is $\tfrac{1}{2}u^2$ only in the small oscillations approximation). With this potential the ϕ-four has stable equilibrium points at $\phi = \pm 1$. Its kink solution takes ϕ from -1 to $+1$, the antikink from $\phi = +1$ to $\phi = -1$. Experts in phase transition theory or familiar with work on the transition of a laser through threshold (see the article on synergetics by H Haken *Physics Bulletin* September 1977 p412) will see in this potential a familiar model. Figure 6a illustrates it: figure 6b shows the cosine potential for the sG.

The ϕ-four has been used as a model field theory. Another particular application has been to the study of displacive phase transitions in one-dimensional crystals. A form of it in two space dimensions models the configuration of oxygen atoms in strontium titanate. This crystal undergoes an antiferrodistortive phase transition in which a unit cell contains four oxygen atoms rather than the two appropriate to higher temperatures. Atoms with the same configuration tend to surround each other in the displacive phase forming clusters. The kinks bound such clusters just as the Bloch wall bounds clusters of similar spins. These kinks act like particles and may explain unusual features ('central peaks') observed in the dynamical form factor $S(k,w)$ of SrTiO$_4$ close to the transition temperature.

We learn a little more about the nature of the soliton if we contrast the kinks of the ϕ-four with those of the sG. The ϕ-four has two stable equilibrium points ± 1: the sG has an infinity ($u = 0, \pm 2\pi, \pm 4\pi,$ etc). Kinks or antikinks join these points. Thus the kink of the ϕ-four must be followed by the antikink, and vice versa: if they collide they bump. The kink of the sG can however be followed by an antikink or another kink. Hence the sG kinks can pass through each when they collide and so act like solitons. This is what they do.

The kinks of the ϕ-four are physically important quasiparticles. They are not true solitons but it is not useful to extend the definition of a soliton and include them because no exact analytical machinery has yet been developed which can handle them. They do not simply bump in collision: they radiate oscillatory phonon-like modes and have weaker stability than the solitons.

Notional spins

Two important applications of the sG arise in physical situations where the interacting spins are notional rather than real as in a ferromagnet. Consider first a short intense optical pulse (10^{-9} s, 100 W cm^{-2} at peak) with electric field envelope ε modulating a carrier wave at optical frequency (10^{15} Hz). The pulse traverses a medium and the carrier is on resonance with an atomic transition within the medium. The resonant transition will be strongly excited and we can ignore all other nonresonant atomic transitions. If the resonant transition is not degenerate it concerns two atomic states only. These can be mapped on to a 'spin-down' state and 'spin-up' state – a spin-$\tfrac{1}{2}$ system. The motion of an atom in the field ε is now described by the motion of this 'spin'. The atoms in the medium couple through Maxwell's equations, the atomic pseudospins act like pendulums, and the motion is described by the sG!

One finds that $\varepsilon \propto u_t$ and u satisfies a form of the sG. From the kink solution (1) one finds $\varepsilon \propto \mathrm{sech}(x-Vt)/(1-V^2)^{1/2}$. This optical pulse has time 'area' $|\int_{t=-\infty}^{t=+\infty} u_t \, dt|$ $= 2\pi$ and the pulse is called a 2π pulse. Since this pulse travels without change of shape or energy the medium is *transparent* to it. This is the phenomenon of self-induced transparency (SIT). Figure 7 shows the break up of larger area resonant pulses in ^{87}Rb vapour into two or three 2π pulses as well as the reshaping of a single 2π pulse. They are optical solitons and the medium is transparent to them. It is interesting to note that as the pulses have finite transverse profiles self-focusing also occurs. Break up takes place but one can measure intensity transmission factors of *several hundred per cent* at the centre of the intensity profile! These pulses are another example of soliton-like objects in more than one space dimension.

Another application of the pseudospin formalism is to large area Josephson junctions – 1 mm^2 pieces of superconductor sandwiching an oxide layer 10^{-6} cm thick. Voltage *pulses* satisfy the Josephson relation $V = (\hbar/2e)\sigma_t$ in this junction: V is the voltage and σ is the phase difference between the two sides of the junction and satisfies $\sigma_{xx} - \sigma_{tt} = \sin \sigma$. The sine term is just the Josephson current. Voltage pulses therefore break up into voltage solitons. They carry a transverse magnetic field and the total flux threading the junction proves to be an integral multiple of the single 'fluxon' $hc/2e$ because it is carried by kinks or antikinks. Work at Bell Laboratories has shown that these solitons can change the voltage/current characteristics. Figure 8 shows extra branches for the I/V curves – assigned by the workers at Bell to the presence of pairs of kinks and antikinks. Kinks can be pinned by micro-shorts or external magnetic fields and perturbation theory about the kink solutions of the SG describes this. A fast shift register, the flux shuttle, based on this has been proposed at Bell. Access time is 10^{-11} s (1 mm × c^{-1}; c is the velocity of light).

Kinks have been invoked to describe the electrical conductivity of one-dimensional conductors like TTF-TCNQ. Depending on the electron–phonon coupling constant a one-dimensional lattice of positive ions surrounded by electrons can lower its energy by a periodic displacement of the lattice positions. The electrons lock to the displaced ions forming a charge density wave (CDW), the configuration of the one stabilising the other. The system forms a 'condensate' (Fröhlich–Peierls condensate) with some of the properties of a superfluid. If the periodicities of lattice and displacement are incommensurate the system can translate without energy change. The translating CDW constitutes a current since electrons translate whilst the ions merely oscillate about average positions. In practice the condensate is pinned by impurity centres, lattice commensurability, interchain coupling, or other factors. Local distortions of the phase σ of the CDW then induce a local charge proportional to σ_x. These can move as kinks or antikinks. If in particular a pinning potential of period $2\pi/a$ is well approximated by $1 - \cos(a\sigma)$, the kinks and antikinks are sine-Gordon kinks and antikinks and carry charges proportional to $\pm 2\pi/a$. At thermal energies below the pinning potential these kinks or antikinks carry the current. Agreement between this theory and experiments in the range 1·6–4·2 K looks promising. It is fair to add that enhanced conductivity is also predicted from a one-dimensional model involving electron–electron rather than electron–phonon interactions. Remarkably this theory maps on to the *quantised* sine-Gordon equation which we discuss below in the context of particle physics.

Yet another application of the SG has been to discern a Josephson junction in liquid ³He. This isotope of helium undergoes a superconducting-type phase transition at 2·6 mK and 35 atm to the so called A-phase. This is believed to involve the pairing of the uncompensated nuclear spins of two ³He atoms. To prevent close approach the pairs take on unit orbital angular momentum. The spin state is then the triplet state $S = 1$ (the metallic superconductor has $l = 0$ and $S = 0$). In the A-phase spin fluctuations ensure that only spin-up and spin-down states play any role. The system now consists of two interpenetrating superfluids labelled by spin-up and spin-down. These couple through the weak spin–dipole interaction. The result is the SG for the phase difference between spins in the two states. A-phase spin waves thus become kinks of the SG.

These kinks consist of the rotation of a spin vector v against the supposedly uniform direction l of the orbital angular momentum. A more general kink is a composite kink involving changes in the directions of both l and v. Such a composite kink may be at rest in equilibrium and oscillations of v against its equilibrium position can then be excited: satellite frequencies in the nuclear magnetic resonance (NMR) spectrum have been predicted from such motions and certainly satellite frequencies have recently been observed. This evidence from composite kinks at rest is perhaps the best evidence for spatially extended kinks in ³He A at this time, although, in the time domain, Wheatley has made extensive observations of NMR signals characteristic of the inverted pendulum solution of the pendulum equation which I mentioned earlier.

These results and properties apply to the A-phase. At lower temperatures and pressures ³He enters the B-phase. Spin waves satisfy a 'double SG' $\sigma_{xx} - \sigma_{tt} = -(\sin \sigma + \frac{1}{2}\sin \frac{1}{2}\sigma)$ with remarkable solutions. Figure 9 shows the collision of a kink–antikink pair each of area $4\pi - 2\delta$ ($\delta = 2\cos^{-1}(-\frac{1}{4})$). In the long trough the pair changes to a 2δ–*anti*kink–kink pair, this loses energy by radiation and the pair reverts back to a $4\pi - 2\delta$ pair. These kinks are not solitons in the way that ϕ-four kinks are not solitons but they are remarkable enough. Interestingly SIT obeys the double SG with positive sign on the right side in a case when the resonant atomic transition is degenerate. Optical pulses now prove to be *bound pairs* of 2π-hyperbolic secant pulses and these can wobble. Pictures of this wobbler appear in the article 'Solitons' referenced below. It has been observed in degenerate D_1 transitions in sodium vapour.

Particle physics

I turn finally to particle physics. Kinks are particles: their energy is localised at a point, they are stable and have finite self-energy. When the velocity $V = 0$ the rest energy of the kink solution of the SG $u_{xx} - u_{tt} = m^2 \sin u$ is $8m\gamma^{-1}$ (γ is a coupling constant: it appears in the Hamiltonian but not in the equation). There is an important solution of the SG I have not mentioned. This is the 'breather'. The breather is a bound kink–antikink pair and its exact analytical form is known. It acts like a soliton and passes through other solitons – kinks or breathers. Its mass is $16m\gamma^{-1}\sin\mu$. The parameter μ is associated with an internal oscillatory degree of freedom of the breather. For $\gamma \ll 1$ kinks and breathers become massive particles much heavier than the 'meson' mass m.

Theoreticians impose commutation relations on the field $u(x,t)$ and 'quantise' it. The usual Hilbert space of vacuum plus meson states is now extended by the soliton states. The soliton states are not accessible by perturbation theory and the soliton masses, of order γ^{-1}, show this. The quantised SG in particular has a mass spectrum consisting of mesons, mass m, kinks and antikinks, mass $8m\gamma^{-1}$, and a *discrete* breather spectrum $16m\gamma^{-1}\sin(n\gamma/16)$ where $n = 1, 2 \ldots, N$ and N is that integer just below $8\pi\gamma^{-1}$. If $\gamma \ll 1$ the lowest states have masses nm and can be thought of as states made up of n mesons.

This beautiful model must be a stimulus for further work since it has no known relevance to the distribution of particle

Figure 8 Extra branches on the I/V characteristics of a large area Josephson junction believed due to sine-Gordon kink–antikink pairs (*from Fulton T A and Dynes R C 1973 Sol. Stat. Commun.* **12** 57)

Figure 9 Computer simulation of a colliding spin wave kink–antikink pair of area $4\pi - 2\delta$ in the B-phase of ³He below 2·6 mK

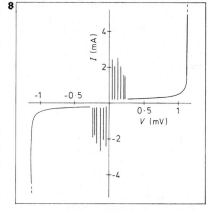

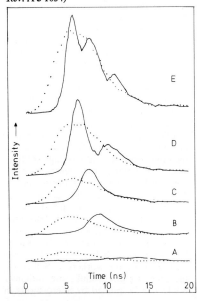

Figure 7 Self-induced transparency in ⁸⁷Rb vapour: dotted (solid) lines show the optical pulse profile ε^2 at input (output). The weak pulse A attenuates. Stronger pulses reshape to solitons (B and C), and still more intense ones break up into several solitons (D and E) (*from Slusher R E and Gibbs H M 1972 Phys. Rev. A* **5** 1634)

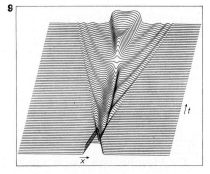

masses which have been observed. One extension must be to three space dimensions, but we know that Derrick's theorem then excludes kink-like solutions of the SG although certain time dependent solutions like the breather remain possible. Derrick's theorem is one reason for exploring relativistically invariant classical gauge theories[3]. Equations with soliton solutions are usually 'infinite dimensional completely-integrable Hamiltonian' systems. This means they have an infinite number of constants of the motion over and above those of energy and momentum. A completely integrable infinite-dimensional relativistically invariant equation in three space dimensions is not ruled out. It could provide a lot of answers if and when it can be found ∎

3 There are no problems, except increasing difficulty of the mathematics, in finding examples of localised multisoliton solutions of nonrelativistic wave equations in more than one space dimension. The Kadomtsev–Petviashvili equation $u_{yy} - (u_t + 6uu_x + u_{xxx})_x = 0$ has such solutions.

Further reading
A short biography of Russell appears in the *Dictionary of National Biography* 1897 **49** 465–6. He was an outstanding example of the Victorian entrepreneur. In middle life he built the *Great Eastern* for Brunel in his own dockyard on the Thames and was held to have tried to steal more credit for the vessel than was his due. The new biography *John Scott Russell* by G S Emmerson (London: Murray 1977) presents a very different picture. *The Wave of Translation* was first published by Teubner in 1878 with a new, London, edition in 1885. The 5 miles estimate agrees with the actual equivalent depth at uniform density; the 5×10^{17} miles is out by at least five orders of magnitude and in any case apparently relies on a g reduced by a factor 10^{-5}.
    Review or semireview articles are 'The soliton: a new concept in applied science' by A C Scott, F Y F Chu and D W McLaughlin (*Proc. IEEE* 1973 **61** 1443–83); 'Solitons' by the author in *Interaction of Radiation with Condensed Matter* Vol I (IAEA-SMR-20/51, Vienna: International Atomic Energy Agency 1977); 'Quantum meaning of classical field theory' by R Jackiw (*Rev. Mod. Phys.* 1977 **49** 681–706). Jackiw also discusses 'instantons'. These multidimensional solutions (of classical gauge theories) are not solitons and not particles; but they connect two vacuum states (rather like those of the ϕ-four shown in one dimension in figure 6a) and tell us about the possibility of penetrating the potential barrier between them. They may be able to say something on the problem of quark confinement.
    Note that techniques developed in soliton theory may also solve some of the problems Haken discusses in his article on synergetics in the September 1977 issue of *Physics Bulletin*.
    I am grateful to J E Balmer for figure 4 and P W Kitchenside for figure 9. The other figures are taken from the published literature.
    This article is dedicated to friends and colleagues in the soliton community. May it serve to indicate their very exacting standards!

R K Bullough, F Inst P, is Professor of Mathematical Physics at the University of Manchester Institute of Science and Technology

Soliton Propagation in Liquid Crystals

Lin Lei,[a] Shu Changqing, and Shen Juelian

Institute of Physics, Chinese Academy of Sciences, Beijing, China

and

P. M. Lam

Institute of Theoretical Physics, Chinese Academy of Sciences, Beijing, China

and

Huang Yun

Department of Physics, Beijing University, Beijing, China

(Received 5 May 1982)

Soliton propagation in nematic liquid crystals under shear is shown to be possible and studied theoretically. Calculations including those pertaining to the modulation of monochromatic or white light passing through such a liquid-crystal cell are presented. Recent experiments are interpreted accordingly and are in good agreement with the theory presented here.

PACS numbers: 61.30.-v, 03.40.Kf, 05.70.Ln, 47.15.-x

Solitons are important and have been found in various objects ranging from celestial bodies to laboratory systems.[1,2] However, unlike the first observation of solitons in shallow water by Scott Russell, many of the recent experimental evidences of solitons in condensed matter are indirect in nature. The experiments[3] on the ordered fluid ^{3}He are no exception. In this regard, we note that in another type of ordered fluid, viz., liquid crystal, because of the strong coupling of the director with light, it may be possible to observe the motion of the molecules and the solitons rather directly.

Discussions of solitons in liquid crystals[4] was first given by Helfrich[5] and subsequently by de Gennes,[6] Brochard,[6] and Leger.[7] In their work in nematics, the solitons (called "walls") are magnetically generated and are small in width (e.g., a few microns). Experimentally, the observation[7] of these solitons is delicate and a polarizing microscope has to be used. Recently, there has been more but still limited attention[8] paid to the role of solitons in the physics of liquid crystals.

In this Letter, we first point out and discuss a new case in liquid crystals, viz., nematics under uniform shear, in which solitons can exist and propagate. In contrast to the magnetic case[5-7]

11.2 Soliton Propagation in Liquid Crystals

the propagation of these solitons can be observed even by the naked eye and easily measured. We then give an analysis and explanations of some recent experiments[9,10] which are found to be in good agreement with our theory.

Let us consider a nematic under uniform shear such that the velocity is given by $\vec{v} = (v(y), 0, 0)$ and $s \equiv \partial v/\partial y = $ const. The incompressibility condition $\nabla \cdot \vec{v} = 0$ is clearly satisfied. Under the assumptions that the director $\vec{n} = (\sin\theta, \cos\theta, 0)$ and $\theta = \theta(x,t)$ we find, according to the Ericksen-Leslie equations, that[11]

$$M\frac{d^2\theta}{dt^2} = K\frac{\partial^2\theta}{\partial x^2} - \gamma_1\frac{d\theta}{dt} + \frac{s}{2}(\gamma_1 - \gamma_2\cos2\theta), \quad (1)$$

where the one-constant assumption ($K_1 = K_2 = K_3 = K$) is used. Here, M is the moment of inertia, K the elastic constant, γ_1 and γ_2 the viscosity coefficients, and $d\theta/dt = \partial\theta/\partial t + v\partial\theta/\partial x \simeq \partial\theta/\partial t$ is assumed. Equation (1) is the damped driven sine-Gordon equation which is known to have soliton solutions.[12] When θ is a traveling wave of velocity c, Eq. (1) becomes

$$m\ddot{\theta} = -\eta\dot{\theta} - \partial U/\partial\theta, \quad (2)$$

where $\theta = \theta(Z)$, $Z \equiv X - \eta T$, $X \equiv x/\lambda$, $T \equiv t/\tau$, $\lambda \equiv (2K/|\gamma_2|s)^{1/2}$, $\tau \equiv 2\gamma/s$, $\eta \equiv c\tau/\lambda$, $m \equiv 1 - Ms\eta^2/(2\gamma^2|\gamma_2|)$, $\gamma \equiv \gamma_1/|\gamma_2|$, $U = \gamma\theta + \frac{1}{2}\sin2\theta$, and $\dot{\theta} \equiv d\theta/dZ$. In (2), the experimental fact that $\gamma_2 < 0$ is adopted and $s > 0$ is assumed. Note that if θ is the soliton for $s > 0$ then $-\theta$ is that for $s < 0$.

Equation (2) describes the damped motion of a particle with mass m in an apparent potential U. The damping coefficient is η and Z plays the role of time. For $0 < \gamma < 1$, U has a series of maxima at $\theta = \theta_0 + k\pi$ and minima at $\theta = -\theta_0 + k\pi$ where k is an arbitrary integer and $\pi/4 < \theta_0 \equiv \frac{1}{2}\cos^{-1}(-\gamma) < \pi/2$. There are only three types of solitons corresponding respectively to the particle starting (with zero velocity) at the maximum at $\theta = \theta_0$ and ending in (A) the adjacent minimum at $\theta = -\theta_0$, (B) the minimum at $\theta = \pi - \theta_0$, or (C) the maximum at $\theta = \theta_0 - \pi$. Type C appears only when $\eta = \eta_c$ but type A (B) is possible for all $\eta > \eta_c$ ($\eta > 0$). Here, η_c is a parameter which increases monotonically with γ from zero at $\gamma = 0$ to $0.84m^{1/2}$ at $\gamma = 1$ (see Fig. 7 of Ref. 12). Note that there is no soliton for $\gamma > 1$. For $\gamma = 1$, type A reduces to type C. With the experiment of Ref. 10 in mind, we will discuss below only the strongly damped case of $\eta \gg 1$.

To this end, (2) is expanded in $1/\eta$ resulting in

$$\dot{\theta} = -(\gamma + \cos2\theta)/\eta - 2(\sin2\theta)\dot{\theta}/\eta^2 + O(1/\eta^3) \quad (3)$$

which has been solved analytically.[13] Numerically, soliton solutions of type A may be approximated very accurately by the more simple expression

$$\theta = \tan^{-1}\{w\tanh[(\gamma - 1)wZ/\eta]\}, \quad (4)$$

where $w \equiv [(1+\gamma)/(1-\gamma)]^{1/2}$, which is actually the solution of (3) to $O(1/\eta)$. As expected, θ decreases monotonically from θ_0 at $Z = -\infty$ to $-\theta_0$ at $Z = +\infty$ ($\gamma < 1$), the two uniform states allowed by the shear flow. In (3) and (4), without loss of generality, $m = 1$ is assumed.

When the shearing nematic is placed between two crossed polarizers which are in the x-z plane with the polarizing direction at 45° with the x axis, the part of the soliton corresponding to $\theta = 0$ will appear as a dark line moving with velocity c in the x direction. The illuminating light is assumed to be in the y direction.

For monochromatic light of wavelength λ_0 the ratio of output to input intensities I/I_0 as a function of Z/η is calculated. It varies from 0 to 1 consisting of a series of minima and maxima. The positions of the points with $I/I_0 = 0.5$ are depicted in Fig. 1. The region between two adjacent points with a minimum in between is painted black. In Fig. 2, the curve I/I_0 for white light is shown. The width of the dark line at the center, Δ, is found to decrease with γ as shown in Fig. 3. In these calculations, (4) is used; thickness of the nematic $2d = 20$ μm, refractive indexes $n_0 = 1.54$ and n_e from Fig. 6 of Ref. 14 for N-[p-methoxybenzylidene]-p-butylaniline (MBBA) are adopted.

What we discussed above is the appearance of the solitons once they are created. There remains the question of how the solitons can be excited. In the experiments of Ref. 10, nematic

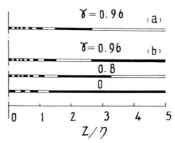

FIG. 1. Theoretical "photograph" of transmitted monochromatic light derived from the calculated I/I_0 vs Z/η curve (see text). The picture is symmetric in Z and $-Z$. (a) $\lambda_0 = 6328$ Å, $\gamma = 0.96$; (b) $\lambda_0 = 6000$ Å, $\gamma = 0.96, 0.8, 0$.

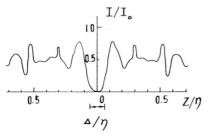

FIG. 2. Transmitted white light intensity I/I_0 vs Z/η. $\gamma = 0.96$. Δ is the width of the dark line defined at the half maximum intensity.

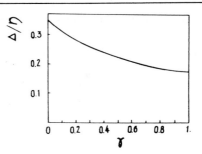

FIG. 3. Dependence of the width of the dark line Δ/η on γ.

MBBA is homeotropic and at rest initially. To excite the solitons, a Mylar plate placed at one end of the liquid-crystal cell is either pushed or pulled steadily along the cell. In our opinion, the movement of this plate creates velocity gradients in the nematic.[15] To first approximation, the velocity profile may be assumed to be steady and of the form shown in Fig. 4. The problem becomes one dimensional and our results presented above may be applied without modification to each of the layers (of thickness d) near the surfaces of the cell.

The good agreement between our theory and the experimental results[10] are evidenced by the following. (i) The behavior of the dark lines in the two cases of pushing and pulling of the plate are similar to each other.[10] Our theory predicts identical behavior when all other conditions are identical. (ii) Experimentally, $c \gg v$. Our theory gives $\delta \equiv \lambda\Delta = (c/s)2\gamma f(\gamma)$ where $f(\gamma)$ is the curve in Fig. 3, implying that a thick dark line (under white light) moves faster than a thin one, in agreement (at least qualitatively) with experiments.[10] (iii) Experiment and our theory both show that the dark line corresponds to molecules normal to the glass plates. (iv) The characteristics of I/I_0 shown in our Fig. 1 are in agreement with that in Ref. 10. In fact, the experimental pattern of the transmitted monochromatic light may be understood as resulting from the overlapping of three patterns of the type similar to Fig. 1 (corresponding to three solitons) as evidenced from theoretical results[13] shown in Fig. 5. (v) For a cell of 30 cm in length, 5 cm in width, and $d = 10$ μm, $K = 10^{-6}$ dyn, $\gamma = 0.96$, $c = 10$ cm/sec, and $V = 0.05$ cm/sec (resulting in $\theta_0 = 81.9°$, $\eta = 1.6 \times 10^3$), the power required to generate and maintain the propagation of one soliton is calculated to be ~ 193 erg/sec. The experimental result (for three solitons)[16] is $\sim 10^2$ erg/sec. With the same set of parameters and from our Fig. 3 we find $\delta = 0.8$ mm while experiment gives $\delta \sim 1$ mm. Note that physically V is always smaller than the velocity of the pushing plate. (vi) The dark line (under white light) is sandwiched between two bright narrow lines (see Fig. 2). This is clearly observed experimentally.[10]

Knowing the temperature dependence[17] of γ, n_0, and n_e, one may obtain δ as a function of temperature. Also, using $s = V/d$ our theory predicts $\delta/c = 2\gamma f(\gamma) d/V$. This can be checked easily by varying the thickness of the cell or of the pushing plate. The occurrence of three dark lines in the experiment[10] is related to the input power and the

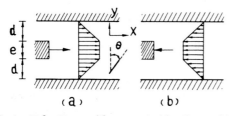

FIG. 4. Velocity profiles created by the pushing (a) or pulling (b) of the Mylar plate at the left of the cell. The maximum velocity in the profile is V.

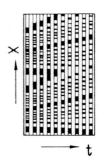

FIG. 5. Modulated monochromatic light pattern corresponding to two solitons (see Ref. 13). $\lambda_0 = 6328$ Å, $\gamma = 0.96$.

shear rate. Only one dark line is observed when a thinner pushing plate is used,[16] or when the soliton is generated by applying pressure gradients.[2,18]

Strictly speaking, those theoretical results obtained above for a traveling wave are applicable only in the time region in which the velocity of the dark lines is almost constant.[13] The major features and conclusions of the above one-dimensional analysis are retained in a more refined two-dimensional study.[13]

Solitons of type B should be observable for nematics initially in planar configuration.[13] With the setup described in Ref. 10 dark lines (under white light) cannot be (and have not been[16]) observed when the homeotropic configuration is replaced by the planar one. For both homeotropic and planar configurations, type C solitons should be observable. Note that the method used in Ref. 10 to excite solitons is not unique. Further discussions and other results can be found in Refs. 2 and 13.

We thank Zhu Guozhen for providing his experimental results of Ref. 10 to us before publication and many helpful discussions.

---

[a] Correspondence address Sept.-Nov. 1982: Laboratoire de Physique des Solides, Université Paris-Sud, Orsay, France.

[1] See, e.g., A. R. Bishop, J. A. Krumhansl, and S. E. Trullinger, Physica (Utrecht) 1D, 1 (1980); *Solitons and Condensed Matter Physics*, edited by A. R. Bishop and T. Schneider (Springer-Verlag, Berlin, 1978); *Solitons in Physics*, edited by H. Wilhelmsson, Phys. Scr. 20, No. 3-4, 289 (1979).

[2] Lin Lei, in Proceedings of the Conference on Statistical Physics and Condensed Matter Theory, Wuhan, China, 8-14 December 1981 (to be published).

[3] T. J. Bartolac et al., Phys. Rev. Lett. 46, 126 (1981); E. Polturak et al., Phys. Rev. Lett. 46, 1588 (1981).

[4] In the soliton literature [e.g., A. Barone et al., Riv. Nuovo Cimento 1, 227 (1971); A. C. Scott et al., Proc. IEEE 61, 1443 (1973)] reference has been made to the work of J. L. Fergason and G. H. Brown, J. Am. Oil Chem. Soc. 45, 120 (1968), on the possibility of the director waves propagating along a liquid-crystalline lipid membrane. It is interesting to note that in this work [and Zhu Guozhen, J. Qinghua Daxue Xuebao 21(4), 83 (1981)] in which damping terms are ignored, only a linear wave equation is presented and there is no mention of any nonlinear wave or soliton (which is impossible within the theory given there).

[5] W. Helfrich, Phys. Rev. 21, 1518 (1968).

[6] P. G. de Gennes, J. Phys. (Paris) 32, 789 (1971); F. Brochard, J. Phys. (Paris) 33, 607 (1972).

[7] L. Leger, Solid State Commun. 10, 697 (1972), and 11, 1499 (1972).

[8] J. Prost, in *Liquid Crystals of One- and Two-Dimensional Order*, edited by W. Helfrich and G. Heppke (Springer-Verlag, Berlin, 1980); E. F. Carr and R. W. H. Kozlowski, in *Liquid Crystals*, edited by S. Chandrasekhar (Heyden, London, 1980); R. Pindak et al., Phys. Rev. Lett. 45, 1193 (1980); M. Yamashita et al., Mol. Cryst. Liq. Cryst. 68, 79 (1981).

[9] Zhu Guozhen, in Proceedings of the Chinese Liquid Crystal Conference, Guilin, China, 20-25 October 1981 (to be published).

[10] Zhu Guozhen, preceding Letter [Phys. Rev. Lett. 49, 1332 (1982)].

[11] Notations used here are those in L. Lam, Z. Phys. B 27, 349 (1977).

[12] See, e.g., M. Büttiker and R. Landauer, Phys. Rev. A 23, 1397 (1981).

[13] Lin Lei and Shu Changqing, to be published.

[14] Y. Takahashi, T. Uchida, and M. Wada, Mol. Cryst. Liq. Cryst. 66, 171 (1981).

[15] Lin Lei and Shen Juelian, in Proceedings of the Chinese Liquid Crystal Conference, Guilin, China, 20-25 October 1981 (to be published).

[16] Zhu Guozhen, private communication.

[17] Ch. Gähwiller, Phys. Rev. Lett. 28, 1554 (1972); S. Meiboom and R. C. Hewitt, Phys. Rev. Lett. 30, 261 (1973).

[18] Shu Changqing, Zhu Guozhen, and Lin Lei (unpublished).

ERRATUM

Soliton Propagation in Liquid Crystals. LIN LEI, SHU CHANGQING, SHEN JUELIAN, P. M. LAM, and HUANG YUN [Phys. Rev. Lett. **49**, 1335 (1982)].

On p. 1336, the seventh line above Eq. (3), $0.84\ m^{1/2}$ should be replaced by $1.68\ m^{1/2}$.

Possible relevance of soliton solutions to superconductivity

T. D. Lee

Department of Physics, Columbia University, 538 West 120th Street, New York, New York 10027, USA

The newly discovered high-temperature superconductors[1] indicate that there might exist another underlying mechanism for superconductivity, different from the BCS theory[2]. Recent suggestions of bipolarons[3,4] and resonant valence bonds[5], among others (see, for example, refs 6,7), point to the importance of an effective complex boson field $\phi(\mathbf{x})$. For example, $\phi(\mathbf{x})$ could be formed from a pair of electron fields $\psi_\uparrow$ and $\psi_\downarrow$ of opposite spin: $\phi(\mathbf{x}) \equiv \langle \psi_\uparrow(\mathbf{x}+\mathbf{l})\psi_\downarrow(\mathbf{x}-\mathbf{l})\rangle_l$ where $\langle \rangle_l$ denotes the appropriate average over the spacing $\mathbf{l}$. The corresponding phenomenological Lagrangian density may be written as

$$\mathcal{L} = \frac{\partial \phi^\dagger}{\partial t}\frac{\partial \phi}{\partial t} - \nabla\phi^\dagger \cdot \nabla\phi - u(\phi^\dagger \phi) \quad (1)$$

where as $|\phi| \to 0$,

$$u \to m^2 \phi^\dagger \phi + g(\phi^\dagger \phi)^2 + \cdots \quad (2)$$

with m acting as the effective mass and g as a nonlinear coupling. The replacement of ∂_μ (where μ denotes the space–time indices) by $\partial_\mu - 2eA_\mu$ gives the modification when there is an electromagnetic field A_μ. In models that assume a relatively tight binding between the electron pair, one might expect the coefficient m^2 to be positive and insensitive to the temperature T (for T less than the transition temperature T_c), in contrast to that in the standard Ginzburg–Landau equation[8] (which has m^2 negative and proportional to $T - T_c$). This difference changes the character of the solution, making the customary acquisition of a non-zero $|\phi|$ through the Higgs mechanism inapplicable. The purpose of this note is to call attention to the property that, when $m^2 > 0$ and if there exists some attractive nonlinear interaction, the low-energy states of equation (1) may be dominated by solitons, instead of by the usual plane-wave solutions of particles of mass m.

Define $\varepsilon \equiv u - m^2 \phi^\dagger \phi$, which contains all the nonlinear interactions of ϕ. We assume that, at least for a limited range of $|\phi|$, ε is negative; that is,

$$\varepsilon < 0 \text{ for } \sigma_1 < |\phi| < \sigma_1 + \delta \quad (3)$$

where σ_1 and δ are both positive. (If $g < 0$ in equation (2), then $\sigma_1 = 0$.) As we shall see, when the space-dimension D is 1, independently of the behaviour of ε outside the range given in equation (3) and no matter how small ε and δ are within that range, there always exist non-topological soliton solutions. Such solutions can also exist in any higher dimension $D > 1$, but there is a condition on ε and δ; this condition becomes more stringent as D increases. Furthermore, these solutions are valid in classical physics as well as in quantum mechanics[9].

Consider first $D = 1$ and assume $\phi = \sigma(x)e^{-i\omega t}$. From equation (1) it follows that $\sigma(x)$ satisfies $\omega^2 \sigma + d^2\sigma/dx^2 - \frac{1}{2}du/d\sigma = 0$, which, together with the boundary condition $\sigma = 0$ at $|x| = \infty$, gives

$$\left(\frac{d\sigma}{dx}\right)^2 - u + \omega^2 \sigma^2 = 0 \quad (4)$$

This is analogous to the equation of energy conservation for a non-relativistic particle of 'position' σ and 'time' x moving in a 'potential' $v(\sigma) \equiv \omega^2 \sigma^2 - u(\sigma)$. Both $u(\sigma)$ and $v(\sigma)$ are illustrated in Figs 1 and 2. Because of expression (3), there exists an $\bar{\omega}^2 < m^2$ such that the parabola $\bar{\omega}^2 \sigma^2$ is tangent to $u(\sigma)$, as shown in Fig. 1. Hence, for any ω^2 between $\bar{\omega}^2$ and m^2, there exists a soliton solution given by

$$x - x_0 = \pm \int_{\sigma_0}^{\sigma} v(\sigma)^{-1/2} d\sigma \quad (5)$$

as shown in Fig. 3. When $x \to \pm\infty$, $\sigma \approx \exp[-(m^2-\omega^2)^{1/2}|x|]$; this gives a correlation length $(m^2 - \omega^2)^{-1/2}$ which approaches ∞ when $\omega \to m$. As a result, there could be a phase transition even in one dimension.

If for small σ, $\varepsilon(\sigma) = g\sigma^4 + g'\sigma^6 + \cdots$ with $g < 0$, then $\sigma_1 = 0$ in equation (3). In this case, solitons exist for any particle number N, up to a maximum value $\bar{N}$ (corresponding to $\omega = \bar{\omega}$); it can be shown[9] that because $dM/dN = \omega < m$, the soliton energy $M(N)$ is always less than Nm, insuring its stability. Thus, at low temperature, the thermodynamics of the system becomes dominated by the soliton phase, which resembles that of a Bose gas of rest mass $M(N)$, where N refers to the average particle number $N(T)$ varying with T. Because $M(N_1 + N_2) < M(N_1) + M(N_2)$ for any N_1 and N_2, we have at $T = 0$, $N = \bar{N}$ (depending on the model, $\bar{N}$ can be quite large). As T increases, $N(T)$ decreases, and it becomes of order unity near T_c; corre-

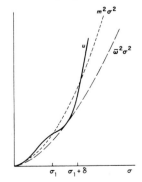

Fig. 1 The solid line gives an example of $u(\sigma) = m^2\sigma^2 + \varepsilon(\sigma)$, where $\varepsilon(\sigma) < 0$ for $\sigma_1 < \sigma < \sigma_1 + \delta$; the short-dashed line is $m^2\sigma^2$ and the long-dashed line is $\bar{\omega}^2\sigma^2$, which is tangent to $u(\sigma)$.

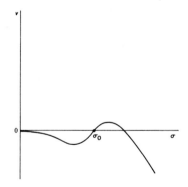

Fig. 2 Schematic plot of $v(\sigma) = \omega^2\sigma^2 - u(\sigma)$, for $\bar{\omega} < \omega < m$.

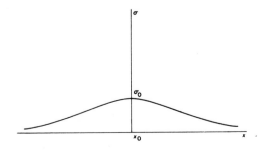

Fig. 3 Example of a non-topological soliton, thus called because as $|x| \to \infty$, $|\phi| \to 0$ (in any dimension), where $\phi = \sigma e^{-i\omega t}$.

spondingly, ω increases from $\bar{\omega}$ to m, accompanied by the related change of correlation length from $(m^2 - \bar{\omega}^2)^{-1/2}$ to ∞. For $D = 1$, the specific heat is proportional to $T^{1/2+n}$ where $n > 0$. The value of n depends on the nonlinear interaction ε, and is associated with the gradual change of $N(T)$.

For $D > 1$, the radial part of ∇^2 is $\partial^2/\partial r^2 + (D-1)r^{-1}\partial/\partial r$; the latter gives a dissipative 'friction' to the corresponding 'energy' expression in the mechanical analogue given by equation (4). This sets a lower limit on the magnitude of the negative part of ε. In order to have solitons, ε cannot be arbitrarily small. Furthermore, stable solitons (that is, for $M(N) < Nm$) now exist only within a relatively limited range of N between, say, $\bar{N} - \Delta$ and $\bar{N}$. Again, the low-lying states are those of solitons, not particles of mass m. The corresponding specific heat is proportional to $T^{D/2+n}$. If Δ is narrow, then n is positive but small; this leads to, when $D = 2$, a nearly linear dependence on T for the specific heat.

A system of these solitons, like that of any bosons, can undergo Bose-Einstein condensation and thereby may exhibit superconductivity, in accordance with the work of M. R. Schafroth[10].

In view of the rapid accumulation of new experimental results in this field and the apparent inadequacy of our present theoretical understanding, exploration of different theoretical avenues may be necessary. It is in this spirit that this note is written.

I thank J. M. Luttinger for helpful discussions. This research was supported in part by the US Department of Energy.

Received 30 October; accepted 12 November, 1987.

1. Bednorz, J. G. & Müller, K. A. *Z. Phys.* **B64**, 189-193 (1986).
2. Bardeen, J., Cooper, L. N. & Schrieffer, J. R. *Phys. Rev.* **108**, 1175-1204 (1957).
3. Alexandrov, A. S., Ranninger, J. & Robaszkiewicz, S. *Phys. Rev. Lett.* **56**, 949-952 (1987).
4. Teller, E. *Proc. 1987 Erice Summer School* (ed. Zichichi, A.) (Academic, New York, in the press).
5. Anderson, P. W. *Science* **235**, 1196-1198 (1987).
6. Lundqvist, S., Tosatti, E., Tosi, M. & Yu Lu (eds) *Proc. Adriatico Res. Conf. High Temperature Superconductors* (World Scientific, Singapore, 1987).
7. Gan, Z. Z., Cu, G. J., Yang, G. Z. & Yang, Q. S. (eds) *Proc. Beijing int. Workshop High Temperature Superconductivity* (World Scientific, Singapore, 1987).
8. Ginzburg, V. L. & Landau, L. D. *Zh. éksp. teor. Fiz* **20**, 1064-1082 (1950).
9. Lee, T. D. *Particle Physics and Introduction to Field Theory*, Ch. 6 (Harwood, London, 1981).
10. Schafroth, M. R. *Phys. Rev.* **100**, 463-475 (1955).

Dendrites, Viscous Fingers, and the Theory of Pattern Formation

J. S. Langer

There has emerged recently a new theoretical picture of the way in which patterns are formed in dendritic crystal growth and in the closely analogous phenomenon of viscous fingering in fluids. Some interesting questions that arise in connection with this theory include: How broad is its range of validity? How do we understand dynamic stability in systems of this kind? What is the origin of sidebranches? Can weak noise, or even microscopic thermal fluctuations, play a role in determining the macroscopic features of these systems?

THE THEORY OF PATTERN FORMATION IN NONLINEAR DISSIpative systems has taken some surprising turns in the last several years. One of the most interesting developments has been the discovery that weak capillary forces act as singular perturbations which lead to beautifully delicate and very nearly identical selection mechanisms both in dendritic crystal growth and in the fingering patterns which emerge when a viscous fluid is displaced by a less viscous one. It now appears likely that important progress has been made, but pieces of the puzzle still seem to be missing.

For most of us, dendritic crystal growth brings to mind pictures of snowflakes. Materials scientists may think also about metallurgical microstructures, which provide very practical reasons for research in this field; but it is the snowflake that most quickly captures our imaginations. Kepler's 1611 monograph "On the Six-Cornered Snowflake" (1) is often cited as the first published work in which morphogenesis—the spontaneous emergence of patterns in nature—was treated as a scientific rather than a theological topic. At a time in which the existence of atoms was merely speculation, Kepler mused about hexagonal packings of spheres, but concluded that the

The author is at the Institute for Theoretical Physics, University of California, Santa Barbara, CA 93106.

problem was beyond his reach. Its solution would have to be left for future generations. In fact, scientists have waited more than three centuries before finding much hint of an answer to the question that Kepler posed.

One part of the answer, of course, is the understanding of crystalline symmetries and their relation to atomic structure. Another part is our modern statistical theory of the fluctuations and dissipative processes that ultimately govern pattern formation. But it is only in very recent years that we have begun to understand how these irreversible processes can amplify weak anisotropies and even very small noisy fluctuations in such a way as to produce intricate patterns in ostensibly featureless systems.

In the pattern-forming systems of interest here, we are dealing with dynamic processes, not just molecular structures or macroscopic forms. Unlike D'Arcy Thompson (2) (who could describe and measure but not explain) or Nakaya (3) (who produced one of the world's most complete and beautiful catalogues of snowflakes), we now have the experimental and analytic tools that we need to find out, for example, how the growth rate of a dendrite and the spacing between its sidebranches are determined by the temperature and composition of the solidifying substance. We may even have most of the tools—if not yet the information—that we need to understand the growth of biological forms. Of the analytic tools, the two which seem most essential are the theory of morphological instabilities in systems far from equilibrium, and the computer, which enables us to explore quantitatively the nonlinear behavior of such systems. [For both of these we must pay tribute to the remarkable insights of Turing (4).] The work to be described here arises largely from the modern interplay between physical insight, mathematical analysis, and numerical methods.

In this article, I shall review briefly the recent history of the dendrite (5) and viscous fingering problems and shall attempt to communicate at least the general flavor of recent developments, specifically, the so-called "solvability theory" (6). As an illustration of this theory, I shall describe Couder's remarkable bubble effect, which, by seemingly turning fingers into dendrites, provides an excellent illustration of the singular perturbation in action. I shall conclude with some conjectures about the range of validity of the solvability theory and its implications for our understanding of more complex dynamical effects such as sidebranching.

Dendritic Solidification of a Pure Substance

In the conventional thermodynamic model of the solidification of a pure substance from its melt, the fundamental rate-controlling mechanism is the diffusion of latent heat away from the interface between the liquid and solid phases. The latent heat that is released in the transformation warms the material in the neighborhood of the solidification front and must be removed before further solidification can take place. This is a morphologically unstable process which characteristically produces dendrites, that is, treelike or snowflake-like structures. In a typical sequence of events, an initially featureless crystalline seed immersed in an undercooled melt develops bulges in crystallographically preferred directions. The bulges grow into needleshaped arms whose tips move outward at constant speed. These primary arms are unstable against sidebranching and the sidebranches, in turn, are unstable against further sidebranching, so that each outward growing tip leaves behind itself a complicated dendritic structure like that shown in Fig. 1.

The dimensionless thermal diffusion field in this model for convenience is chosen to be

$$u = \frac{T - T_\infty}{(L/c)} \quad (1)$$

where T_∞ is the temperature of the liquid infinitely far from the growing solid, and the ratio of the latent heat L to the specific heat c is an appropriate unit of undercooling. The field u satisfies the diffusion equation

$$\frac{\partial u}{\partial t} = D\nabla^2 u \quad (2)$$

where D is the thermal diffusion constant, which can be taken to be the same in both liquid and solid phases. The remaining ingredients of the model are the boundary conditions imposed at the solidification front. First, there is heat conservation:

$$v_n = -[D\hat{n}\cdot\nabla u] \quad (3)$$

where $\hat{n}$ is the unit normal directed outward from the solid, v_n is the normal growth velocity, and the square brackets denote the discontinuity of the flux across the boundary. In these units, the left-hand side of Eq. 3 is the rate at which latent heat is generated at the boundary and the right-hand side is the rate at which it is being diffused away. The physically more interesting boundary condition is the statement of local thermodynamic equilibrium, which determines the temperature u_s at the two-phase interface:

$$u_s = \Delta - d_0\kappa \quad (4)$$

where

$$\Delta = \frac{T_M - T_\infty}{(L/c)} \quad (5)$$

and T_M is the melting temperature. Δ is the dimensionless under-

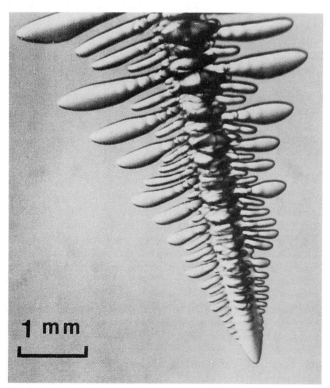

Fig. 1. Primary dendrite of succinonitrile (a transparent plastic crystal with cubic symmetry) growing in its undercooled melt. Note the smooth paraboloidal tip, the secondary sidebranching oscillations emerging behind the tip, and the beginnings of tertiary structure on the well-developed secondaries. (Photograph courtesy of M. E. Glicksman.)

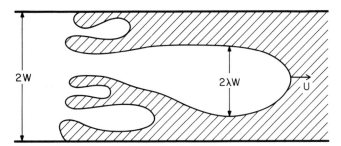

Fig. 2. Schematic illustration of a Hele-Shaw experiment adapted from a photograph by J. Maher. The inviscid fluid is entering from the left and the viscous fluid (shown shaded) is being pushed to the right. The motion is effectively two-dimensional, constrained by narrowly separated glass plates in the plane of the figure. As shown here, the flow takes place in a channel of width $2W$. The initially irregular pattern formed by the instability of the interface between the two fluids is developing into a single finger of width $2\lambda W$.

cooling, a measure of the driving force for the processes that we are considering. The second term on the right-hand side of Eq. 4 is the Gibbs-Thomson correction for the melting temperature at a curved surface: κ is the sum of the principal curvatures and $d_0 = \gamma c T_M / L^2$ is a length, ordinarily of order angstroms, which is proportional to the solid-liquid surface tension γ. The latter quantity and, accordingly, d_0 may be functions of the angle of orientation of the interface relative to the axes of symmetry of the crystal. In particular, for a cubic crystal in the (1, 0, 0) plane, d_0 is proportional to $(1 - \alpha \cos 4\theta)$, where θ is the angle just mentioned and α is a measure of the strength of the anisotrophy.

Viscous Fingering

The hydrodynamic analog of dendritic solidification is the fingering instability that occurs when one causes an inviscid fluid to drive a viscous one through a porous medium. The two-dimensional version of this situation (7) is a Hele-Shaw cell in which the two immiscible fluids are constrained to move between narrowly separated parallel plates. The configuration is shown schematically in Fig. 2. The invading inviscid fluid can be visualized as playing the role of the growing solid, and the more viscous fluid that is being pushed away is like the melt. The analog of the thermal field u is the pressure P, which can be taken to be constant in the "solid" and to satisfy Laplace's equation in the "melt." Here is the main difference between fingering and solidification; the Laplace equation is not the diffusion equation. The velocity of the viscous fluid in the porous medium is given by Darcy's law to be simply proportional to $-\nabla P$; thus the expression for the velocity of the interface between the two fluids is precisely the analog of the conservation law, Eq. 3. Finally, the interfacial tension γ causes the pressure at the interface to be reduced by an amount proportional to $\gamma\kappa$, in exact analogy to the thermodynamic boundary condition, Eq. 4. There is, however, no crystalline anisotrophy associated with this γ. Directional information can be provided only by the interaction between the long-ranged pressure field and the walls of the container, or else by adding to the model—"by hand," so to speak—some anisotropy of the medium through which the fluids are moving.

Pattern Selection

There are sharply defined problems of pattern selection associated with both of these models. In solidification, it is known that the growth rate v and the tip radius ρ of a dendrite are determined uniquely by the undercooling Δ. In the hydrodynamic case, specifically, the two-dimensional Saffman-Taylor (7) experiment in which a steady-state finger forms in a long channel, the ratio λ of the width of the finger to the width of the channel is determined uniquely by the flow speed. In both cases, surface tension appears at first glance to be a negligible perturbation; the length d_0 is orders of magnitude smaller than other characteristic lengths. However, the omission of surface tension in either problem leads to continuous families of solutions and, thus, to no explanation whatsoever of the experimentally observed selection principles. It turns out that surface tension is playing an especially subtle role in these processes.

In the case of the dendrite, if one neglects surface tension altogether, one arrives at Ivantsov's paradox (8). Instead of there being a unique growth velocity v and tip radius ρ at a fixed Δ, as required by experiment, there exists a continuous family of steady-state, shape-preserving solidification fronts—paraboloids of revolution—that satisfy the Ivantsov relation

$$\Delta = p\, e^p \int_p^\infty dy\, \frac{e^{-y}}{y} \tag{6}$$

where $p = \rho v / 2D$ is the thermal Péclet number. The tips of dendrites often do look very paraboloidal, and quantitative experiments generally indicate that the Ivantsov relation, Eq. 6, is satisfied. But obviously some essential ingredient of the theory is missing.

Over a decade ago, Müller-Krumbhaar and I (9) explored the idea [originally suggested by Oldfield (10)] that the missing element of the theory might have something to do with stability of the growth form. We performed many complicated calculations, but what was left in the end was a relatively simple conjecture that has since been confirmed remarkably well by experiment. In the simplest possible terms, our conjecture was that the tip radius ρ might scale like the Mullins-Sekerka (11) wavelength $\lambda_s = 2\pi(2Dd_0/v)^{1/2}$. Note that λ_s is the geometric mean of the microscopic capillary length d_0 and the macroscopic diffusion length $2D/v$; it is of roughly the right magnitude to characterize dendritic structures. A planar solidification front moving at speed v is linearly unstable against sinusoidal deformations whose wavelengths are larger than λ_s. Therefore, we reasoned, a dendrite with tip radius ρ appreciably greater than λ_s must be unstable against sharpening or splitting. The dynamical process that leads to the formation of the dendritic tip might naturally come to rest at a state of marginal stability, that is, at a state for which the dimensionless group of parameters

$$\sigma = \frac{2Dd_0}{v\rho^2} = \left(\frac{\lambda_s}{2\pi\rho}\right) \tag{7}$$

is a constant, independent of Δ. Moreover, if we take the idea literally and set ρ equal to λ_s, then the value of this constant should be $\sigma^* \cong (1/2\pi)^2 \cong 0.025$. The assumption $\sigma = \sigma^* =$ constant is consistent with a wide range of experimental observations (12) (when convective effects are eliminated or otherwise taken into account) and the specific value $\sigma^* \cong 0.0195$ for succinonitrile—by far the most carefully studied material—is quite close to the naïve prediction.

What, then, is wrong with the marginal stability theory? It seems that its mathematical foundation has been knocked from under it by the discovery that the Ivantsov family of solutions does not survive in the presence of surface tension (13–17). A nonvanishing d_0, no matter how small, reduces the continuum of solutions to, at most, a discrete set; and the existence of any solution whatsoever depends on there being some angular dependence of the surface tension, that is, a nonvanishing anisotropy strength α. Thus, the stability calculation that Müller-Krumbhaar and I thought we were performing was unfounded because the family of steady-state solutions whose stability we supposedly were testing did not exist.

All is not lost, however, because the mathematics immediately suggests an alternative selection mechanism, albeit one that has little of the intuitive appeal of marginal stability. A natural guess is that the selected dendrite is the one for which a stable solution exists. In more formal language, we guess that the condition for solvability of the steady-state equations is equivalent to a condition for the existence of a stable fixed point with a large basin of attraction in the space of configurations of this dynamical system. If this conjecture is correct, orderly, steady-state dendritic growth does not occur at all in isotropic materials. In suitably anisotropic systems, a growing body of analytic and computational evidence suggests that there is a denumerably infinite set of solutions, and that only the fastest (and thus sharpest) of these solutions can be dynamically stable. The hypothesis that this unique solution exists and that it describes the tip of a dynamically selected dendrite has come to be known as the "solvability theory."

Special Features of the Solvability Theory

This is not the place for a detailed exposition of the mathematics of the solvability theory, but there are several features that do need to be mentioned. In the limit of small Péclet number p, the controlling group of parameters in the theory is the same quantity σ, defined in Eq. 7, that appeared in the stability analysis. This happens because one is looking for a small surface tension–induced correction to the shape of the Ivantsov parabola and, in computing this correction, one encounters an equation quite similar to the one which arises in linear stability theory. [As shown by Pomeau and coworkers (15), linearization is not a necessary ingredient of the argument for solvability.] To be precise, σ enters the theory as a singular perturbation; it describes the strength of the curvature effect in Eq. 4 and, accordingly, multiplies the highest derivative in the equation for the shape correction once one has reduced this equation to dimensionless form.

There is a very nice way to visualize the effect of this perturbation. In practical numerical calculations (18–20), and also in the analytic approaches that have been applied successfully to this problem (14, 17), one can generally assure the existence of some kind of solution by relaxing a boundary condition—most commonly the condition of smoothness at the tip. Suppose one allows the tip to have a cusp of outer angle Θ and then, either numerically or analytically, computes what value Θ must have in order to achieve a solution at a given value of σ. Because Θ must vanish for a physically acceptable solution, a formally exact statement of the solvability condition is

$$\Theta(\sigma, p, \alpha) = 0 \tag{8}$$

We may think of Θ as a measure of how close we have come to finding a solution at an arbitrary value of σ. The special values of σ for which Eq. 8 is satisfied are denoted $\sigma^*(p, \alpha)$.

If one tries to compute Θ by expanding it in powers of σ, one finds that Θ vanishes at all orders, a result that would be consistent with the original expectation of a continuous family of solutions. If the calculation is performed more carefully, however, the answer—at small p and zero anisotropy α—has the form

$$\Theta(\sigma, 0, 0) \propto \exp\left(-\frac{\text{constant}}{\sqrt{\sigma}}\right) \tag{9}$$

This function has an essential singularity at $\sigma = 0$ and no possible expansion about that point. It is extremely small for small σ, but it does not vanish exactly unless $\sigma = 0$. Thus, an arbitrarily small amount of isotropic surface tension destroys all solutions. For small, positive anisotropy α, however, the function $\Theta(\sigma)$ has the same form as Eq. 9 for large σ but oscillates rapidly in the limit $\sigma \to 0$. The largest value of σ at which Θ passes through zero occurs at $\sigma = \sigma^* \propto \alpha^{7/4}$, the latter approximation being valid only in the limit of very small α.

The solvability theory for the Saffman-Taylor (21–23) problem is strikingly similar to the analysis for the dendrite. In this case, the system is automatically in the limit $p \to 0$ because $p = \nu \rho/2D$ and the diffusion equation, Eq. 2, reduces to the Laplace equation in the limit $D \to \infty$. The parameter σ is replaced by the dimensionless group of parameters σ_{ST}:

$$\sigma_{ST} = \frac{\gamma b^2 \pi^2}{12\mu U W^2 (1-\lambda)^2} \tag{10}$$

where γ is the surface tension, b the spacing between the plates, μ the viscosity, U the speed of the finger, $2W$ the width of the channel, and $2\lambda W$ the width of the finger. All other essential ingredients of the solvability function $\Theta_{ST}(\sigma, \lambda)$ defined in analogy to $\Theta(\sigma, p \to 0, \alpha)$ are the same except that the function $(1 - \alpha \cos 4\theta)$ is replaced by a function of θ and λ. It then turns out that the boundary-related quantity $\lambda - 1/2$ plays a role in this problem that is closely analogous to that played by the anisotropy strength α for the dendrite. For $\lambda < 1/2$, Θ_{ST} looks like Θ in Eq. 9 and there are no solutions of $\Theta_{ST} = 0$. For $\lambda < 1/2$, on the other hand, Θ_{ST} oscillates for small values of σ_{ST}, and the physically meaningful solution of the solvability condition has the form $\sigma_{ST}^* \propto (\lambda - 1/2)^{3/2}$. The convergence of λ to the value 1/2 at small σ_{ST} (large U) is consistent with experiment (7).

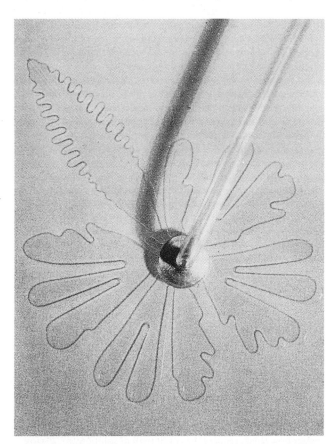

Fig. 3. Hele-Shaw experiment in which the inviscid fluid enters through a central orifice and forms fingers that move radially outward. One of these fingers has trapped a small bubble at its tip. As a result, it is growing stably at constant speed and is emitting sidebranches. (Photograph courtesy of Y. Couder.)

The picture summarized above—an apparently accurate analytic description of a mechanism for selecting steady-state configurations in two different nonlinear dissipative systems—seems elegant and mathematically compelling. The mathematics looks especially sound in view of recent work of Combescot et al. (22) in which a nonlinear formulation originally suggested by Kruskal and Segur (24) has been developed into an amazingly complete solution of the viscous fingering problem. There remains the possibility, however, that the solvability theory might be mathematically correct but physically irrelevant—that real systems might simply ignore these steady-state solutions and find other, perhaps oscillatory or even irregular, states of motion. With this possibility in mind, let us consider some of the evidence regarding the validity—or lack thereof—of the solvability theory.

Dendritic Growth Rates

First, there is the question of whether the solvability theory really agrees with experiment for the dendrites and, if so, what is its range of validity? Experiments (12) indicate that the parameter σ, as predicted, is a Δ-independent constant at small Δ (apart from corrections at very small Δ where convection in the melt becomes important). The solvability theory also provides a natural explanation for the previously unexplained fact that free dendrites grow only in directions parallel to crystalline axes of symmetry; lack of symmetry precludes the existence of solutions in other directions.

The trouble is that we do not know yet whether the values of σ^* predicted by the theory agree quantitatively with those found experimentally. At the moment, the available evidence seems inconclusive, and we are waiting both for new measurements and for more extensive, three-dimensional calculations. A particularly worrisome aspect of the situation is that the theory predicts a strong dependence of σ^* on the anisotropy strength α; specifically, σ^* is predicted to be proportional to $\alpha^{7/4}$ in the limit $\alpha \to 0$ and to be roughly linear in α for most of its accessible range of values (19, 25, 26). No such strong dependence on anisotropy has so far been confirmed experimentally.

In my opinion, it is most likely that the solvability theory will turn out to be a correct description of a large but limited class of relatively simple dendritic phenomena. It may break down in complex situations where competing processes such as thermal and solutal diffusion might produce time-dependent behavior that would be invisible in the present steady-state theory. It may also break down at large crystalline anisotropies where the solvability calculations become extremely difficult and perhaps intrinsically impossible. Almost certainly, the solvability theory will fail at small anisotropies, and the $\alpha^{7/4}$ law will turn out not to be physically meaningful. The last conjecture is based on considerations of stability that deserve a few paragraphs of their own. I shall return to that topic shortly.

Couder's Bubbles

A second category of evidence regarding validity of the solvability theory is the bubble effect discovered by Couder and co-workers (27), which indicates that something very much like solvability is occurring in variants of the Saffman-Taylor problem. For both the dendrite and the Saffman-Taylor finger, physically acceptable solutions require $\Theta = 0$, that is, the structure is not allowed to have a cusplike discontinuity at its tip. However, if one were able to perturb the system in such a way as to fix Θ at some nonvanishing positive value, then the mathematics tells us that dendrites should exist in the absence of anisotropy and that viscous fingers should occur with relative widths λ less than 1/2. Couder et al. have produced such perturbations of the fingers by attaching small bubbles to their tips, and in this way have succeeded in observing anomalously small values of λ. Their results for the dependence of λ on the channel width W (a function whose form should not depend on details of the flow in the neighborhood of the tip) are in excellent agreement with the solvability theory (23). Moreover, in the circularly symmetric geometry where radial fingers ordinarily suffer tip-splitting instabilities, they have shown that fingers with bubbles at their tips behave very much like dendrites, complete with sidebranches! A picture of such a finger—behaving like a dendrite—is shown in Fig. 3.

Stability and Sidebranching

Perhaps the most dramatic of the conceptual developments stemming from the solvability theory is a growing understanding of the dynamics of pattern-forming systems. In particular, we are beginning to understand the stability of dendritic tips and the manner in which perturbations of these tips may be amplified to form complex arrays of sidebranches.

Note the following apparent paradox. It has been known for some time that, in the absence of surface tension ($\sigma = 0$), Ivantsov's needlelike solutions of the solidification problem are manifestly unstable (9). In fact, the $\sigma = 0$ problems for both the dendrite and the viscous finger are not even dynamically well defined because interfaces destabilized arbitrarily rapidly at arbitrarily short-length scales. On the other hand, the most complete stability analyses performed to date (28) indicate that the tips of fingers and dendrites remain linearly stable at all nonzero values of $\sigma = \sigma^*$. How can it happen that an indefinitely small amount of surface tension can so completely change the behavior of this system?

The answer to this question, and to several others of related interest, can be seen in the result of a simple calculation. It will be convenient to describe this calculation in terms appropriate to the dendrite; the analogous result for the viscous finger is slightly different in technical aspects that need not concern us here.

In principle, the correct way to study stability of a moving, open-ended system like the dendrite is to look at its response to a localized perturbation, for example, a short pulse of heat applied near the tip. The analysis that is needed for this purpose is similar to that used by Zel'dovich and colleagues to study the stability of flame fronts (29). To linear order in the deviation from a steady-state solution determined by solvability, we find that this pulse generates a wavepacket-like deformation whose center moves away from the tip as shown schematically in Fig. 4. More precisely, the center of the wavepacket stays at a fixed position along the side of the dendrite as viewed in the laboratory frame of reference, while the tip grows at constant speed away from the perturbation.

This wavepacket has several important properties (30–32). First, its amplitude $A(s)$ continues to grow as its center moves away from the tip. More specifically,

$$A(s) \approx \exp\left[\frac{0.647}{(\sigma^*)^{1/2}}\left(\frac{s}{\rho}\right)^{1/4}\right] \quad (11)$$

where s is the distance measured along the front from the tip of the dendrite to the center of the packet. Equation 11 is an asymptotic estimate valid for $s \gg \rho$. Second, the packet spreads and stretches in such a way that, as it grows, it acquires a sharply defined wavelength that increases slowly with distance from the tip. Finally, although this deformation grows as it moves, it leaves the tip of the dendrite unchanged after a sufficiently long time. That is, any point on the

Fig. 4. Schematic illustration of two stages in the growth of a localized sidebranching deformation. An initial noisy pulse is indicated at the tip of an otherwise unperturbed parabolic needle crystal. At a later time, the tip has regained its shape and has moved beyond the point of perturbation, leaving behind it a smooth wavepacket that will grow into sidebranches.

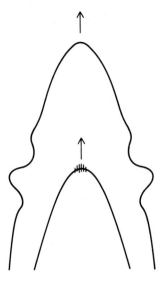

solidification front at a fixed distance from the tip ultimately will return to its original position after the wavepacket has passed. This is the sense in which the front is stable despite the fact that sidebranches continue to grow.

A crucial aspect of Eq. 11 is its singular dependence on σ^*, which is reminiscent of the σ-dependence of the solvability function in Eq. 9. We immediately can see from this how the crossover occurs from stability at $\sigma^* > 0$ to manifest instability at $\sigma^* = 0$; as σ^* becomes small, perturbations become increasingly amplified before leaving the neighborhood of the tip. We also can see why the solvability theory is likely to break down at small anisotropy strength α. If taken literally at arbitrarily small α, the theory predicts arbitrarily small values of σ^* and, according to Eq. 7, tip radii ρ that are much larger than the stability length λ_s. Intuition tells us that such tips should be unstable. According to Eq. 11, the linear instability is controlled because the perturbation moves away from the tip but not until it has grown by an amount which may be large enough to carry it beyond the limits of validity of the linear theory. It seems likely, therefore, that dendrites with small crystalline anisotropies are nonlinearly unstable. Whether or not dendritic behavior occurs in such situations is unknown at present. Perhaps such systems find stable oscillatory modes of growth, or perhaps sufficiently isotropic materials always form chaotic patterns when they solidify in undercooled melts.

The above considerations lead naturally to a theory of sidebranching. Until quite recently, most workers in this field had assumed that the tip of a real dendrite must be weakly—perhaps marginally—unstable against some oscillatory mode of deformation, and that this oscillation must generate the train of sidebranches that seems always to be observed in these systems. Couder's fingers with bubbles at their tips, when driven fast enough, quite definitely do oscillate and emit coherent trains of sidebranches. As mentioned above, however, neither the theorists nor the experimentalists have found any evidence for oscillatory tip modes in the purely thermal dendrites that we have been considering.

One possibility that is suggested by the properties of the wavepacket described above is that dendritic sidebranches are generated by the selective amplification of noise (30, 32–34). In order to construct a satisfactory theory of sidebranching, it seems that we need only to identify the pulses that generate wavepackets with the ambient noise—perhaps just the thermal fluctuations—in the solidifying material. If we look at some fixed distance behind the tip, say, at the point where initially very small deformations have grown out of the linear regime and are big enough to be visible, then it turns out that only a relatively narrow band of wavelengths has been selected from the original broad-band perturbation. This is what is meant by selective amplification; small, noisy perturbations near the tip produce large deformations away from the tip that look very much like sidebranches. One can even estimate the noise temperature required to generate the sidebranches that are seen experimentally. Purely thermal noise seems too small according to present best estimates, but only by about one order of magnitude. The important lesson is that the dendrite is an extremely sensitive and selective amplifier of weak fluctuations in its environment.

Snowflakes

In conclusion, let us return to Kepler and ask what we now might tell him about snowflakes. We know that snowflakes, at least those that seem aesthetically attractive to us, are flat, hexagonal, ice crystals that have grown under conditions in which dendritic instabilities have taken place at the six corners. (The actual growth mechanisms for real ice crystals are more complex than any I have described in this article, but I do not think that these technical differences are relevant to the main points that I want to make.) We understand why these dendritic arms of snowflakes can grow only along the six preferred crystalline axes, and we know that their precise behavior—their growth rates, their thicknesses, the spacings of their sidebranches, and so on—are extremely sensitive to small changes in the temperature and humidity of the vapor out of which they are being formed. Because these conditions are very nearly uniform across the millimeter or less that is occupied by a growing snow crystal, the six branches of a single snowflake will be nearly—but usually not quite—identical to one another. On the other hand, because the atmosphere in a snowstorm is generally turbulent on scales of meters and more, each tiny crystal encounters a different sequence of growth conditions. Thus, no two crystals, not even if they have started from neighboring seeds, are likely to be identical to one another. Of all the new ideas we have learned recently about pattern formation, I think it may be this quantitative understanding of the close relationship between instability and diversity that will turn out to be the most important.

REFERENCES

1. J. Kepler, *The Six-Cornered Snowflake* (Clarendon Press, Oxford, 1966) [translated by C. Hardie, originally published as *De Nive Sexangula* (Godfrey Tampach, Frankfurt am Main, 1611)].
2. D'Arcy Wentworth Thompson, *On Growth and Form* (Cambridge University Press, Cambridge, 1944).
3. U. Nakaya, *Snow Crystals* (Harvard University Press, Cambridge, MA, 1954).
4. A. Hodges, *Alan Turing, the Enigma* (Simon and Schuster, New York, 1983).
5. A review of the physics of pattern formation in crystal growth can be found in J. S. Langer, *Rev. Mod. Phys.* **52**, 1 (1980).
6. For more detailed reviews of recent developments, see J. S. Langer, in *Chance and Matter* (Lectures in the Theory of Pattern Formation, Les Houches Summer School, 1986), J. Souletie, J. Vannimenus, R. Stora, Eds. (North-Holland, New York, 1987), pp. 629–711; D. Kessler, J. Koplik, H. Levine, *Adv. Phys.* **37**, 255 (1988); in *Dynamics of Curved Fronts*, P. Pelcé, Ed. (Academic Press, New York, 1988).
7. P. G. Saffman and G. I. Taylor, *Proc. Roy. Soc. A* **245**, 312 (1958); J. W. McLean and P. G. Saffman, *J. Fluid Mech.* **102**, 455 (1981). For reviews of modern developments in the viscous fingering problem, see D. Bensimon, L. P. Kadanoff, S. Liang, B. Shraiman, C. Tang, *Rev. Mod. Phys.* **58**, 977 (1986) or Kessler *et al.* (6).
8. G. P. Ivantsov, *Dokl. Akad. Nauk SSSR* **58**, 567 (1947).
9. J. S. Langer and H. Müller-Krumbhaar, *Acta Metall.* **26**, 1681; 1689; 1697 (1978); H. Müller-Krumbhaar and J. S. Langer, *ibid.* **29**, 145 (1981).
10. W. Oldfield, *Mater. Sci. Eng.* **11**, 211 (1973).
11. W. W. Mullins and R. F. Sekerka, *J. Appl. Phys.* **34**, 323 (1963); *ibid.* **35**, 444 (1964).
12. M. E. Glicksman, R. J. Schaefer, J. D. Ayers, *Metall. Trans. A* **7**, 1747 (1976); S. C. Huang and M. E. Glicksman, *Acta Metall.* **29**, 701 and 717 (1981); M. E. Glicksman, *Mater. Sci. Eng.* **65**, 45 (1984); A. Dougherty and J. P. Gollub, *Phys.*

Rev. A **38**, 3043 (1988). For a counterexample, see J. H. Bilgram, M. Firmann, W. Känzig, *Phys. Rev. B* **37**, 685 (1988).
13. E. Ben-Jacob, N. Goldenfeld, J. S. Langer, G. Schön, *Phys. Rev. A* **29**, 330 (1984).
14. J. S. Langer, *ibid.* **33**, 435 (1986).
15. P. Pelcé and Y. Pomeau, *Stud. Appl. Math.* **74**, 245 (1986); M. Ben Amar and Y. Pomeau, *Europhys. Lett.* **2**, 307 (1986).
16. D. Kessler, J. Koplik, H. Levine, in *Proceedings of the NATO Advanced Research Workshop on Patterns, Defects, and Microstructures in Non-Equilibrium Systems* (Austin, Texas, March 1986), ASI Series E, 121, D. Walgraef, Ed. (Nijhoff, Dordrecht, 1987).
17. A. Barbieri, D. C. Hong, J. S. Langer, *Phys. Rev. A* **35**, 1802 (1987).
18. D. Meiron, *ibid.* **33**, 2704 (1986).
19. M. Ben-Amar and B. Moussallam, *Physica D* **25**, 7 (1987); *ibid.*, p. 155.
20. D. A. Kessler, J. Koplik, H. Levine, *Phys. Rev. A* **33**, 3352 (1986); D. A. Kessler and H. Levine, *Phys. Rev. B* **33**, 7867 (1986).
21. B. I. Shraiman, *Phys. Rev. Lett.* **56**, 2028 (1986).
22. R. Combescot, T. Dombre, V. Hakim, Y. Pomeau, A. Pumir, *ibid.*, p. 2036; *Phys. Rev. A* **37**, 1270 (1988).
23. D. C. Hong and J. S. Langer, *Phys. Rev. Lett.* **56**, 2032 (1986); *Phys. Rev. A* **36**, 2325 (1987).
24. See the discussion of the work of M. Kruskal and H. Segur, in (*6*).
25. A. Barbieri and J. S. Langer, *Phys. Rev. A*, in press.
26. D. Kessler and H. Levine, *Phys. Rev. A* **36**, 4123 (1987).
27. Y. Couder, N. Gerard, M. Rabaud, *ibid.* **34**, 5175 (1986); Y. Couder, O. Cardoso, D. Dupuy, P. Tavernier, W. Thom, *Europhys. Lett.* **2**, 437 (1986).
28. D. Kessler and H. Levine, *Phys. Rev. A* **33**, 2621 (1986); *ibid.*, p. 2634 (1986); *Europhys. Lett.* **4**, 215 (1987).
29. Ya. B. Zel'dovich, A. G. Istratov, N. I. Kidin, V. B. Librovich, *Combust. Sci. Technol.* **24**, 1 (1980).
30. R. Pieters and J. S. Langer, *Phys. Rev. Lett.* **56**, 1948 (1986); R. Pieters, *Phys. Rev. A.* **37**, 3126 (1988).
31. M. Barber, A. Barbieri, J. S. Langer, *Phys. Rev. A* **36**, 3340 (1987).
32. J. S. Langer, *ibid.*, p. 3350.
33. R. Deissler, *J. Stat. Phys.* **40**, 371 (1985).
34. The relation between noise and sidebranching has been examined experimentally by A. Dougherty, P. D. Kaplan, J. P. Gollub, *Phys. Rev. Lett.* **58**, 1652 (1987); see also A. Dougherty and J. P. Gollub, *Phys. Rev. A* **38**, 3043 (1988).
35. This article is based in large part on the text of the author's Marian Smoluchowski Memorial Lecture presented in Warsaw, Poland, on 17 March 1988. I thank the Polish Academy of Sciences for its hospitality. I also thank Y. Couder, M. Glicksman, and J. Maher for providing photographs and information. The research described here was supported by U.S. Department of Energy grant DE-FG03-84ER45108 and by National Science Foundation grant PHY 82-17853, supplemented by funds from the National Aeronautics and Space Administration.

Tip splitting without interfacial tension and dendritic growth patterns arising from molecular anisotropy

Johann Nittmann* & H. Eugene Stanley†

*Etudes et Fabrication Dowell Schlumberger, 42003 St. Etienne, France
† Center for Polymer Studies and Department of Physics, Boston University, Boston, Massachusetts 02215, USA

Two growth mechanisms of considerable recent interest are related to a single statistical mechanical model. Tip splitting without interfacial tension occurs when a fluid pushes into another miscible fluid of higher viscosity. Dendritic growth occurs when anisotropic molecules aggregate—a common example is the snowflake. We find that both structures are fractal objects, and can be obtained from a single statistical mechanical model, implying that there is a relation between the underlying physical processes involved.

GROWING structures have fascinated mankind for centuries, and today the field of growth phenomena elicits interest from many disciplines, ranging from medicine and biology to fluid mechanics. Two growth forms that have attracted recent interest are the following:

(1) Dendritic growth[1-9]. No two snowflakes are identical; each is assembled by the random aggregation of water molecules. Yet every child can distinguish a snowflake from other growth forms. The key scientific question is by what mechanism the anisotropy of a water molecule becomes amplified from its weak 'local' effect at the molecular level to its pronounced 'global' effect at the macroscopic level of the snowflake.

(2) Tip splitting[10-20]. A classic experiment in fluid mechanics concerns the splitting of a low-viscosity body of fluid which results when it is forced under pressure into a high-viscosity fluid. If the two fluids are immiscible, then the interfacial tension between them serves to establish a length scale at which tip splitting occurs. When the two fluids are miscible there is no interfacial tension, yet tip splitting nonetheless occurs. Thus an important question concerns the physical mechanism which determines the point at which the finger splits.

The scientific questions in (1) and (2) have been the object of research for many years, in part because our present state of understanding is so incomplete[21] that even a little progress would be valuable. The two categories of growth mechanism (1) and (2) have been considered to be quite different, in the sense that the physical basis for one has no relation to that of the other. Here we develop a statistical mechanical model which incorporates both dendritic growth and tip-splitting, thereby relating two disparate fields of enquiry.

Relation between noise and tip splitting

The model is most clearly explained if we begin with the dielectric breakdown model (DBM) of Niemeyer et al.[22] on, for example, a triangular lattice. We first place a seed particle at the origin of a large circular domain of radius R. If we think of this seed particle as being the source of a fluid of infinitesimally small viscosity, which is being forced under pressure to displace a fluid with much higher viscosity[23,24], then the interface must move according to Darcy's law:

$$v_n = -\mathbf{n} \cdot \nabla P \qquad (1)$$

Here v_n is the velocity component normal to the interface, $\hat{\mathbf{n}}$ is the normal unit vector and P is the pressure field. P is constant in the less viscous fluid and, because $\nabla \cdot \mathbf{v} = 0$, P satisfies the Laplace equation

$$\nabla^2 P = 0 \qquad (2)$$

in the more viscous fluid. Hence the relevant boundary conditions are $P(r, \theta) = 1$ anywhere in the low-viscosity body of fluid, and $P(R, \theta) = 0$ along a circle of radius R.

In a perfect medium with radial symmetry and no pressure fluctuations, the interface will spread out in concentric circles. However, because there is always some noise in the system, a fluid-dynamical instability[25] will occur and irregularities in the interface will grow. This noise phenomenon is reproduced in the DBM, which includes fluctuations by means of the following algorithm. First, ∇P is calculated at every perimeter site of the cluster; this is done by solving equation (2) with an overrelaxation technique. At step 1 there is a single seed on a triangular lattice with six perimeter sites. As all sites have equal values of ∇P, the first perimeter site is mapped to the numerical interval $[0, \frac{1}{6}]$ the second to $[\frac{1}{6}, \frac{2}{6}]$, the third to $[\frac{2}{6}, \frac{3}{6}]$ and so forth. Next, a random number generator is used to choose a number in the interval $[0, 1]$. Suppose that this random number is 0.2603238: the second perimeter site is then occupied, and the procedure iterated. For this two-site cluster, ∇P is calculated at the eight perimeter sites, the values are normalized to unity, a new random number is chosen, and one of the eight sites occupied. Such a DBM cluster is characterized by a high degree of noise: as each growth step is determined by only one random number, it is always possible that the random number chosen corresponds to a perimeter site with an extremely small value of ∇P, which by equation (1) should almost never grow. Thus the DBM violates the fundamental Darcy law due to the noise inherent in the algorithm.

We now describe a procedure whereby this noise can be systematically reduced in a controllable fashion. Clearly we need an algorithm such that perimeter sites with extremely small values of ∇P are extremely unlikely to be chosen. This is accomplished by advancing to a new perimeter site only after it has been chosen s times, where s is a parameter which can be tuned. Each perimeter site has a counter which registers how many times that particular site has been chosen. As $s \to \infty$, the growth of the interface will approach Darcy-law growth, in which any point of the interface grows according to the true local pressure gradient. In the Darcy 'zero-fluctuation' or 'mean-field' limit[26] $(s \to \infty)$, the interface would be a perfect circle if there were no underlying lattice.

Figure 1a-c shows the results of calculations for successive values of s. We find that Fig. 1b and c resemble tip splitting as observed in the viscous fingering of both newtonian (refs 19, 27; J. D. Chen, personal communication; R. Lenormand, personal communication) and non-newtonian[10,12,18] fluids. When $s = 2$ (Fig. 1a), the structure resembles the DBM both qualitatively (although the branches look thicker) and quantitatively

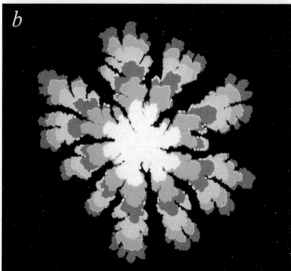

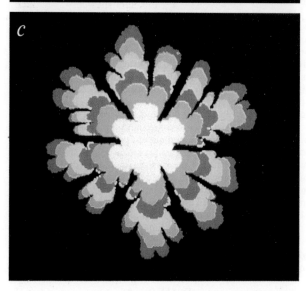

Fig. 2 A typical fractal structure on a square lattice with $s = 50$ and a microscopic anisotropy (defined by equations (6), (7)) of $k - 1 = 10$. The colour coding is the same as in Fig. 1.

($d_f = 1.7$). Here d_f is the fractal dimension obtained, for example, from the slope of a log–log plot of the mass against the caliper diameter. For large s (for example, $s = 20$; Fig. 1b), the qualitative appearance appears to differ: the system appears at first sight to cross over to a new 'universality class', with a larger value of d_f. However, when we extrapolate the apparent fractal dimension to large cluster sizes we find that $d_f = 1.7$ for all values of s; that is, the growth forms are quantitatively identical, independent of the degree of noise reduction.

Note that tip splitting always occurs by the same mechanism. First a cluster grows 'smoothly', without tip splitting. However, as the radius of curvature increases, the interface becomes 'rough', with both positive (outward) and negative (inward) fluctuations. The positive fluctuations are not significant, as they are soon damped out; however, the negative fluctuations persist (Fig. 3). This is because, for a charged fractal object, the electric field inside a single notch is very small, and the equation relating the electric field to the gradient of the potential is formally identical to the Darcy law relating growth velocity to the gradient of the pressure. Hence, the tiny notch is not likely to be filled in so quickly as one would expect if interfacial tension were present (Fig. 3d). The tiny protrusions on both sides of the notch see a much larger field than does the notch, so they attract mass. The tiny notch thus becomes the terminus of a long fjord (Fig. 3e). A fjord is almost perfectly screened, and so is almost never filled in. In Fig. 3, $s = 50$. If $s > 50$ (less noise), then the same tip-splitting mechanism will apply but a negative fluctuation (notch) will decay more efficiently: the system is less susceptible to negative fluctuations and a fjord is formed only when the cluster has reached a larger radius of curvature.

Although the asymptotic fractal dimension d_f is independent of s, the finger thickness W_f clearly increases with s. Moreover, our model explains the existence of a well-defined W_f: the less the noise, the thicker the finger (see Fig. 1). We find the quantitative law:

$$W_f \simeq 4.5 \log s + 2 \qquad (3)$$

Fig. 1 Examples of fractal structures generated when the anisotropy parameter k is held fixed at unity, but the noise parameter $1/s$ is decreased. In a, b and c, $s = 2, 20$ and 200, respectively. For all finite values of s, we find that the fractal dimension is equal to the DBM value, $d_f = 1.7$, providing we take care to extrapolate the apparent mass dependence of d_f to its asymptotic limit. The colour coding is as follows: the first one-sixth of the sites are white, the next sixth are blue, followed by magenta, yellow, green and red.

Fig. 3 Schematic illustration of the difference between an outward ('positive') and an inward ('negative') interface fluctuation. A positive fluctuation tends to be damped out rather quickly, as mass quickly attaches to the side of the extra site that is added. On the other hand, a negative fluctuation grows, in the sense that mass accumulates on both sides of the tiny notch. The notch itself has a lower and lower probability of being filled in, as it becomes the end of a longer and longer fjord. This is the underlying mechanism for the tip-splitting phenomenon when no interfacial tension is present. *a* shows the advancing front (row α) of a cluster with $s = 50$. The heavy line separates the cluster sites (all of which were chosen 50 times) from the perimeter sites (all of which have counters registering less than 50). In *a*, no fluctuations in the counters of these three sites have occurred yet, and all three perimeter counters register 49. *b* shows a negative fluctuation, in which the central perimeter site is chosen slightly less frequently than the two on either side; the latter now register 50, and so they become cluster sites in row β. The perimeter site left in the notch between these two new cluster sites grows much less quickly because it is shielded by the two new cluster sites. For the sake of concreteness, let us assume it is chosen 10 times less frequently. Hence by the time the notch site is chosen one more time, the two perimeter sites at the tips have been chosen 10 times (*c*). The interface is once again smooth (row γ), as it was before, except that the counters on the three perimeter sites differ. After 40 new counts per counter, the situation in *d* arises. Now we have a notch whose counter lags behind by 10, instead of by 1 as in *b*. Thus the original fluctuation has been amplified, due to the tremendous shielding of a single notch. Note that no new fluctuations were assumed: the original fluctuation of 1 in the counter number is amplified to 10 solely by electrostatic screening. This amplification of a negative 'notch fluctuation' has the effect that the tiny notch soon becomes the end of a long fjord. To see this, note that *e* shows the same situation after 50 more counts have been added to each of the two tip counters, and hence (by the 10:1 rule) 5 new counts to the notch counter. The tip counters therefore become part of the cluster, but the notch counter has not yet reached 50 and remains a perimeter site. The notch has become an incipient fjord of length 2, and the potential at the end of this fjord is now exceedingly low. Indeed it is quite possible that the counter will never pass from 45 to 50 in the lifetime of the cluster. In our simulations we can see tiny notch fluctuations become the ends of long fjords, and all of the above remarks on the time-dependent dynamics of tip splitting are confirmed quantitatively.

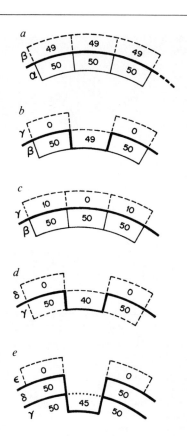

We also find that W_f is independent of the magnitude of the pressure field. To see this, we varied the global pressure gradient by changing the size of our computational grid from 200 to 25 units and found no variation of the finger thickness. This discovery is explained if one considers that the ratio of the local pressure gradient between a site at a finger tip and a site within a fjord does not change if the global pressure gradient is changed: this in turn is a direct consequence of the fact that the pressure field satisfies a Laplace equation.

Previous work on tip-splitting phenomena has focused on explaining the non-zero value of the finger thickness W_f as arising from the presence of interfacial tension σ (ref. 28; see also ref. 29). However this explanation cannot be applied to miscible fluids, such as those used in recent experiments[10,12,18,27,28], because in this case, by definition, $\sigma = 0$. In our model interfacial tension does not exist (that is, σ acts only on the length scale of a single fluid element or 'pixel'); our observed finger thickness is thus related solely to the concept of noise.

Before proceeding further we note that in the limiting case $s = 1$ the DBM is equivalent to diffusion-limited aggregation[30-32] (DLA). The diffusion analogue of the DBM for $s > 1$ is a DLA-type model in which growth occurs only after a perimeter site has been hit by s random walkers. If all the counters are reset to zero after each growth step, then we have the Meakin model[33] or the Kertész–Vicsek model[8], in which growth of the positive fluctuations (the tips) is amplified because a tip of size 1 pixel is more likely to experience the next growth event—so the apparent value of d_f decreases toward unity as the cluster grows. The DLA-type analogue of our DBM-type model in which the counters are not reset to zero after each growth event is the Tang model[26], for which it is not the positive fluctuations (the tips) that display amplified growth but the negative fluctuations (the notches). Amplified growth of negative fluctuations is the characteristic feature of DLA, explaining our result that d_f has its DLA value for all finite values of s.

Thus we conclude that noise reduction—arising from suppression of fluctuations—does not change the overall 'universality class', but does introduce a characteristic finger thickness.

Local anisotropy and dendritic growth

Real growth phenomena are never perfectly isotropic. In fact, anisotropy appears to dominate dendritic crystal growth; thus, for example, a snowflake is recognized by its six-fold anisotropy, although the noise is also reflected in the variability from one snowflake to another[9]. No two are alike, although the eye immediately recognizes the pattern of a snowflake.

The problem of understanding the growth of a snowflake has a rich history. A large class of models has focused on introducing anisotropy in a 'global' or macroscopic fashion by introducing angular variables and assuming that the growth depends sensitively upon these variables[2-4]. Although the resulting patterns have, by virtue of their rules of construction, the requisite six-fold symmetry, their resemblance to real snowflakes is not striking. Moreover, they lack the random variations that seem to characterize real snowflakes and also fractal objects.

A snowflake grows by successive landings of water molecules, and we have therefore focused our attention on how microscopic irregularities in the landing surface can be translated into the macroscopic structure of the snowflake. To reflect the presence of these microscopic irregularities, we must incorporate into our model the essential fact that the landing sites seen by an incoming molecule are not all equivalent. Hence we replace equation (1) by

$$v_n = -\mathbf{n} \cdot (k \nabla P) \qquad (4)$$

where the conservation of mass condition $\nabla \cdot \mathbf{v} = 0$ implies that equation (2) is replaced by

$$\nabla \cdot (k \nabla P) = 0 \qquad (5)$$

with the same boundary conditions as for $k = 1$. Here the anisotropy parameter $k = k(x, y)$ would be the permeability in a fluid problem.

Consider a square lattice. One simple choice for $k(x, y)$ is (see Fig. 2)

$$k(x, y) = 1 \quad (6)$$

for x or y even,

$$k(x, y) = k > 1 \quad (7)$$

otherwise equations (6) and (7) express mathematically the fact that the surface affinity for incoming water molecules depends on the spatial coordinate: the incoming particles do not see a perfectly smooth and homogeneous 'landing surface'.

Moreover, our anisotropy is fundamentally different from that considered in, for example, refs 4 and 11. Schematically, in these models the interface is moved according to the rule

$$u = f(\kappa) - f(\theta)u_n \quad (8)$$

where u is the growth velocity, $f(\kappa)$ is an interfacial tension term and u_n is essentially the local pressure (or temperature, or concentration) gradient at the interface. The function $f(\theta)$ indicates the extent to which the growth is enhanced along directions separated by an angle θ. In marked contrast, our model assumes that the anisotropy is present on a molecular level at the interface. We assume that along the interface, the affinity for an incoming water molecule alternates from site to site:

$$u = -f(x, y)u_n \quad (9)$$

We believe that our model is more realistic, as an incoming water molecule in snowflake formation cannot possibly sense the angle $\theta = \arctan(y/x)$, but does see a 'landing surface' whose 'attraction' fluctuates from point to point.

Next we consider the effect of tuning the anisotropy parameter k. Figure 5a-d shows structures grown with a succession of increasing values of k, ranging from 1.1 to 11. We hold s fixed at the value $s = 50$; if s were too small, then noise effects would complicate visualization of the effect of anisotropy. Figure 5 is for a triangular lattice, for which equations (6) and (7) are replaced by a different rule: we set $k > 1$ for every fifth row of the three principal directions of the lattice (E-W, NE-SW, NW-SE). We see from Fig. 5 that as k increases there is a pronounced change from the isotropic case $k = 1$, and the resulting growth (see, for example, Fig. 5d) resembles a 'snowflake' for reasons more subtle than merely the characteristic 6-fold axis of rotation[34]. Using standard methods (for example, all three methods of ref. 18), we measured d_f for this 'snowflake' and found values that decrease with the number of particles used in the calculation. Extrapolating to infinite size, we find[35,36] $d_f = 1.5 \pm 0.1$.

Although the structure at first sight appears to be somewhat ordered, we realize that this is a trick played by the 6-fold axis. In fact, an individual branch is quite disordered, with side branches of all sizes extending from it. The reason $d_f > 1$ is that the side branches occur with many different length scales. This is especially apparent from Fig. 5d, where we see from the colour coding that the latest particle to arrive can attach to the side branches as well as to the tip. Figure 5e shows real snowflakes with side branches, which show a striking resemblance to the anisotropic simulations of Fig. 5d. The differences between Fig. 5d and e are the subject of current investigation.

We now address the actual structure of the fractal objects in the presence of anisotropy. It is important to note that there are distinct effects that cooperate to generate the final structure obtained. The first effect is the fine structure of the side branches (see Fig. 2), consisting of a set of 'trees' of varying height, as shown schematically in Fig. 4a. The trees are mainly without branches, as the anisotropy favours growth only in even-numbered rows or columns of the lattice. However the height of a tree varies widely from one tree to the next, due to the tendency of tall trees to screen shorter trees. An analogous variation in the height of trees has been found by Meakin[37] in his classic studies of DLA on a planar substrate: he found that the resulting fractal structure is a 'forest' of trees, with fewer but taller trees surviving at large times due to their tendency to shield the shorter trees.

The main difference between our work and the Meakin (planar substrate) DLA simulations is our lattice anisotropy (parameterized by $k-1$) and our noise reduction (parameterized by $1/s$), which have the effect of making the trees tall and straight instead of ramified. Consider now the overall profile for the height of the trees in the side branches. This profile can be understood mathematically as arising from the product of two functions. The first, a decreasing function from origin to tip, is related to the fact that the regions of the branches that were formed at early times tend to be larger than the regions of the branches that were formed at late times (Fig. 4b). The second, an increasing function, is related to 'screening'; that is, to ∇P, which is larger near the tips and smaller near the origin (Fig. 4c). As $v \propto \nabla P$, the growth rate is larger near the tips. The product of the increasing and decreasing functions gives the characteristic profile for the height of the trees in the side branches (Fig. 4d).

We also measured as a function of cluster mass: (1) the caliper width of the side branches of Fig. 2, and (2) the caliper diameter of the entire cluster. Both log-log plots are parallel, with slope $1/d_f$.

Discussion

We have shown that two fundamental physical phenomena that are not yet understood, dendritic growth and tip splitting in the absence of interfacial tension, can be related in that both arise from the same statistical mechanical model—a generalization of the DBM[22]. This means that there are physical features common to both phenomena: they differ only in parameter values. In our model we can incorporate in a direct and systematic fashion the crucial role played by fluctuation phenomena

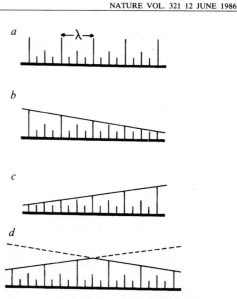

Fig. 4 Schematic illustration to explain the characteristic shape of the four arms of the 'snowflake' cluster in Fig. 2; a recalls the fundamental structure of aggregation onto an equipotential surface, first studied by Meakin[37]. For simplicity, the 'trees' are drawn as straight line segments, and the hierarchical or fractal distribution of tree height is indicated by a difference of a factor of 2 between successive sizes, together with a spacing, $\lambda(M)$, which increases as M^{1/d_f}, where M is the total cluster mass. b shows the modification expected from the fact that the regions of an arm near the centre have more time to accumulate mass than the regions near the tip. c shows the effect of the fact that ∇P is much larger near the tip; d shows the result of combining $a-c$, and resembles the overall shape observed in Figs 2 and 5d.

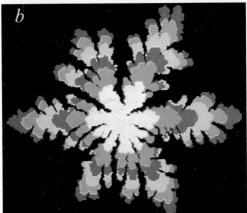

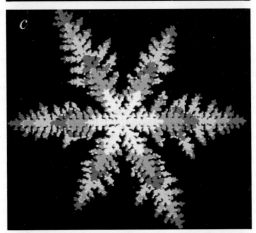

Fig. 5 a–d, Examples of fractal structures formed when the noise parameter s is held constant at $s = 50$, but the anisotropy parameter k is chosen to be, respectively, $k - 1 = 0.1$, 0.31622, 1.0 and 10.0 (0.31622 interpolates logarithmically between 0.1 and 1.0). The limiting fractal dimension (as mass $\to \infty$) is $d_f \simeq 1.5$, independent of k, for all $k > 1$. The colour coding is the same as in Fig. 1. e, Examples of real snowflakes (reproduced with permission from ref. 9) which show a striking resemblance to d.

and anisotropy. The physical picture that we have proposed is embodied in two fundamental equations, (4) and (5) (or (1) and (2) for $k = 1$). The second equation describes the spatial change of the pressure field which drives the instability; the first represents the 'growth law', which relates the growth rate of the interface to the pressure field. We have used a generalized Darcy-type law, which enables us to selectively tune both noise and anisotropy.

The overall physical picture that emerges is as follows: Tip-splitting phenomena in the absence of interfacial tension are triggered by microscopic fluctuations (that is, noise). Although positive and negative fluctuations of the interface occur symmetrically, the stability (and hence the subsequent growth) of positive and negative fluctuations are totally different: tip splitting is the direct consequence of this asymmetry in the stability of positive and negative fluctuations. A small protrusion of size 1 pixel is much less long-lived than a small notch of the same size; in fact, it is remarkably difficult to fill even the shallowest notch. Zero noise ($s = \infty$) results in a compact (non-fractal) circular object. A very low noise level (large s) has little effect when a cluster is small, but its effect becomes much more pronounced as the cluster grows larger. In the limit of infinite cluster size, an arbitrarily small but non-zero amount of noise is sufficient to make the cluster fractal. The measured fractal dimension is identical to that of DBM and DLA, two models designed to describe phenomena in the limit in which there is a very high noise level.

The tip-splitting phenomena that occur in the case of zero anisotropy are generalized into a fractal hierarchy of side branches in the presence of anisotropy. In the limit of infinite cluster size even a tiny degree of anisotropy changes the fractal dimension from the DLA value of 1.7 to the value 1.5.

Thus, the complete phase diagram has $1/s$ on the abscissa and $(k-1)$ on the ordinate. Asymptotically we find that d_f is constant, at the DBM value of ~ 1.7, everywhere on the x-axis, and d_f is also constant, at the value 1.5, everywhere else in the phase diagram except on the y-axis (zero noise), where $d_f = 1$. Thus noise reduction is not a sufficient perturbation to change d_f from its DBM value, because the negative fluctuations persist for all values of s, and these negative fluctuations control the value of d_f. On the other hand, anisotropy at the microscopic level does change d_f. Further details of this phase diagram suggest an intriguing analogy to critical point phenomena, and this will be the subject of future investigation.

Finally, we return to the question posed in the introduction, of how a tiny anisotropy can become 'amplified' from its local effect at the molecular level to a global effect at the macroscopic

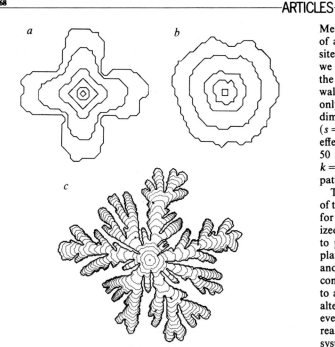

Fig. 6 *a*, Initial growth of an interface on a square lattice for the case $s = 500$. Growth can occur in any of the four space directions (four nearest neighbours). The interface is shown after 5, 21, 85, 200, 500 and 1,000 growth steps. *b*, As in *a*, except that here growth can occur into eight directions (four nearest neighbours and four next-nearest neighbours). The interface is shown after 9, 21, 85, 200, 500 and 1,000 steps. *c*, Viscous fingering structure for $s = 50$, after 15,000 growth steps for eight-fold coordination on a square lattice. The first four contour lines are drawn after 100, 300, 650 and 1,000 steps; subsequent lines are drawn at intervals of 1,000 steps.

level. Our model directly demonstrates this fact: we have shown that in the presence of anisotropy, the resulting fractal dimension is not the DBM value of 1.7, but rather tends asymptotically toward 1.5, a new 'universality class'. Our result is supported by Meakin's very recent calculations for DLA[38] without any local anisotropy, except for that arising from the square-lattice substrate. For this classic and well-studied system, 'pure DLA',

Meakin finds the same behaviour that we find in the presence of anisotropy, but only after the cluster size reaches 5 million sites—almost three orders of magnitude larger than the clusters we study! Thus Meakin's local anisotopy, arising solely from the effect of the square lattice itself on the trajectories of random walkers, can lead to a pronounced global effect, altering not only the overall appearance of the cluster but also the fractal dimension itself. As pure DLA is the limit of maximum noise ($s = 1$), we have to wait for an extremely large cluster to see its effect. To support this idea, we systematically reduced s from 50 to 1 while keeping the local anisotropy fixed at the value $k = 2$. When $s = 50$ it is easy to see the snowflake anisotropy pattern, but as s decreases the snowflake vanishes.

To understand better the subtle role played by the anisotropy of the square lattice, we show in Fig. 6a the initial growth events for the case $s = 500$, $k = 1$. The early stage of growth is characterized by the competition of 'lattice anisotropy', which attempts to pull the interface into the four principal directions of the plane, and 'interface smoothing' (due to the decay of positive and negative fluctuations), which initially prevents splitting. The competition between these two contradicting tendencies leads to an oscillation of the interface: the structure of the interface alternates between a circle and a diamond-shaped cusp until, eventually, the weak anisotropy of the lattice dominates. As in real systems, no cusp singularities[39] occur: the noise in our systems smooths the sharp corners of the initial cusp as interfacial tension would do. A well-defined finger thickness has developed. The larger the value of s, the smaller is the noise and the larger is W_f. To weaken the lattice anisotropy, which results from the rule that growth is possible only in one of the four space directions (nearest neighbours), we can also allow growth into the four diagonal directions (next-nearest neighbours). Figure 6b,c shows that such growth is initially almost circular until it reaches a critical radius, after which negative fluctuations are no longer filled in. This gives rise to the characteristic fingering structure shown in Fig. 6c. Thus this model seems to represent both qualitatively and quantitatively the viscous fingering phenomenon for the case of miscible fluids (zero interfacial tension).

We thank G. Daccord, R. Lenormand and J. D. Chen for sharing their experimental results with us before publication, and we also thank them and F. Rondelez for helpful discussions. We thank David Kamins for generous and patient assistance with colour computer graphics, and the Boston University Academic Computer Center for generous computer time on the mainframe IBM-3090 computer.

Received 19 December 1985; accepted 21 April 1986.

1. Langer, J. S. *Rev. mod. Phys.* **52**, 1 (1980).
2. Ben-Jacob, E., Goldenfeld, N., Langer, J. S. & Schön, G. *Phys. Rev.* **A29**, 330–340 (1984).
3. Brower, R. C., Kessler, D. A., Koplik, J. & Levine, H. *Phys. Rev.* **A29**, 1335–1342 (1984).
4. Kessler, D. A., Koplik, J. & Levine, H. *Phys. Rev.* **A30**, 2820–2823 (1984).
5. Honjo, H., Ohta, S. & Sawada, Y. *Phys. Rev. Lett.* **55**, 841–844 (1985).
6. Vicsek, T. *Phys. Rev. Lett.* **53**, 2281–2284 (1984).
7. Szép, J., Cserti, J. & Kertész, J. *J. Phys.* **A18**, L413–L416 (1985).
8. Kertész, J. & Vicsek, T. *J. Phys. A* (in the press).
9. Bentley, W. A. & Humphreys, W. J. *Snow Crystals* (Dover, New York, 1962).
10. Nittmann, J., Daccord, G. & Stanley, H. E. *Nature* **314**, 141–144 (1985).
11. Sander, L. M., Ramanlal, P. & Ben-Jacob, E. *Phys. Rev.* **A32**, 3160–3165 (1985).
12. Van Damme, H., Obrecht, F., Levitz, P., Gatineau, L. & Laroche, C. *Nature* **320**, 731–733 (1986).
13. DeGregoria, A. J. & Schwartz, L. W. *J. Fluid Mech.* **164**, 383–400 (1986).
14. Bensimon, D. *Phys. Rev.* **A33**, 1302–1308 (1986).
15. Lenormand, R. & Zarcone, C. *Phys. chem. Hydrodyn.* **6**, 497–506 (1985).
16. Chen, J. D. & Wilkinson, D. *Phys. Rev. Lett.* **55**, 1892–1895 (1985).
17. Måløy, K. J., Feder, J. & Jøssang, T. *Phys. Rev. Lett.* **55**, 2688–2691 (1985).
18. Daccord, G., Nittmann, J. & Stanley, H. E. *Phys. Rev. Lett.* **56**, 336–339 (1986).
19. Ben-Jacob, E. *et al. Phys. Rev. Lett.* **55**, 1315–1318 (1985).
20. Paterson, L. *J. Fluid. Mech.* **113**, 513–529 (1981).
21. Maddox, J. *Nature* **313**, 93 (1985).
22. Niemeyer, L., Pietronero, L. & Wiesmann, H. J. *Phys. Rev. Lett.* **52**, 1033–1036 (1984).
23. Paterson, L. *Phys. Rev. Lett.* **52**, 1621–1624 (1984).
24. Sherwood, J. D. & Nittmann, J. *J. Phys., Paris* **47**, 15–21 (1986).
25. Saffman, P. G. & Taylor, G. I. *Proc. R. Soc.* **A245**, 312–329 (1958).
26. Tang, C. *Phys. Rev.* **A31**, 1977–1979 (1985).
27. Paterson, L. *Physics Fluids* **28**, 26–30 (1985).
28. Chuoke, R. L., Van Meurs, P. & Van der Pol, C. *Trans. Am. Inst. Min. Engrs* **216**, 188–194 (1959).
29. Mullins, W. W. & Sekerka, R. F. *J. appl. Phys.* **34**, 323–329 (1963).
30. Witten, T. A. & Sander, L. M. *Phys. Rev. Lett.* **47**, 1400–1403 (1981).
31. Witten, T. A. & Sander, L. M. *Phys. Rev.* **B27**, 5686–5697 (1983).
32. Meakin, P. in *On Growth and Form: Fractal and Non-Fractal Pattern in Physics* (eds Stanley H. E. & Ostrowsky, N.) (Nijhoff, Dordrecht, 1985).
33. Meakin, P. *Phys. Rev.* (submitted).
34. Mason, B. J. *Scient. Am.* **204**, No. 1, 120–130 (1961).
35. Jullien, R., Kolb, M. & Botet, R. *J. Phys., Paris* **45**, 395–399 (1984).
36. Ball, R. C., Brady, R. M., Rossi, G. & Thompson, B. R. *Phys. Rev. Lett.* **55**, 1406–1409 (1985).
37. Meakin, P. *Phys. Rev.* **A27**, 2616–2623 (1983).
38. Meakin, P. *Pap. presented at int. Conf. Fragmentation, Form and Flow in Fractured Media* Neve Ilan, 6–9 January 1986.
39. Shraiman, B. I. & Bensimon, D. *Phys. Rev.* **A30**, 2840–2844 (1984).

Oblique Roll Instability in an Electroconvective Anisotropic Fluid

R. Ribotta and A. Joets

Laboratoire de Physique des Solides, Université de Paris-Sud, 91405 Orsay, France

and

Lin Lei

Queensborough Community College of the City University of New York, Bayside, New York 11364
(Received 4 February 1986)

> We have experimentally discovered that in a nematic liquid crystal subjected to an ac electric field, the first convective structure at low frequencies is in fact a set of oblique rolls, contrary to the accepted picture. We show that it is a new structure with a helical flow motion, and thus lower in symmetry than the usual normal rolls recovered at high frequencies. Besides indicating the limitations of the available theoretical models, these results clearly show that the highest-symmetry flow structure corresponds to the normal rolls.

PACS numbers: 61.30.−v, 47.20.Tg, 47.65.+a

In order to study experimentally the disorganization of a convective flow inside an extended layer of fluid, it is preferable to start from an ordered flow with a well-defined wave vector. This means that the ordered structure must have an orientation fixed in space. In fully isotropic convection, i.e., when the fluid is isotropic and when there is no coupling to an aligning external field, the flow usually appears disordered at threshold (e.g., Rayleigh-Bénard[1]). In some cases the system may be rendered anisotropic and the flow oriented along a given direction by the coupling to a magnetic field[2] (magnetohydrodynamic convection). Another possibility is to use an anisotropic fluid, for instance, a nematic liquid crystal—hereafter referred to simply as a nematic. It has recently been found[3] that in a nematic subjected to an increasing ac electric field there exists a complete sequence of prechaotic stationary structures. These structures are ordered and spontaneously oriented with respect to the initial average molecular direction **n** (the anisotropy axis). However, up to now, the essential features of the different flows were not recognized, and more importantly, the identification of the flow of highest symmetry at the first threshold could not be determined.

It is presently well known that a layer of a nematic subjected to a transverse ac electric field undergoes a transition to a convective flow when the voltage reaches some threshold. After the first observation by Williams[4] of an ordered spatial structure, a one-dimensional (1D) electrohydrodynamical model was constructed.[5] In this model, the convective flow is made of parallel rolls oriented perpendicularly to the initial direction of the molecular axis. The frequency of the field is an additional parameter and it was established that the rolls would appear from dc to some cutoff frequency f_c, in the so-called "conduction regime." However it was often experimentally found that the ordering was less effective at low frequencies and indeed it is clear that Williams's observations are not accounted for by the model since his results show tilted domains of parallel rolls.[4] Such a discrepancy was either disregarded[6] or attributed to "inhomogeneities" in the alignment.

In this Letter, we present experimental results which show, in fact, that in a nematic under an ac electric field the convective structure is, at threshold and depending on the frequency, either a set of rolls perpendicular to the molecular axis, or domains of parallel rolls oriented obliquely to this axis, contrary to the widely accepted description. Our purpose is to identify the essential features of each flow in order to determine the highest-symmetry one.

The experimental procedure is the usual one and special care is taken to ensure a correct homogeneous molecular alignment along **x**. The nematic is sandwiched between two glass plates coated with semitransparent electrodes which are rubbed along **x** for planar alignment. The liquid crystal of negative dielectric anisotropy used here is a Merck Phase-V compound. Similar results were also obtained with N-(p-methoxybenzylidene)-p-butylaniline, but are not reported here. The experiments are started with the frequency set at 60 Hz, well below the cutoff frequency f_c ($\simeq 120$ Hz). At rest the sample is uniformly transparent. The voltage is increased by steps of 25 mV every minute. At $V_r = 14$ V, a static periodic bending of the molecular axis appears along **x** as a set of bright parallel lines on a dark background [Fig. 1(a)]. It corresponds to parallel rolls uniformly oriented perpendicular to **x**. This structure is consistent with the 1D model and was named the "Williams domains"; however, we shall refer to it as the normal-rolls (NR) structure. As the voltage is further increased, the NR structure becomes undulatory along the roll axis **y** at a

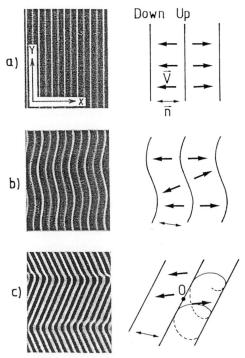

FIG. 1. Plan form of a sample observed under a microscope, and sketch of the velocity field. (a) Normal rolls. The bright lines are focal lines for the up and down flows. The up plane is a symmetry plane. (b) The undulatory rolls. (c) The oblique rolls, with symmetry about a point O.

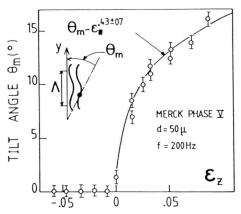

FIG. 2. Maximum tilt angle θ_m of the undulation as a function of $\epsilon_z = V^2 - V_{uz}^2/V_{uz}^2$, where V_{uz} is the voltage threshold value.

well-defined voltage $V_{uz} = 15.5$ V. The deformation is static and has a sinusoidal shape with a spatial period $\Lambda = 2\pi/q_y$ of order 5 to 7 times the roll diameter d [Fig. 1(b)]. In order to describe the undulatory roll we shall measure the local tilt angle along $\mathbf{y}$, $\theta = \theta_m \sin(q_y y)$. The maximum tilt angle θ_m over $\mathbf{y}$, measured at the inflection points, increases with the voltage (Fig. 2), while Λ remains almost constant. Similar to the case of the NR, a control parameter ϵ_z is defined as $\epsilon_z = (V^2 - V_{uz}^2)/V_{uz}^2$ and we have found that $\theta_m \sim \epsilon_z^{0.43}$. Such a behavior is characteristic of a direct bifurcation. At higher voltage, θ_m remains constant while Λ increases sharply. The deformation is no longer sinusoidal but becomes angular while the rolls straighten [Fig. 1(c)]. The increase in Λ is limited by the defects which nucleate more easily and pile up along $\mathbf{y}$ in order to form grain boundaries. The final stage of evolution of the system is an ensemble of domains of parallel rolls, tilted symmetrically with respect to $\mathbf{y}$. We shall refer to this last state as the oblique-roll or zigzag structure. To our knowledge, the undulation instability has never been previously identified. Only the final state (the zigzag) has already been reported[7] as a "modified Williams domain." In order to track the streamlines of the flow we immerse small glass spheres (3–5 μm in diameter) in the

nematic. In the NR the convective motion is a pure rotation around the roll axis $\mathbf{y}$ and has a tangential velocity $v_t \simeq 5$ μm/s at 2% above the threshold. In the oblique roll, there exists, in addition, a small axial component v_a of the velocity, of order $0.1v_t$. The v_a component changes sign from one roll to the next, and the continuity of the flow along the roll axis is ensured through the grain boundaries.[8] In order to describe completely the state of the nematic, it is necessary to determine accurately the molecular axis orientation inside the flow. While any inclination over the horizontal xOy plane is easily detected by the birefringence effects, the azimuthal tilt out of a vertical plane xOz cannot be measured by a change in the state of polarization of an outgoing light wave. This restriction is due to the Mauguin condition[9] which is always fulfilled in our experiments. Up to now, we have not been able to measure accurately this azimuthal deviation α. However, an estimation from the intensity ratios between the polarized and the depolarized intensities in a diffraction experiment would lead to $\alpha < \theta$. The experiments are repeated at a lower frequency value of 10 Hz. At a threshold $V_{zz} \simeq 7$ V a new instability occurs. The resulting structure consists of domains of parallel straight rolls tilted symmetrically over $\mathbf{y}$, by a fixed angle $\theta \simeq 30°$. The motion of the glass spheres indicates an axial component for the velocity, as in the oblique rolls which were obtained at higher frequencies, and for higher voltages. We plot the different thresholds as a function of the frequency f and obtain a structure diagram with a triple point M (Fig. 3). Typically the point M occurs at a frequency $f_M = 30$–40 Hz which increases with the conductivity σ. It decreases by 40% when a stabilizing magnetic field $H \simeq 5$ kG is applied along $\mathbf{x}$. It is clear that the oblique rolls, which are consistent with Williams's observations[4] rather than with the 1D model, correspond to a flow field with helicoidal streamlines, while in the normal rolls the

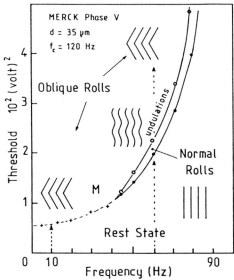

FIG. 3. Structure diagram. Below the triple point M the transition is direct to the oblique roll with a finite angle θ_m ($f = 10$ Hz). Beyond M, the normal rolls are met first, followed by the undulatory rolls which then change continuously into the oblique rolls ($f = 60$ Hz).

motion is a pure rotation around the roll axis. It is also clear that the symmetry in a vertical mirror of the normal rolls is replaced in the oblique rolls by a symmetry about a point [Figs. 1(a) and 1(c)].

In the following, we would like to suggest some elements and outline some conditions for an anisotropic mechanism of the undulation instability. We start from the stable convective flow of the NR where the molecule is subjected to two simple shears: $s_x = dv_x/dz$ along x and $s_z = dv_z/dx$ along z. In the initial 1D model only s_z was considered. However, we notice that s_x can be destabilizing, i.e., if s_x were alone, a small fluctuation δn_y of $\mathbf{n}$ out of the xOz plane by an azimuthal angle θ could be amplified. Such a condition is similar to that of Pikin's instability,[10] although the initial conditions are different here (the Leslie coefficient α_3 is negative in our experiment). Once the molecular axis deviates from the xOz plane the ionic charges also deviated. If we suppose a fluctuation of splay of $\mathbf{n}$ such that $\delta n_y \sim \delta n_{y0} \sin k_y y$, then the charges are also periodically focused along $\mathbf{y}$. This focusing induces a transverse component E_y for the total electric field $\mathbf{E}_T$ responsible for the dielectric torque which acts on the molecule. This torque adds up to the viscous torque exerted by the drag of charges in the destabilizing process. The result is a static undulation of the rotation axis of the flow, i.e., an undulatory roll. Such a mechanism would be a mere extension of the 1D mechanism.[5] However, here no characteristic length appears along $\mathbf{y}$ in order to impose a spatial period Λ at threshold. A homogeneous solution corresponds to a uniform value for the deviation angle θ along $\mathbf{y}$, in the absence of boundary conditions in this direction. The final state is then a domain of parallel straight rolls at an angle θ with $\mathbf{y}$: the oblique-rolls structure. The misalignment of the molecular axis with respect to the rotation plane implies transverse forces $f_j = \partial_i t_{ij}$, where t_{ij} is the complete viscous-stress tensor. These forces induce transverse components of the flow which create an axial component of the velocity of reversed sign from one roll to the next one. A three-dimensional linear stability analysis has recently been given by Zimmermann and Kramer.[11] This analysis confirms our previous observations regarding the stability domain for the normal rolls, but it does not account either for the undulation instability, or for the essential difference in the flow symmetry between the NR and the oblique rolls.

In conclusion, we find that in laterally extended containers, two intrinsically different structures may occur at the first convective threshold, depending on the frequency. At high frequencies, beyond a triple point it is the usual normal-roll structure consistent with the 1D theoretical model. At low frequencies the normal rolls are not a stable solution as shown by Zimmermann and Kramer and the 1D model is no longer valid here. Then the normal rolls are replaced by oblique rolls rather consistent with Williams's first observations. In the oblique rolls the velocity field is helicoidal. The normal rolls are of highest symmetry and, therefore, a hydrodynamic description of the evolution to the chaotic state must start from this structure. A purely anisotropic mechanism that could be based upon our suggestions remains to be built for the undulation which is the basic instability to the oblique rolls.

---

[1]R. Krishnamurti, J. Fluid. Mech. **42**, 295 (1970); J. P. Gollub and J. F. Steinman, Phys. Rev. Lett. **47**, 505 (1981).

[2]S. Chandrasekhar, *Hydrodynamic and Hydromagnetic Stability* (Clarendon, Oxford, 1961).

[3]A. Joets and R. Ribotta, in *Cellular Structures in Instabilities*, edited by J. E. Wesfreid and S. Zaleski (Springer-Verlag, New York, 1984), p. 294, and J. Phys. (Paris) (to be published).

[4]R. Williams, J. Chem. Phys. **39**, 384 (1963).

[5]E. Dubois-Violette, P. G. de Gennes, and O. Parodi, J. Phys. (Paris) **32**, 305 (1971); I. W. Smith, Y. Galerne, S. T. Lagerwall, E. Dubois-Violette, and G. Durand, J. Phys. (Paris), Colloq. **36**, C1-237 (1975).

[6]Orsay Liquid Crystal Group, Mol. Cryst. Liq. Crystal, **12**, 251 (1971).

[7]C. Hilsum and F. C. Saunders, Mol. Cryst. Liq. Cryst., **64**, 25 (1980).

[8]A. Joets, X. D. Yang, and R. Ribotta, to be published.

[9]R. Cano, Bull. Soc. Fr. Miner. Cristallogr. **90**, 333 (1967).

[10]S. A. Pikin, Zh. Eksp. Teor. Fiz. **65**, 2495 (1973) [Sov. Phys. JETP **38**, 1246 (1974)].

[11]W. Zimmermann and L. Kramer, Phys. Rev. Lett. **55**, 402 (1985).

Oblique Roll Instability in an Electroconvective Anisotropic Fluid. R. RIBOTTA, A. JOETS, and LIN LEI [Phys. Rev. Lett. **56**, 1595 (1986)].

In the caption of Fig. 2, add "These data are obtained from a sample having a cutoff frequency $f_c \simeq 500$ Hz (f_c is known to increase with the ionic charge density of the sample, e.g., when the sample ages)."

In the byline after Lin Lei, add "Department of Physics, City College, City University of New York, New York, New York 10031."

The work at City University of New York was supported in part by a grant from the Professional Staff Congress–City University of New York Research Award Program.

Critical Behavior in the Transitions to Convective Flows in Nematic Liquid Crystals

A. Joets and R. Ribotta

Laboratoire de Physique des Solides, Université de Paris-Sud, 91405 ORSAY, France

and

L. Lam

Department of Physics, San Jose State University, San Jose, California, 95192

ABSTRACT

Experimental results on the undulatory rolls and oblique rolls in the stability diagram of stationary electroconvective structures in nematic liquid crystals are presented. There exists a multicritical point which features a Lifshitz point. We find an anology with the NAC thermodynamic phase diagram of liquid crystals and we present a Landau-type functional with two order parameters, which enables us to describe quantitatively the stability diagram and other properties of the rolls.

PACS numbers : 61.30.-v, 47.65.+a, 47.20.Hw, 64.60.Kw

12.4 Critical Behavior in the Transitions to Convective Flows ...

Analogy between bifurcations in nonequilibrium systems and phase transitions in equilibrium systems has been noted before.[1] For example, in the language of critical phenomena, the divergence of the coherence length,[2] critical slowing down[3] and tricritical point[4] have been observed experimentally in the flow of either liquid crystals[2] or simple liquids[3,4]. All these results can be described by a Landau "free energy" functional with <u>one</u> order parameter, in exact analogy to the case of phase transitions. Even though there are extensive investigations on multicritical phenomena in nonequilibrium systems (relating mostly to codimension-2 points[5,6]), to our knowledge, the case of a Landau model with <u>two</u> order parameters has not been established, either theoretically or experimentally. This raises the question of how far the analogy between nonequilibrium systems and equilibrium systems can be extended.

Recently, a stability structure diagram in the (V^2,f) plane describing the different stationary patterns obtained in a nematic liquid crystal subjected to an ac electric field was established experimentally.[7] Here V is the voltage and f is the frequencey of the electric field, respectively. It was found that at low frequency, there is a <u>discontinous</u> transition from the rest state (RS) to the oblique roll (OR) structure at threshold V_{zz}, while at high frequency the RS transforms <u>continuously</u> to a normal roll (NR) structure at threshold V_r and at a second threshold V_{uz} ($> V_r$) the NR transforms <u>continuously</u> to undulatory rolls (UR) which then evolve gradually to OR as the voltage is increased further (Fig. 1). The three thresholds, V_{zz}, V_r and V_{uz}, are functions of f and intersect at a triple point M at $f = f_M$. If we identify the continuous transitions as second-order transitions and the discontinuous ones as first-order transitions, then the resemblance of the structure diagram with phase transition diagram in equilibrium systems is very striking.

In this Letter, the properties of the UR and the OR are investigated in detail. Specifically the wavelength Λ of the UR and the tilt angle θ_l of the OR for $f \geq f_M$, and the tilt angle θ_o of the OR at threshold for $f < f_M$ are measured. The UR is shown to be described rather well by the behavior of a θ^4 nonlinear oscillator. A Landau-type functional with two order parameters is next constructed, resulting in a "phase" diagram in excellent agreement with the structure diagram[7]. The behavior of Λ, θ_l and θ_o can also be understood within our theory.

The experimental procedure is the same as in Ref. 7. Merck Phase-V compound in the nematic phase was used. For $f > f_M$, sine-like curves of the UR first appear at the threshold V_{uz}, which then become more angular in shape with increasing wavelength until finally a zigzag structure composed of rectilinear rolls (the OR) is reached[7]. In Fig. 2, the wavevector of the UR at V_{uz}, $q_o = 2\pi/\Lambda_o$, where Λ_o is the corresponding wavelength, and the tilt angle θ_l of the OR as functions of f are shown. Both q_o and θ_l decrease with f. For fixed f, as V is increased, the measured Λ of the UR as a function of θ_m^2/θ_l^2 is plotted in Fig. 3. Here θ_m is the maximum tilt angle of the UR and is found to increase with V [7]. All the tilt angles are measured with respect to the y axis, the axis of the NR, which is perpendicular to both the cell normal and the aligning direction of the molecules at the cell surfaces (i.e., the x axis). As seen from Fig. 3, Λ increases slowly and then very rapidly as θ_m (or V) is increased. At high voltage, θ_m

tends to θ_ℓ and Λ seems to diverge.

For $f < f_m$, at the threshold V_{zz} the RS directly bifurcates to the OR with tilt angle θ_o. The angle θ_o is found to decrease with f (Fig. 4). Note that the values for f_M in Figs. 1,2 and 4 differ from each other because of the use of different samples. While the purity or aging of the sample is found to have marked effect on f_M, as usually is the case in liquid crystal experiments, the general trend of our results is not affected.

To shed light on these experimental results let us note that phenomenologically the various type of rolls observed can be described by two quantities, φ and $\tilde{\theta}$, the amplitude and phase of the deviation of the director from the x axis, respectively. Let this deviation angle be ψ. We then have

$$\psi(x,y) = \varphi \sin[q_x x - \tilde{\theta}(y)]. \tag{1}$$

Here φ is independent of x and y, but φ, q_x and $\tilde{\theta}$ are functions of V and f. A more convenient but equivalent quantity to $\tilde{\theta}$ is θ, the local tilt angle. The two are related by $\theta = \tan^{-1}(q_x^{-1} d\tilde{\theta}/dy)$. The RS corresponds to $\varphi = 0$. For $\varphi \neq 0$, $\theta(y) = $ const. corresponds to the NR while a y-dependent θ corresponds to the UR and OR.

For fixed f and $V > V_{uz}$, $\theta(y)$ is governed by a nonlinear equation. The most simple one consistent with the physical symmetry $\theta \to -\theta$ of the system is that derived from the variation of the functional,

$$F_o = \int dy \, [-\tfrac{1}{2} A\theta^2 + \tfrac{1}{4} \theta^4 + \tfrac{1}{2} L(d\theta/dy)^2], \quad A, L > 0. \tag{2}$$

The corresponding Lagrange equation is that of a θ^4 nonlinear oscillator. The oscillating solutions are the Jacobian elliptic functions, $\theta = \theta_m sn(u|k^2)$,[8] with $u = yL^{-1/2} \theta_\ell (1+k^2)^{-1/2}$, $\theta_\ell = A^{1/2}$, and $k = [2(\theta_\ell/\theta_m)^2 - 1]^{-1/2}$. The period of $\theta(u)$ is

$4K(k^2)$ where $K(k^2)$ is the complete elliptic integral of the first kind. For $0 \leq \theta_m \leq \theta_\ell$ we have $0 \leq k \leq 1$, and $-\theta_m \leq \theta \leq \theta_m$ for each k. We note that for $k \to 0$ ($\theta_m \to 0$), one has $sn(u|k^2) \to \sin(u)$; for $k \to 1$ ($\theta_m \to \theta_\ell$), $sn(u|k^2) \to \tanh(u)$. The latter is a soliton-like solution with infinite wavelength corresponding to infinitely large domains of rectilinear rolls tilted by $\pm\theta_\ell$. These shapes are exactly those observed for the UR and the OR if we identify the θ_m and θ_ℓ here with that defined in the observed experimental patterns. More quantitatively, we may transform θ from the u space into the physical y space and obtain for the wavelength of the UR,

$$\Lambda/\Lambda_o = (1 + k^2)^{1/2} K(k^2)/K_o, \quad (3)$$

where $K_o \equiv K(0)$ and $\Lambda_o \equiv \Lambda(0)$. By Eq. (3), for the OR ($\theta_m \to \theta_\ell$), Λ diverges logarithmically. The theoretical universal curve, Λ/Λ_o vs θ_m^2/θ_ℓ^2, is plotted in Fig. 3 and is in good agreement with the experiments. The curve should be valid for different samples and different materials, and is independent of f. Note that

$$\theta_\ell = 4K_o L^{1/2} \Lambda_o^{-1} = (2K_o/\pi) L^{1/2} q_o. \quad (4)$$

Consequently, Λ_o is finite at the threshold V_{uz} ($\theta_m \to 0$), in complete agreement with the experiments (Fig. 2). Also, if we assume L to be independent of f we should have θ_ℓ proportional to q_o. This seems to be the case experimentally. In view of the large uncertainty in the experimental data we have refrained from a quantitative check of this prediction.

To describe the structure diagram we use φ and θ_m as the two order parameters and construct a Landau-type "free energy"[9]

$$F = a\varphi^2 + \tfrac{1}{2}b\varphi^4 + \tfrac{1}{3}c\varphi^6 + A\theta_m^2 + \tfrac{1}{2}B\theta_m^4 - \delta\varphi^2\theta_m^2, \quad (5)$$

where all the coefficients may vary with V and f, in principle. However, for our purpose here we allow a to change sign while all the other coefficients are restricted to be positive. The RS pattern corresponds to $\varphi = 0 = \theta_m$; the NR to $\varphi \neq 0$, $\theta_m = 0$; the UR and OR to $\varphi \neq 0 \neq \theta_m$. The stable pattern is the one having the minimum free energy, in exact analogy to the case of phase transitions in equilibrium systems. Analytic results are obtained. For the particular choice of parameters, $b = 94.77(ec)^{1/2}$, $\delta^2 B = 2.31(ec)^{1/2}(80.026 - f)$ and $a/e = -1 + 10.71[-v^2 + 86.81 + 0.124(f - 29.826)^2]$, a structure diagram in excellent agreement with the experiment[7] is shown in Fig. 1. Here $e \equiv \delta A/B$; both e and c are arbitrary positive constants ; V is in volt and f in hertz. In Fig. 1 the first-order line (RS-OR) at low frequency actually passes the triple point M at $f = 38$ Hz and meets smoothly at a tricritical point at $f = 39$ Hz with the second-order line (NR-UR). Our experimental uncertainty near M precludes a precise comparison with the theory.

Our theory also gives θ_m as a function of V and f in the UR and OR region, i.e., $\theta_m = \theta_m(V,f)$. By definition, θ_o in Fig. 4 for $f \leq f_M$ is given by $\theta_o(f) = \theta_m(V_{zz}, f)$. We predict that $\theta_o(f_M)$ is non zero ; for the parameters used in Fig. 1, $\theta_o(f)/\theta_o(f_M)$ is a monotonic decreasing function of f ($f < f_M$) with $\theta_o(0)/\theta_o(f_M) = 8.64$. All these results are consistent with the data in Fig. 4.[10]

For $f > f_M$ and f fixed, our theory gives θ_m as an increasing function of V^2 with an effective exponent[11] β varying between 0.25 and 0.5. $\beta = 0.5$ when ε is extremely small. Here β is defined by $\theta_m \sim \varepsilon^\beta$, with $\varepsilon \equiv (V^2 - V_{uz}^2)/V_{uz}^2$. This is consistent with the experimental result of $\beta = 0.43 \pm 0.07$.[7] With the parameters used in Fig. 1, $f = 60$ Hz and $0 \leq \varepsilon \leq 1$, we obtain $\beta = 0.43$.[12]

The general agreement between experimental data and the theory gives us the confidence that, at least for the electroconvective nematic system under consideration and for the first few patterns (on the route to chaos[13]) discussed here, the Landau-type "free energy" description with two order parameters is valid and useful. However, we have not been able to give here a general functional, in the form of $\int dxdy F(x,y)$, say, from which Eqs.(2) and (5) can be derived in a unified way. Yet, this is not at all impossible. In this regard, the amplitude-equation approach[14] may also be helpful.

The structure diagram in Fig. 1 contains a "modulated phase", the UR's. The point M may also be interpreted as a Lifshitz point.[15-17]. However, the wavevector in classical modulated phases increases from zero beyond the Lifshitz point,[15] in contrast to the trend shown in Fig.2. In their stability analysis of the OR, Zimmermann and Kramer[18] found a nonlinear dependence of θ_o of f near f_M, while our Fig 4 shows a almost linear dependence. Our experimental uncertainties on the angle measurements are large (± 3°) and do not allow us to conclude either way. Pesch and Kramer[19] have recently proposed an anisotropic model (not specific to the electroconvection) which is derived from a functional. Their phenomenological model accounts for the main features of these experimental results (e.g., the existence of a triple point). Unfortunately, we cannot compare their model to ours concerning the existence of the UR and the variation of the wavelength.

In conclusion, we have experimentally found that the structure diagram of the convective flows in a nematic liquid crystal subjected to an ac electric field presents a multicritical triple point which may appear as a Lifshitz point by analogy with the NAC transition point in equilibrium liquid crystals and the magnetic modulated phases. The main results can be described by the use of a Landau-type functional with two coupled order parameters. We have demonstrated here that in such a nonequilibrium system where there does not exist yet any tractable model (a Landau functional), the use of an ad hoc functional may reveal itself to be useful even though it cannot be a priorily justified.

We thank P.G.de Gennes, P.C. Hohenberg, P. Pfeuty and G. Xu for useful discussions. This work was supported by the Direction des Recherches et Etudes Techniques (DRET) under contract No. 84 049.

REFERENCES

1. J. Gea-Banacloche, M. O. Scully and M. G. Velarde, in "Nonequilibrium Cooperative Phenomena in Physics and Related Fields", edited by M.G. Velarde (Plenum, New York, 1984).
2. R. Ribotta, Phys. Rev. Lett. 42, 1212 (1979)
3. J. Wesfreid, Y. Pomeau, M. Dubois, C. Normand and P. Bergé, J. Phys. (Paris) 39, 725 (1978).
4. A. Aitta, G. Ahlers and D.S. Cannell, Phys. Rev. Lett. 54, 673 (1985).
5. H.R. Brand and B.J.A. Zielinska, Phys. Rev. Lett. 57, 3167 (1986) ; and references therein.
6. I. Rehberg and G. Ahlers, Phys. Rev. Lett. 55, 500 (1985) ; G. Ahlers and I. Rehberg, Phys. Rev. Lett. 56, 1373 (1986).
7. R. Ribotta, A. Joets and Lin Lei (L. Lam), Phys. Rev. Lett. 56, 1595 (1986) ; 56, 2335(E) (1986).

8. "Handbook of Mathematical Functions", edited by M. Abramowitz and I.A. Stegun (Dover, New York, 1972).

9. Such a free energy was used by L. Benguigui [J. Phys.(Paris), Colloq. 40, C3-419 (1979)] in describing the NAC phase diagram of liquid crystals. Our results differ from his in the location of the tricritical point.

10. The number 8.64 here refers to the sample used in Fig. 1, which is different from that in Fig. 4. However, it is interesting to note that the data in Fig. 4 is consistent with this number too. In Fig. 4, there seems to be a drop of $\theta_o(f)$ at f_M, as predicted by our theory.

11. Lin Lei, J. Phys. (Paris) 43, 251 (1982).

12. Note that β may depend on f, the range of ϵ and the sample. The number 0.43 refers to the sample in Fig. 1, which differs from that used in Ref. 7. The excellent agreement between the theory and experiment obtained here may be a pure coincidence.

13. A. Joets and R. Ribotta, J. Phys. (Paris) 47, 595 (1986).

14. A.C. Newell and J.A. Whitehead, J. Fluid Mech. 38, 279 (1969); C.Q. Shu and L. Lin, Mol. Cryst. Liq. Cryst. 146, 97 (1987)

15. R.M. Hornreich, M. Luban and S. Shtrikman, Phys. Rev. Lett. 35, 1678 (1975) ; A. Michelson, Phys. Rev. Lett. 39, 464 (1977).

16. J.H. Chen and T.C. Lubensky, Phys. Rev. A14, 1202 (1976) ; T.C. Lubensky, Mol. Cryst. Liq. Cryst. 146, 55 (1987).

17. A. Joets, Thèse de 3eme Cycle, Paris VII (1984)

18. W. Zimmermann and L. Kramer, Phys. Rev. Lett. 55, 402 (1985).

19. W. Pesch and L. Kramer, Z. Phys. B63, 121 (1986).

FIGURE CAPTIONS

Fig. 1. Stability structure diagram. The points are experimental results from Ref.7. The theoretical curves are represented by the solid lines (second-order transition) and the broken line (first-order transition).

Fig. 2. As a function of the excitation frequency are represented: (●) the wavevector q_o of the Undulatory Rolls at threshold above the triple point M ($f \gtrsim f_M$) and (+) the tilt angle θ_l of the Oblique Rolls below M ($f \lesssim f_M$).

Fig. 3 Reduced spatial wavelength Λ/Λ_o of the Undulatory Rolls as a function of the reduced maximum tilt angle θ_m/θ_l. The solid line represents the theoretical curve.

Fig. 4. Tilt angle θ_o at threshold, of the Oblique Rolls ($f \lesssim f_M$), as a function of the reduced frequency f/f_M for two different samples [(●): $f_M = 120$ Hz;(+): $f_M=155$ Hz]. Above f_M the Undulatory Rolls obviously appear with a tilt angle $\theta_o = 0$.

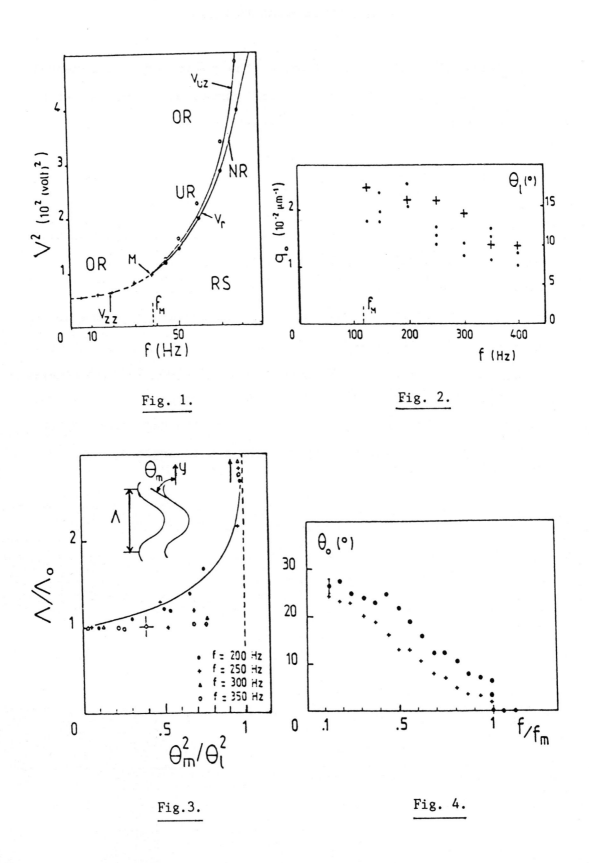

Fig. 1.

Fig. 2.

Fig.3.

Fig. 4.

Chemical Waves

JOHN ROSS, STEFAN C. MÜLLER, CHRISTIAN VIDAL

Spatial structures may occur in nonlinear systems far from equilibrium. Chemical waves, which are concentration variations of chemical species propagating in a system, are an example of such structures. A survey is given of some experiments on chemical waves by spectroscopic and microphotographic techniques, arranged according to different types of waves, different geometries, and various properties.

CHEMICAL FRONTS AND WAVES (1) ARE VARIATIONS IN concentrations of chemical species, or possibly in other state variables such as temperature, which travel in space and occur in nonlinear reactive systems far from equilibrium. Waves in the so-called Belousov-Zhabotinsky (2–8) reaction were first reported by Zhabotinsky (9). The overall reaction is the oxidative bromination by acidic bromate of an organic substrate such as malonic acid; it can be approximated by

$$2BrO_3^- + 3CH_2(COOH)_2 + 2H^+ \rightarrow$$
$$2BrCH(COOH)_2 + 3CO_2 + 4H_2O \quad (1)$$

The reaction is catalyzed by an oxidation-reduction couple such as Ce^{3+}/Ce^{4+} or ferroin-ferriin, $Fe(phen)_3^{2+}/Fe(phen)_3^{3+}$, where phen is phenanthroline. The reaction can be oscillatory, in which case the concentrations of some chemical intermediates vary in time periodically. The period of such oscillations is typically about 40 seconds, and the concentration of Br^- may vary by over five orders of magnitude.

There are many reports of visual observations of waves in this and a few other chemical systems [see (7, 8) and citations therein]. Detailed studies by spectroscopic and photographic methods of different types of waves, the structure of the fronts of waves, and other properties such as dispersion relations have appeared only in the last 2 years (10–19). In this article we present a brief survey of such experiments on chemical waves considered as reaction-diffusion processes but largely omit theory, calculations, and the interaction of such processes with convection. We begin with a categorization of waves: kinematic, trigger, and phase diffusion waves. We then discuss different geometries: plain, circular, spiral, multiarmed vortices, and scroll waves. Then we consider different properties of waves: the amplitude, the velocity, the front structure, dispersion relations, and the relation of curvature and velocity.

Chemical waves are described by solutions of reaction-diffusion equations

J. Ross is in the Department of Chemistry, Stanford University, Stanford, CA 94305. S. C. Müller is at the Max-Planck Institut für Ernährungsphysiologie, Rheinlanddamm 201, D 4600 Dortmund 1, West Germany. C. Vidal is in the Centre de Recherche Paul Pascal, Domaine Universitaire, F33405 Talence Cedex Bordeaux, France.

$$\frac{\partial \Psi}{\partial t} = D\nabla^2 \Psi + F[\Psi] \quad (2)$$

where Ψ is a vector of state variables, such as the concentrations of chemical species and possibly temperature, each dependent on space and time; D is a matrix of transport (diffusion) coefficients, each assumed to be constant; and $F[\Psi]$ represents the variations in time that arise from the chemical reactions. These nonlinear equations can seldom be solved in closed form; a variety of approximations and numerical techniques lead to useful solutions. Reviews on the theory of reaction-diffusion equations are available (8, 20–23).

The field of temporal and spatial structures in nonlinear chemical systems far from equilibrium, such as chemical systems with multiple stationary states, with oscillations in chemical intermediates and products, with chaotic variations of concentrations and other state

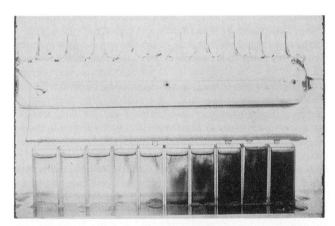

Fig. 1. Kinematic wave in a row of ten adjacent cells. An initial phase (or frequency) gradient imposed on the oscillatory Briggs-Rauscher reaction leads to the apparent propagation of colored fronts. [From (39), reprinted with permission, copyright (1982) *Actual. Chim.*]

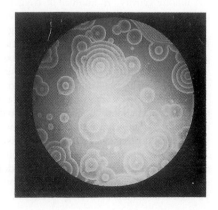

Fig. 2. Two-dimensional target patterns in the Belousov-Zhabotinsky reaction in a layer of solution 1 mm deep. The scale of the photograph is about 5 cm. [From (13), reprinted with permission, copyright (1986) *J. Phys. (Paris)*]

variables, and with spatial structures (*24*) such as chemical waves is receiving significant attention. Systems far from equilibrium are less well understood, both experimentally and theoretically, than systems at equilibrium; yet most natural phenomena are indeed far from equilibrium.

In the past 20 years many examples of oscillatory chemical, biochemical, and biological (both in vivo and in vitro) reactions have been reported (*7, 25*). Waves and fronts in biological systems have been studied for a long time: in signal propagation in nerves (*26–28*); in peristaltic motion (*29*); in the development of embryos (*30, 31*); in phage-bacterium systems (*32*); in the aggregation step of the life cycle of slime mold cells (*33*); in waves across the chambers of the heart (*34, 35*); in pulses of pheromone emission (odor song) (*36*); and in spreading depression in the cerebral cortex (*37*), among others. All these waves can only be electrochemical in origin, and this fact provides additional motivation for the study of the physical chemistry of chemical waves.

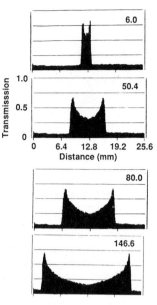

Fig. 3. Wave propagation sequence from initiation at a single point; transmission of argon ion laser pulse at 488 nm as a function of distance. Temperature is 18.0°C. Time (in seconds) since wave initiation is shown in the upper right corner of each scan. [From (*10*), reprinted with permission, copyright (1985) American Institute of Physics]

Types of Waves

Kinematic waves. Another oscillatory chemical system is the Briggs-Rauscher reaction (*38*). Consider such a solution placed in the long upper tube shown in Fig. 1 (*39*). On turning the tube so that the many sidetubes face downward, we observe that the vessels below the long tube become filled. If there exists initially either a small variation in frequency of oscillation from one end of the long tube to the other, induced, for example, by a temperature gradient imposed on that tube, or if there exists a phase gradient in the oscillations, then one will see the passage of waves across the separate beakers. Such a wave, which is an optical illusion, is called a kinematic wave (*40, 41*). It clearly does not involve mass transfer from one beaker to another. If an initial phase gradient or frequency gradient is set up in the long tube, then kinematic waves will also sweep from one end of the long tube to the other, initially without mass transfer. However, after a time, owing to concentration differences in neighboring spatial elements of the tube along its axis, diffusion will occur and different types of waves will appear.

Trigger waves. Let us return to the Belousov-Zhabotinsky reaction in an oscillatory state (*42–49*), say, a solution of that reaction in a shallow (1 mm or less) layer in a petri dish. The solution is first well stirred and then left quiescent. Homogeneous concentration oscillations in the ratio of the oxidation reduction couple, say, Fe^{2+}/Fe^{3+}, are observed with an appropriate indicator ferroin [tris(1,10-phenanthroline) ferrous sulfate] [$Fe(phen)_3^{2+}$ is pink, $Fe(phen)_3^{3+}$ is blue]; the color changes from pink to blue and back to pink, repeatedly. After a while inhomogeneities appear, frequently with a gas bubble or dust particle at the center of the inhomogeneity. In time, a chemical front proceeds radially outward from the center of the inhomogeneity. A number of trigger waves are shown in a photograph (*13*) of such a solution in a petri dish (Fig. 2).

The transmission profile of a propagating trigger wave (*10*) in the Belousov-Zhabotinsky reaction in an excitable stationary state is given in Fig. 3. An appropriate perturbation from such a stationary state, of a threshold magnitude, leads to large concentration variations, just as in an oscillation, before return to the stationary state. The measurements were made by shining a diffuse laser beam onto the surface of the solution, perpendicular to the plane of the layer of the solution. The transmitted light is measured on a diode array system which consists of 1024 photosensitive spots located on a line 25 μm apart. The array can be scanned about once a second, and the data are stored in a computer system and displayed on a screen. By repeating such measurements in time we obtain the concentration of Fe^{2+} (the dominant absorbing species) as a function of space and time and hence a quantitative analysis of the front as it propagates in space and time.

Trigger waves have sharp fronts, 0.1 to 0.5 mm in width [concentration gradients up to 17 mM/mm of ferroin (*18*)] that arise from a phase in the oscillatory Belousov-Zhabotinsky reaction during which rates are very rapid and hence concentration variations are steep in time. Both reaction and diffusion play crucial roles in the propagation of trigger waves. Trigger waves do not interpenetrate each other but annihilate each other on contact, when concentration gradients are equal and opposite in the two colliding waves; unlike kinematic waves, they are blocked by walls.

Trigger waves in an excitable medium are induced by a disturbance of the right direction and magnitude at one location, which leads to sharp concentration differences (gradients) with neighboring locations. Diffusion occurs, which then sets off the same disturbance in the neighboring locations with consequent propagation of a front. The mechanism of wave propagation in an oscillatory reaction is the same.

Phase diffusion waves. Consider a solution of the Belousov-Zhabotinsky reaction, in a petri dish, which oscillates autonomously. We impose a well-focused laser beam, of wavelength absorbed by Fe^{2+}, on a small area (0.2 mm^2) and irradiate the solution for a limited time (about 12 seconds). The photons absorbed are converted into heat, which raises the temperature of the volume beneath the small area and brings about a perturbation that sets off a trigger wave (see Fig. 4, first cycle, graphs at 108.1 to 113.3 seconds) (*11*). That wave, with sharp concentration fronts, propagates into the unperturbed medium. When the phase of the oscillation of the concentration of Fe^{2+} in the unperturbed medium matches that of the wave front, the front disappears (Fig. 4, first cycle, graph at 130.3 seconds). However, because of the temperature difference between the irradiated spot and the remainder of the solution, there exists a difference in frequency of oscillation and after a passage of a cycle the wave reappears. With the elapse of some time, the energy deposited by the light, turned into heat in the solution, diffuses and the temperature gradient becomes shallower. As that happens, a different wave appears (Fig. 4, fourth cycle, graphs at 449.0 to 453.0 seconds), called a phase diffusion wave (*41, 50*). The velocity of a phase diffusion wave is defined as the ratio of the variation of the phase with time divided by the variation of the phase in space

$$v = -\frac{(\partial\phi/\partial t)}{(\partial\phi/\partial r)} \qquad (3)$$

Hence, as the heat in the solution at the irradiated spot diffuses away, the phase gradient in space becomes smaller and hence the velocity of the phase wave increases. When the velocity of the phase diffusion wave exceeds that of the trigger wave (of essentially constant velocity), then a phase wave and not a trigger wave propagates in the solution. Velocities of phase diffusion waves are nearly constant for a limited time interval and become very large as the phase difference between the wave and the solution disappears (see Eq. 3). The concentration gradients in a phase diffusion wave are much shallower than those in a trigger wave; compare, for instance, the gradients at 451.6 seconds with those at 110.7 seconds in Fig. 4. In an autonomous oscillatory reaction, phase diffusion waves appear only if concentration gradients are small; for large concentration gradients, trigger waves appear.

Fronts in bistable systems. Chemical reactions far from equilibrium may have multiple, stable stationary states. On transition from one such stable state to another, a front of concentration variations travels through space. Such fronts have been observed in the iodate–arsenous acid reaction (*51*) and others.

Geometric Forms of Waves

The fronts of waves of chemical activity may assume various shapes depending on the geometry and volume of the container. All the characteristic features of these shapes are best realized in quiescent solutions in which trigger waves can be excited.

One-dimensional waves. These waves are formed in a confined space, as in a narrow (a few millimeters) test tube, and consist of a train of fronts following each other at a distance. Measurements in such simple geometries are performed, for instance, in order to study the effect of an externally applied electrical field on wave properties (*52, 53*). The dependence of the velocity of chemical waves on electric field intensity is nonlinear. Waves can be slowed with appropriate polarity of the field; at high fields, waves may be split and the direction of propagation may be altered.

Two-dimensional waves. In a thin layer (about 1 mm^2) of reactive solution in a petri dish there occur several types of two-dimensional wave forms. An apparatus (*14–19*) for computerized digital spectrophotometry of structures in two dimensions consists of ultraviolet (UV) optical components mounted on a vibration-isolated table for illumination and imaging purposes, a UV-sensitive video camera serving as the two-dimensional intensity detector, and a fast, large-memory computer for storage of the digitized data and further data processing. The sample layer in an optically flat petri dish is illuminated from above with a parallel, spatially homogeneous light beam emerging from a 300-W xenon short-arc lamp (Cermax) that has high temporal stability. Square sections (1 by 1 mm^2 to 15 by 15 mm^2) are imaged by a UV photo lens on the target of the video camera with an image raster resolution of 512 by 512 picture elements (pixels). The video signal is converted to digital data with 256 digital units (gray levels) intensity resolution. The apparatus combines spatial, temporal, and intensity resolution satisfactory for the analysis of chemical patterns and their temporal evolution. A comprehensive software package for the presentation of two-dimensional data arrays includes extraction of profiles of transmitted light intensity, logarithmic conversion of intensities into concentrations, pseudocolor and three-dimensional perspective graphical presentations, and fitting procedures for specific isointensity or isoconcentration lines.

1) Distributed sets of concentric annuli (*42–49*) are frequently also called "target" patterns (Fig. 2) and have been discussed in the section on trigger waves. One important property of these waves is their mutual annihilation upon collision, which leads to typical cusplike structures in the vicinity of the area of collision between two annuli.

2) Spiral-shaped waves are formed by the disruption of an expanding circular wave front, as can be done in a controlled manner with a gentle blast of air from a pipette onto the surface of the reacting solution (*15, 43, 45*). The irregularly shaped open ends of the circular wave are then the starting points of a rapid evolution toward a structure composed of a pair of counterrotating spiral waves with highly regular geometry. The tips of these spiral-shaped vortices turn inward with a rotation period of 17 seconds, whereas the fronts move in the outward direction. Spiral waves have been studied in some detail (*14–17*). Figure 5A shows the digital image of a pair of spiral waves in three-dimensional perspective presentations with a specific isoconcentration level of the light-absorbing catalyst marked in black. Outside a small region surrounding the center of rotation, the so-called spiral core, the structure follows in good approximation an Archimedean geometry, but the involute of a circle fits the structure equally well. These two curves are asymptotically identical and differ only slightly in the immediate neighborhood of the core, where the resolution of the measured data points is not yet sufficient to distinguish between them. Theoretical analysis of simplified models of this reaction yielded results close to the shape of the involute (*54*). The three-dimensional

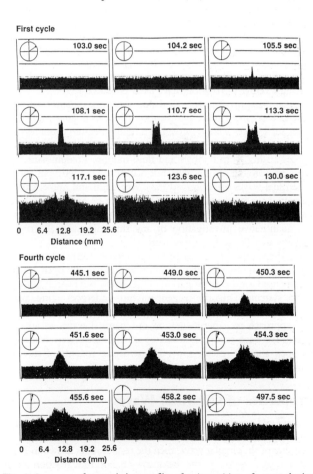

Fig. 4. Sequence of transmission profiles after imposition of a perturbation on the oscillatory Belousov-Zhabotinsky reaction at a given point at time t_0. The perturbation consists of a laser pulse of intensity 90 μW. Each profile is marked by the time $(t - t_0)$ and by the corresponding phase of the unperturbed oscillatory reaction. [From (*11*) reprinted with permission, copyright (1987) American Institute of Physics]

Fig. 5. (**A**) Symmetric pair of counterrotating spiral waves in a thin layer of an excitable Belousov-Zhabotinsky reaction. The digital image shows an 8.2 mm by 8.2 mm area and is composed of 410 × 410 picture elements. Each element has one out of 256 possible gray levels of transmitted light intensity, which is a measure of the local concentration of the catalyst ferroin. One concentration level is enhanced in black. (**B**) The inner section of (A) is rendered as a three-dimensional surface image by interpolation of the measured pixel values and subsequent projection on perspective. The level enhanced in black is the same as that in (A). [From (*18*), reprinted with permission, copyright (1987) *Biophys. Chem.*]

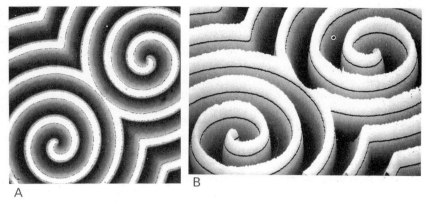

perspective image in Fig. 5B indicates the asymmetric shape of the wave profiles. The gradients of ferroin are about 10 m*M*/mm at the steep fronts and 1 m*M*/mm at the gentler sloping backs. The rotation center is a singular site in that it is the only location in the system where the chemistry remains quasi-stationary; in the areas outside the center periodic redox transitions take place that are correlated with the outward propagation of the wave front. The size of the site is smaller than 10 μm. The core region is just that portion of the layer where a transition takes place from the spiral center to the outer area where the solution is excited to full-amplitude waves. This transition gives the core a "tornado"-like structure, as shown in the cover.

3) Multiarmed vortices have been seen in a shallow layer of an excitable medium. They are obtained by the controlled addition of a drop of a chemical close to the center of rotation of a spiral wave. Two-, three-, and four-armed vortices, that is, spiral waves having the same center of rotation, have been produced (*55*).

Three-dimensional waves. Evidence for three-dimensional wave spheres first came from experiments with stacks of filter paper (*45*). Later such waves were observed in a liquid medium contained in a cylinder (8 mm), whereby special care was taken to avoid disturbances arising from bubbles and fluid motion. Predominant wave types are toroidal scroll waves, but occasionally other types such as spheroidal waves are observed (*56*). Analytical and numerical predictions have been made for the possibility of twisted and linked scroll rings (*57*). Their experimental realization, however, as well as the quantitative investigation of any three-dimensional structure remains a challenge for future work.

All these experiments to produce different forms of waves are carried out in closed (batch) systems and not in the open reactors usually used for the investigation of oscillatory behavior in homogeneous systems. Thus the reagent mixture undergoes aging, which results in a slow drift of the reacting system toward equilibrium. A first approach to experimentation on chemical waves in an open system has been reported; sustained wave patterns were realized in an open annular reactor (*58*). The structure consists of traveling azimuthal wave pieces which look very much like pinwheels.

Properties of Chemical Waves

Shape and amplitude. If we select a wave with a single bright (blue-oxidizing) wave front and measure the transmission profile along a diameter of the circular wave, then the results are as shown in Fig. 3 for a trigger wave in the Belousov-Zhabotinsky reaction in an excitable stationary state. A three-dimensional representation of photographs of the same reaction in oscillatory conditions is shown in Fig. 6. The photographs are subjected to a quantitative analysis by first recording the images from a TV camera on video tape and then digitizing selected frames of pictures. With the techniques of geometrical corrections, background subtraction, enhancement of pattern by filtering and thresholding, among others, characteristic features of the patterns can be obtained. The oxidation domain grows from a small area around the center. A flat cylinder appears with radius and height (amplitude) increasing monotonically. Beyond a critical value of the radius, the reduction phase then begins at the center and the amplitude of the cylinder at the center decreases.

The constancy of the front and shape of a trigger wave (*10*) is shown in Fig. 7. Figure 7A gives a measurement of the transmission profile of Fe^{2+} in space at a given time. Figure 7B is the transmission profile at a given point in space as a function of time. Figure 7C shows a superposition of the measurements in (A) and (B) translated by the measured velocity of propagation of the front according to the equation $\phi = X - vt$, where ϕ is the phase, X is position, and t is time. The constancy of the shape in time is quantitatively confirmed. A wave initiated in water when a rock is dropped into it propagates with constantly decreasing amplitude due to the viscosity of the medium. The comparison, however, is not fair since the water wave propagates into a medium at equilibrium. In the case of chemical trigger waves the disturbance propagates into a medium that is far from equilibrium. In spite of the fact that diffusion removes concentration gradients, the combination of reaction and diffusion propagates the wave with constant shape. The energy necessary for this process comes from the Gibbs free energy change of the reaction. With respect to the constancy of wave profile, energetics, and nonlinear origin, there are interesting similarities between these waves and nerve conduction; the velocity of chemical waves,

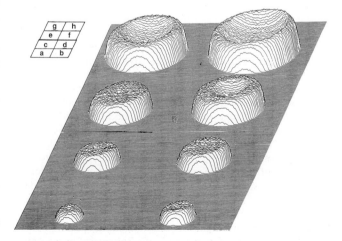

Fig. 6. Three-dimensional perspective representation of light transmission by a layer of oscillatory Belousov-Zhabotinsky reaction. The eight successive snapshots, taken at 1-second intervals, show the buildup of the oxidizing wave front. [Courtesy of C. Vidal and A. Pagola, in preparation]

however, is considerably slower (by about a factor of 400,000).

The range of variation in the concentration of Fe^{3+} in an oscillation is a factor of 2 or 3 (from $1 \times 10^{-4} M$ to $3 \times 10^{-4} M$). The distance in which this sharp variation occurs, the front width, varies with initial concentration in the system in a manner that has not yet been explored.

Velocity. At a given temperature the velocity of a trigger wave depends on the initial concentrations of reactants, 4.0 mm/min for the wave shown in Fig. 3. The velocity has a square root dependence (*10, 46, 47*) on the concentrations of sulfuric acid and bromate ion, $v \sim k[H_2SO_4]^{1/2}[BrO_3^-]^{1/2}$, in a limited range of initial concentrations, and is much less sensitive to the concentrations of the other chemical species. The variation of the coefficient k has been studied so far only over a narrow range of temperature (284 to 318 K) and is reasonably given by an Arrhenius law with an apparent activation energy of about 35 kJ mol^{-1}.

Dispersion of chemical waves. Dispersion is the variation of the velocity of wave propagation with the period (of oscillation in an oscillatory chemical system). Consider the emission of a wave from a center with a period T into a medium of oscillatory period T_0. The first wave emitted from the center propagates into the medium with a velocity that is described by the concentrations in the medium. If the period of emission is of the same order as T_0, then the second and successive waves propagate into a medium that is essentially relaxed after the passage of the prior wave, and the velocity of the propagation is little affected. However, if the period of emission is below $T_0/2$, then successive waves propagate into a medium that is incompletely relaxed and a reduction in the velocity of propagation may be expected. Measurements of the dispersion relation (*12*) in an oscillatory Belousov-Zhabotinsky reaction show that for T/T_0 larger than 0.5, the reduction in velocity in propagation is small, whereas for T/T_0 smaller than 0.5 the reduction in propagation velocity is substantial.

In order to solve reaction-diffusion equations numerically, we must have a reaction mechanism and rate coefficients for the various steps in that mechanism. Much effort has gone into unraveling the mechanism of complex reactions, in particular that of the Belousov-Zhabotinsky reaction, and a number of simplified models of the mechanism have been proposed. Even a two-variable model (*54, 59–61*), which oversimplifies the mechanism and neglects diffusion of all but one of the reacting species (bromous acid), provides a fairly good description of the main features of the wave profile, the velocity of propagation, and dispersion.

Origin of trigger waves. A fundamental point, not yet settled, concerns the origin of trigger waves that appear spontaneously in a thin layer of solution of an oscillatory or excitable chemical system. Do these waves arise spontaneously (*62*) as a result of a symmetry-breaking fluctuation? Or do these waves arise deterministically from a heterogeneous center (a dust particle, an impurity, a gaseous bubble)? Extensive measurements have been made (*13*) on samples of more than 500 centers, and distributions of velocities of wave propagation have been determined, with no conclusive decision as yet. Experiments on thoroughly filtered solutions of the Belousov-Zhabotinsky reaction show that formation of trigger waves can thus be suppressed in an excitable system but not in an oscillatory system. A theory has been proposed for these results, which favors heterogeneous centers (*61*).

Propagation velocity and curvature in spiral waves. An important problem in the study of spiral waves is the relation between the curvature and the propagation velocity of a front (*63*). Quantitative confirmation of the curvature-velocity equation used by Keener and Tyson (*54*) has been obtained (*64*) by measuring the temporal evolution of the cusplike structures that form immediately after wave collision (compare the dip in the wave crests in Fig. 5B). Here one can easily produce areas of extremely high curvature and follow the very rapid changes with a video movie.

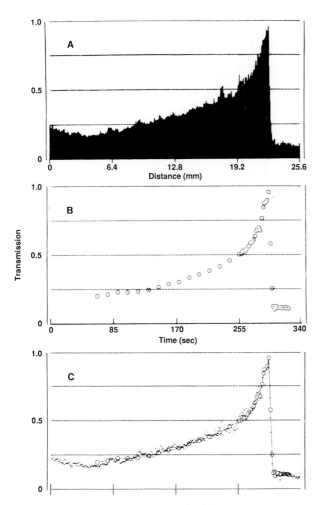

Fig. 7. Transmission profile of a trigger wave: (**A**) sequence similar to Fig. 3, as a function of distance on a reticon; (**B**) sequence as a function of time on one photosensitive spot on the reticon; (**C**) superposition of the measurements in (A) and (B) as described in text. The superposition shows the constancy of the wave profile in time. [From (*10*), reprinted with permission, copyright (1985) American Institute of Physics]

Conclusions

Chemical waves are an interesting phenomenon that characterizes many nonlinear reaction systems far from equilibrium. The quantitative investigation of these waves is in the initial stage. Much work needs to be done including the study of their characteristics in the limit of zero velocity, that is, stable spatial (Turing) structures, and their application to biological systems.

REFERENCES AND NOTES

1. The earliest report of a chemical wave appears to be that of R. Luther [*Z. Elektrochem.* **12**, 596 (1906)].
2. B. P. Belousov, in *Sb. Ref. Radiat. Med.* (collection of abstracts on radiation medicine) (Medgiz, Moscow, 1959), p. 145; A. M. Zhabotinsky, *Biofizika* **9**, 306 (1964).
3. For references on this and other oscillatory reactions, see (*4–8*).
4. R. M. Noyes and R. J. Field, *Annu. Rev. Phys. Chem.* **25**, 95 (1974).
5. J. J. Tyson, *The Belousov-Zhabotinsky Reaction*, Lecture Notes on Biomathematics, (Springer-Verlag, Berlin, 1976), vol. 10.
6. C. Vidal and A. Pacault, in *Evolution of Order and Chaos*, H. Haken, Ed. (Springer-Verlag, Heidelberg, 1982), p. 74.

7. R. Field and M. Burger, Eds., *Oscillations and Traveling Waves in Chemical Systems* (Wiley, New York, 1985).
8. C. Vidal and P. Hanusse, *Int. Rev. Phys. Chem.* **5**, 1 (1986).
9. A. M. Zhabotinsky, in *Oscillatory Processes in Biological and Chemical Systems*, G. M. Frank, Ed. (Science Publishers, Moscow, 1967), p. 252.
10. P. M. Wood and J. Ross, *J. Chem. Phys.* **82**, 1924 (1985).
11. J. M. Bodet, C. Vidal, J. Ross, *ibid.* **86**, 4418 (1987).
12. A. Pagola, C. Vidal, J. Ross, *J. Phys. Chem.* **92**, 163 (1988).
13. C. Vidal *et al.*, *J. Phys. (Paris)* **47**, 1999 (1986).
14. S. C. Müller, Th. Plesser, B. Hess, *Naturwissenschaften* **73**, 165 (1986).
15. _____, *Science* **230**, 661 (1985).
16. _____, *Physica* **24D**, 71 (1987).
17. _____, *ibid.*, p. 87.
18. _____, *Biophys. Chem.* **16**, 357 (1987).
19. Zs. Nagy-Ungvarai *et al.*, *Naturwissenschaften*, in press.
20. H. Eyring and D. Henderson, Eds., *Theoretical Chemistry, Advances and Perspectives*, vol. 4, *Periodicities in Chemistry and Biology* (Academic Press, New York, 1978).
21. P. Fife, *Mathematical Aspects of Reacting and Diffusing Systems* (Springer-Verlag, New York, 1979).
22. Y. Kuramoto, *Chemical Oscillations, Waves, and Turbulence* (Springer-Verlag, Heidelberg, 1984).
23. P. Hanusse, J. Ross, P. Ortoleva, *Adv. Chem. Phys.* **38**, 317 (1978).
24. A. M. Turing, *Philos. Trans. R. Soc. London Ser. B* **237**, 37 (1952).
25. B. Hess and A. Boiteus, *Annu. Rev. Biochem.* **40**, 237 (1971); P. E. Rapp, *J. Exp. Biol.* **81**, 281 (1979).
26. R. Fitzhugh, *Biophys. J.* **1**, 445 (1961).
27. E. C. Zeeman, in *Towards a Theoretical Biology* (Aldine-Atherton, Chicago, 1972), vol. 4, p. 8.
28. G. Matsumoto and H. Shimizu, *J. Phys. Soc. Jpn.* **44**, 1399 (1978).
29. N. W. Weisbrodt, in *Physiology of the Gastrointestinal Tract*, L. R. Johnson, Ed. (Raven, New York, 1981), vol. 1, pp. 411–433.
30. L. Wolpert, *J. Theor. Biol.* **25**, 1 (1969).
31. A. Gierer and H. Meinhardt, *Kybernetik* **12**, 80 (1972).
32. G. R. Ivanitsky, A. S. Kunisky, M. A. Tzyganov, in *Self-Organization—Autowaves and Structures Far from Equilibrium*, V. I. Krinsky, Ed., vol. 28 of Synergetics Series (Springer-Verlag, Berlin, 1984), p. 214.
33. G. Gerisch, *Curr. Top. Dev. Biol.* **3**, 157 (1968); J. L. Martiel and A. Goldbeter, *J. Biophysics* **52**, 807 (1987), and references therein.
34. M. A. Allessie, F. I. M. Bonke, F. J. G. Schopman, *Circ. Res.* **41**, 9 (1977).
35. A. T. Winfree, *Sci. Am.* **248** (no. 5), 144 (1983).
36. W. H. Bossert, *J. Theor. Biol.* **18**, 157 (1968).
37. J. Bures, V. I. Koroleva, N. A. Gorelov, in (*32*), p. 180.
38. T. S. Briggs and W. C. Rauscher, *J. Chem. Educ.* **50**, 496 (1973).
39. M. Sadoun-Goupil, P. de Kepper, A. Pacault, C. Vidal, *Actual. Chim.* 7, 37 (1982).
40. N. Koppel and L. N. Howard, *Science* **180**, 1171 (1973).
41. P. Ortoleva and J. Ross, *J. Chem. Phys.* **60**, 5090 (1974).
42. A. N. Zaikin and A. M. Zhabotinsky, *Nature (London)* **225**, 535 (1970).
43. A. T. Winfree, *Science* **175**, 634 (1972).
44. A. M. Zhabotinsky and A. N. Zaikin, *J. Theor. Biol.* **40**, 45 (1973).
45. A. T. Winfree, *Sci. Am.* **230** (no. 6), 82 (1974).
46. R. J. Field and R. M. Noyes, *J. Am. Chem. Soc.* **96**, 2001 (1974).
47. T. A. Gribschaw, K. Showalter, D. L. Banville, I. R. Epstein, *J. Phys. Chem.* **85**, 2152 (1981); L. Kuhnert, H.-J. Kong, L. Pohlman, *J. Phys. Chem.* **89**, 2022 (1985).
48. M. Orban, *J. Am. Chem. Soc.* **102**, 4311 (1980).
49. P. De Kepper, I. R. Epstein, K. Kustin, M. Orban, *J. Phys. Chem.* **86**, 170 (1982).
50. There is no unaminity on nomenclature. What we call kinematic waves are sometimes referred to as phase waves.
51. J. Harrison and K. Showalter, *ibid.* **90**, 225 (1986).
52. P. J. Ortoleva and S. L. Schmidt, in (*7*), p. 333.
53. H. Szevcikova and M. Marek, *Physica* **13D**, 379 (1984).
54. J. P. Keener and J. J. Tyson, *ibid.* **21D**, 307 (1987).
55. K. I. Agladze and V. I. Krinsky, *Nature (London)* **296**, 424 (1982).
56. B. J. Welsh, J. Gomatam, A. E. Burgess, *ibid.* **304**, 611 (1983).
57. A. T. Winfree and S. H. Strogatz, *Physica* **13D**, 221 (1983).
58. Z. Noszticzius, W. Horsthemke, W. D. McCormick, H. L. Swinney, W. Y. Tam, *Nature (London)* **329**, 619 (1987).
59. J. D. Dockery, J. P. Keener, J. J. Tyson, in preparation.
60. J. J. Tyson and P. Fife, *J. Chem. Phys.* **73**, 2224 (1980).
61. J. J. Tyson, *ibid.*, in press.
62. D. Walgraef, G. Dewel, P. Borckmans, *ibid.* **78**, 3043 (1983).
63. V. S. Zykov and G. L. Morozova, *Biofizika* **23**, 717 (1979).
64. P. Foerster, S. C. Müller, B. Hess, in *Spatial Inhomogeneities and Transient Behavior in Chemical Kinetics*, G. Nicolis and P. Gray, Eds. (Springer-Verlag, Berlin, in press).
65. This work was supported in part by the National Science Foundation and the Air Force Office of Scientific Research.

More Is Different

Broken symmetry and the nature of the hierarchical structure of science.

P. W. Anderson

The reductionist hypothesis may still be a topic for controversy among philosophers, but among the great majority of active scientists I think it is accepted without question. The workings of our minds and bodies, and of all the animate or inanimate matter of which we have any detailed knowledge, are assumed to be controlled by the same set of fundamental laws, which except under certain extreme conditions we feel we know pretty well.

It seems inevitable to go on uncritically to what appears at first sight to be an obvious corollary of reductionism: that if everything obeys the same fundamental laws, then the only scientists who are studying anything really fundamental are those who are working on those laws. In practice, that amounts to some astrophysicists, some elementary particle physicists, some logicians and other mathematicians, and few others. This point of view, which it is the main purpose of this article to oppose, is expressed in a rather well-known passage by Weisskopf (1):

Looking at the development of science in the Twentieth Century one can distinguish two trends, which I will call "intensive" and "extensive" research, lacking a better terminology. In short: intensive research goes for the fundamental laws, extensive research goes for the explanation of phenomena in terms of known fundamental laws. As always, distinctions of this kind are not unambiguous, but they are clear in most cases. Solid state physics, plasma physics, and perhaps also biology are extensive. High energy physics and a good part of nuclear physics are intensive. There is always much less intensive research going on than extensive. Once new fundamental laws are discovered, a large and ever increasing activity begins in order to apply the discoveries to hitherto unexplained phenomena. Thus, there are two dimensions to basic research. The frontier of science extends all along a long line from the newest and most modern intensive research, over the extensive research recently spawned by the intensive research of yesterday, to the broad and well developed web of extensive research activities based on intensive research of past decades.

The effectiveness of this message may be indicated by the fact that I heard it quoted recently by a leader in the field of materials science, who urged the participants at a meeting dedicated to "fundamental problems in condensed matter physics" to accept that there were few or no such problems and that nothing was left but extensive science, which he seemed to equate with device engineering.

The main fallacy in this kind of thinking is that the reductionist hypothesis does not by any means imply a "constructionist" one: The ability to reduce everything to simple fundamental laws does not imply the ability to start from those laws and reconstruct the universe. In fact, the more the elementary particle physicists tell us about the nature of the fundamental laws, the less relevance they seem to have to the very real problems of the rest of science, much less to those of society.

The constructionist hypothesis breaks down when confronted with the twin difficulties of scale and complexity. The behavior of large and complex aggregates of elementary particles, it turns out, is not to be understood in terms of a simple extrapolation of the properties of a few particles. Instead, at each level of complexity entirely new properties appear, and the understanding of the new behaviors requires research which I think is as fundamental in its nature as any other. That is, it seems to me that one may array the sciences roughly linearly in a hierarchy, according to the idea: The elementary entities of science X obey the laws of science Y.

X	Y
solid state or many-body physics	elementary particle physics
chemistry	many-body physics
molecular biology	chemistry
cell biology	molecular biology
.	.
.	.
.	.
psychology	physiology
social sciences	psychology

But this hierarchy does not imply that science X is "just applied Y." At each stage entirely new laws, concepts, and generalizations are necessary, requiring inspiration and creativity to just as great a degree as in the previous one. Psychology is not applied biology, nor is biology applied chemistry.

In my own field of many-body physics, we are, perhaps, closer to our fundamental, intensive underpinnings than in any other science in which nontrivial complexities occur, and as a result we have begun to formulate a general theory of just how this shift from quantitative to qualitative differentiation takes place. This formulation, called the theory of "broken symmetry," may be of help in making more generally clear the breakdown of the constructionist converse of reductionism. I will give an elementary and incomplete explanation of these ideas, and then go on to some more general speculative comments about analogies at

---

The author is a member of the technical staff of the Bell Telephone Laboratories, Murray Hill, New Jersey 07974, and visiting professor of theoretical physics at Cavendish Laboratory, Cambridge, England. This article is an expanded version of a Regents' Lecture given in 1967 at the University of California, La Jolla.

other levels and about similar phenomena.

Before beginning this I wish to sort out two possible sources of misunderstanding. First, when I speak of scale change causing fundamental change I do not mean the rather well-understood idea that phenomena at a new scale may obey actually different fundamental laws—as, for example, general relativity is required on the cosmological scale and quantum mechanics on the atomic. I think it will be accepted that all ordinary matter obeys simple electrodynamics and quantum theory, and that really covers most of what I shall discuss. (As I said, we must all start with reductionism, which I fully accept.) A second source of confusion may be the fact that the concept of broken symmetry has been borrowed by the elementary particle physicists, but their use of the term is strictly an analogy, whether a deep or a specious one remaining to be understood.

Let me then start my discussion with an example on the simplest possible level, a natural one for me because I worked with it when I was a graduate student: the ammonia molecule. At that time everyone knew about ammonia and used it to calibrate his theory or his apparatus, and I was no exception. The chemists will tell you that ammonia "is" a triangular pyramid

with the nitrogen negatively charged and the hydrogens positively charged, so that it has an electric dipole moment (μ), negative toward the apex of the pyramid. Now this seemed very strange to me, because I was just being taught that nothing has an electric dipole moment. The professor was really proving that no nucleus has a dipole moment, because he was teaching nuclear physics, but as his arguments were based on the symmetry of space and time they should have been correct in general.

I soon learned that, in fact, they were correct (or perhaps it would be more accurate to say not incorrect) because he had been careful to say that no stationary state of a system (that is, one which does not change in time) has an electric dipole moment. If ammonia starts out from the above unsymmetrical state, it will not stay in it very long. By means of quantum mechanical tunneling, the nitrogen can leak through the triangle of hydrogens to the other side, turning the pyramid inside out, and, in fact, it can do so very rapidly. This is the so-called "inversion," which occurs at a frequency of about 3×10^{10} per second. A truly stationary state can only be an equal superposition of the unsymmetrical pyramid and its inverse. That mixture does not have a dipole moment. (I warn the reader again that I am greatly oversimplifying and refer him to the textbooks for details.)

I will not go through the proof, but the result is that the state of the system, if it is to be stationary, must always have the same symmetry as the laws of motion which govern it. A reason may be put very simply: In quantum mechanics there is always a way, unless symmetry forbids, to get from one state to another. Thus, if we start from any one unsymmetrical state, the system will make transitions to others, so only by adding up all the possible unsymmetrical states in a symmetrical way can we get a stationary state. The symmetry involved in the case of ammonia is parity, the equivalence of left- and right-handed ways of looking at things. (The elementary particle experimentalists' discovery of certain violations of parity is not relevant to this question; those effects are too weak to affect ordinary matter.)

Having seen how the ammonia molecule satisfies our theorem that there is no dipole moment, we may look into other cases and, in particular, study progressively bigger systems to see whether the state and the symmetry are always related. There are other similar pyramidal molecules, made of heavier atoms. Hydrogen phosphide, PH_3, which is twice as heavy as ammonia, inverts, but at one-tenth the ammonia frequency. Phosphorus trifluoride, PF_3, in which the much heavier fluorine is substituted for hydrogen, is not observed to invert at a measurable rate, although theoretically one can be sure that a state prepared in one orientation would invert in a reasonable time.

We may then go on to more complicated molecules, such as sugar, with about 40 atoms. For these it no longer makes any sense to expect the molecule to invert itself. Every sugar molecule made by a living organism is spiral in the same sense, and they never invert, either by quantum mechanical tunneling or even under thermal agitation at normal temperatures. At this point we must forget about the possibility of inversion and ignore the parity symmetry: the symmetry laws have been, not repealed, but broken.

If, on the other hand, we synthesize our sugar molecules by a chemical reaction more or less in thermal equilibrium, we will find that there are not, on the average, more left- than right-handed ones or vice versa. In the absence of anything more complicated than a collection of free molecules, the symmetry laws are never broken, on the average. We needed living matter to produce an actual unsymmetry in the populations.

In really large, but still inanimate, aggregates of atoms, quite a different kind of broken symmetry can occur, again leading to a net dipole moment or to a net optical rotating power, or both. Many crystals have a net dipole moment in each elementary unit cell (pyroelectricity), and in some this moment can be reversed by an electric field (ferroelectricity). This asymmetry is a spontaneous effect of the crystal's seeking its lowest energy state. Of course, the state with the opposite moment also exists and has, by symmetry, just the same energy, but the system is so large that no thermal or quantum mechanical force can cause a conversion of one to the other in a finite time compared to, say, the age of the universe.

There are at least three inferences to be drawn from this. One is that symmetry is of great importance in physics. By symmetry we mean the existence of different viewpoints from which the system appears the same. It is only slightly overstating the case to say that physics is the study of symmetry. The first demonstration of the power of this idea may have been by Newton, who may have asked himself the question: What if the matter here in my hand obeys the same laws as that up in the sky— that is, what if space and matter are homogeneous and isotropic?

The second inference is that the internal structure of a piece of matter need not be symmetrical even if the total state of it is. I would challenge you to start from the fundamental laws of quantum mechanics and predict the ammonia inversion and its easily observable properties without going through the stage of using the unsymmetrical pyramidal structure, even though no "state" ever has that structure. It is fascinating that it was not until a couple of decades ago (2) that nuclear physicists stopped thinking of the nucleus as a featureless, symmetrical little ball and realized that while it really never has a dipole moment, it can become football-shaped or plate-shaped. This has observable consequences in the reactions and excitation spectra that are studied in nuclear physics, even though it is much more difficult to demonstrate directly than the ammonia inversion. In my opinion, whether or not one calls this intensive research, it is as fundamental in nature as many things one might so label. But it needed no new knowledge of fundamental laws and would have been extremely difficult to derive synthetically from those laws; it was simply an inspiration, based, to be sure, on everyday intuition, which suddenly fitted everything together.

The basic reason why this result would have been difficult to derive is an important one for our further thinking. If the nucleus is sufficiently small there is no real way to define its shape rigorously: Three or four or ten particles whirling about each other do not define a rotating "plate" or "football." It is only as the nucleus is considered to be a many-body system—in what is often called the $N \to \infty$ limit—that such behavior is rigorously definable. We say to ourselves: A macroscopic body of that shape would have such-and-such a spectrum of rotational and vibrational excitations, completely different in nature from those which would characterize a featureless system. When we see such a spectrum, even not so separated, and somewhat imperfect, we recognize that the nucleus is, after all, not macroscopic; it is merely approaching macroscopic behavior. Starting with the fundamental laws and a computer, we would have to do two impossible things —solve a problem with infinitely many bodies, and then apply the result to a finite system—before we synthesized this behavior.

A third insight is that the state of a really big system does not at all have to have the symmetry of the laws which govern it; in fact, it usually has less symmetry. The outstanding example of this is the crystal: Built from a substrate of atoms and space according to laws which express the perfect homogeneity of space, the crystal suddenly and unpredictably displays an entirely new and very beautiful symmetry. The general rule, however, even in the case of the crystal, is that the large system is less symmetrical than the underlying structure would suggest: Symmetrical as it is, a crystal is less symmetrical than perfect homogeneity.

Perhaps in the case of crystals this appears to be merely an exercise in confusion. The regularity of crystals could be deduced semiempirically in the mid-19th century without any complicated reasoning at all. But sometimes, as in the case of superconductivity, the new symmetry—now called broken symmetry because the original symmetry is no longer evident—may be of an entirely unexpected kind and extremely difficult to visualize. In the case of superconductivity, 30 years elapsed between the time when physicists were in possession of every fundamental law necessary for explaining it and the time when it was actually done.

The phenomenon of superconductivity is the most spectacular example of the broken symmetries which ordinary macroscopic bodies undergo, but it is of course not the only one. Antiferromagnets, ferroelectrics, liquid crystals, and matter in many other states obey a certain rather general scheme of rules and ideas, which some many-body theorists refer to under the general heading of broken symmetry. I shall not further discuss the history, but give a bibliography at the end of this article (3).

The essential idea is that in the so-called $N \to \infty$ limit of large systems (on our own, macroscopic scale) it is not only convenient but essential to realize that matter will undergo mathematically sharp, singular "phase transitions" to states in which the microscopic symmetries, and even the microscopic equations of motion, are in a sense violated. The symmetry leaves behind as its expression only certain characteristic behaviors, for instance, long-wavelength vibrations, of which the familiar example is sound waves; or the unusual macroscopic conduction phenomena of the superconductor; or, in a very deep analogy, the very rigidity of crystal lattices, and thus of most solid matter. There is, of course, no question of the system's really violating, as opposed to breaking, the symmetry of space and time, but because its parts find it energetically more favorable to maintain certain fixed relationships with each other, the symmetry allows only the body as a whole to respond to external forces.

This leads to a "rigidity," which is also an apt description of superconductivity and superfluidity in spite of their apparent "fluid" behavior. [In the former case, London noted this aspect very early (4).] Actually, for a hypothetical gaseous but intelligent citizen of Jupiter or of a hydrogen cloud somewhere in the galactic center, the properties of ordinary crystals might well be a more baffling and intriguing puzzle than those of superfluid helium.

I do not mean to give the impression that all is settled. For instance, I think there are still fascinating questions of principle about glasses and other amorphous phases, which may reveal even more complex types of behavior. Nevertheless, the role of this type of broken symmetry in the properties of inert but macroscopic material bodies is now understood, at least in principle. In this case we can see how the whole becomes not only more than but very different from the sum of its parts.

The next order of business logically is to ask whether an even more complete destruction of the fundamental symmetries of space and time is possible and whether new phenomena then arise, intrinsically different from the "simple" phase transition representing a condensation into a less symmetric state.

We have already excluded the apparently unsymmetric cases of liquids, gases, and glasses. (In any real sense they are more symmetric.) It seems to me that the next stage is to consider the system which is regular but contains information. That is, it is regular in space in some sense so that it can be "read out," but it contains elements which can be varied from one "cell" to the next. An obvious example is DNA; in everyday life, a line of type or a movie film have the same structure. This type of "information-bearing crystallinity" seems to be essential to life. Whether the development of life requires any further breaking of symmetry is by no means clear.

Keeping on with the attempt to characterize types of broken symmetry which occur in living things, I find that at least one further phenomenon seems to be identifiable and either universal or remarkably common, namely, ordering (regularity or periodicity) in the time dimension. A number of theories of life processes have appeared in which regular pulsing in time plays an important role: theories of development, of growth and growth limitation, and of the memory. Temporal regularity is very commonly observed in living objects. It plays at least two kinds of roles. First, most methods of extracting energy from the environment in order to set up a continuing, quasi-stable process involve time-periodic machines, such as oscillators and generators, and the processes of life work in the same way. Second, temporal regularity is a means of handling information, similar to information-bearing spatial regularity. Human spoken language is an example, and it is noteworthy that all computing machines use temporal pulsing. A possible third role is suggested in some of the theories mentioned above: the use of phase relationships of temporal pulses to handle information and control the growth and development of cells and organisms (5).

In some sense, structure—functional structure in a teleological sense, as opposed to mere crystalline shape—must also be considered a stage, possibly intermediate between crystallinity and information strings, in the hierarchy of broken symmetries.

To pile speculation on speculation, I would say that the next stage could be hierarchy or specialization of function, or both. At some point we have to stop talking about decreasing symmetry and start calling it increasing complication. Thus, with increasing complication at each stage, we go on up the hierarchy of the sciences. We expect to encounter fascinating and, I believe, very fundamental questions at each stage in fitting together less complicated pieces into the more complicated system and understanding the basically new types of behavior which can result.

There may well be no useful parallel to be drawn between the way in which complexity appears in the simplest cases of many-body theory and chemistry and the way it appears in the truly complex cultural and biological ones, except perhaps to say that, in general, the relationship between the system and its parts is intellectually a one-way street. Synthesis is expected to be all but impossible; analysis, on the other hand, may be not only possible but fruitful in all kinds of ways: Without an understanding of the broken symmetry in superconductivity, for instance, Josephson would probably not have discovered his effect. [Another name for the Josephson effect is "macroscopic quantum-interference phenomena": interference effects observed between macroscopic wave functions of electrons in superconductors, or of helium atoms in superfluid liquid helium. These phenomena have already enormously extended the accuracy of electromagnetic measurements, and can be expected to play a great role in future computers, among other possibilities, so that in the long run they may lead to some of the major technological achievements of this decade (6).] For another example, biology has certainly taken on a whole new aspect from the reduction of genetics to biochemistry and biophysics, which will have untold consequences. So it is not true, as a recent article would have it (7), that we each should "cultivate our own valley, and not attempt to build roads over the mountain ranges . . . between the sciences." Rather, we should recognize that such roads, while often the quickest shortcut to another part of our own science, are not visible from the viewpoint of one science alone.

The arrogance of the particle physicist and his intensive research may be behind us (the discoverer of the positron said "the rest is chemistry"), but we have yet to recover from that of some molecular biologists, who seem determined to try to reduce everything about the human organism to "only" chemistry, from the common cold and all mental disease to the religious instinct. Surely there are more levels of organization between human ethology and DNA than there are between DNA and quantum electrodynamics, and each level can require a whole new conceptual structure.

In closing, I offer two examples from economics of what I hope to have said. Marx said that quantitative differences become qualitative ones, but a dialogue in Paris in the 1920's sums it up even more clearly:

FITZGERALD: The rich are different from us.

HEMINGWAY: Yes, they have more money.

References

1. V. F. Weisskopf, in *Brookhaven Nat. Lab. Publ. 888T360* (1965). Also see *Nuovo Cimento Suppl. Ser 1* 4, 465 (1966); *Phys. Today* 20 (No. 5), 23 (1967).
2. A. Bohr and B. R. Mottelson, *Kgl. Dan. Vidensk. Selsk. Mat. Fys. Medd.* 27, 16 (1953).
3. Broken symmetry and phase transitions: L. D. Landau, *Phys. Z. Sowjetunion* 11, 26, 542 (1937). Broken symmetry and collective motion, general: J. Goldstone, A. Salam, S. Weinberg, *Phys. Rev.* 127, 965 (1962); P. W. Anderson, *Concepts in Solids* (Benjamin, New York, 1963), pp. 175–182; B. D. Josephson, thesis, Trinity College, Cambridge University (1962). Special cases: antiferromagnetism, P. W. Anderson, *Phys. Rev.* 86, 694 (1952); superconductivity, ———, *ibid.* 110, 827 (1958); *ibid.* 112, 1900 (1958); Y. Nambu, *ibid.* 117, 648 (1960).
4. F. London, *Superfluids* (Wiley, New York, 1950), vol. 1.
5. M. H. Cohen, *J. Theor. Biol.* 31, 101 (1971).
6. J. Clarke, *Amer. J. Phys.* 38, 1075 (1969); P. W. Anderson, *Phys. Today* 23 (No. 11), 23 (1970).
7. A. B. Pippard, *Reconciling Physics with Reality* (Cambridge Univ. Press, London, 1972).

Cellular automata as models of complexity

Stephen Wolfram

The Institute for Advanced Study, Princeton, New Jersey 08510, USA

Natural systems from snowflakes to mollusc shells show a great diversity of complex patterns. The origins of such complexity can be investigated through mathematical models termed 'cellular automata'. Cellular automata consist of many identical components, each simple, but together capable of complex behaviour. They are analysed both as discrete dynamical systems, and as information-processing systems. Here some of their universal features are discussed, and some general principles are suggested.

IT is common in nature to find systems whose overall behaviour is extremely complex, yet whose fundamental component parts are each very simple. The complexity is generated by the cooperative effect of many simple identical components. Much has been discovered about the nature of the components in physical and biological systems; little is known about the mechanisms by which these components act together to give the overall complexity observed. What is now needed is a general mathematical theory to describe the nature and generation of complexity.

Cellular automata are examples of mathematical systems constructed from many identical components, each simple, but together capable of complex behaviour. From their analysis, one may, on the one hand, develop specific models for particular systems, and, on the other hand, hope to abstract general principles applicable to a wide variety of complex systems. Some recent results on cellular automata will now be outlined; more extensive accounts and references may be found in refs 1–4.

Cellular automata

A one-dimensional cellular automaton consists of a line of sites, with each site carrying a value 0 or 1 (or in general $0, \ldots, k-1$). The value a_i of the site at each position i is updated in discrete time steps according to an identical deterministic rule depending on a neighbourhood of sites around it:

$$a_i^{(t+1)} = \phi[a_{i-r}^{(t)}, a_{i-r+1}^{(t)}, \ldots, a_{i+r}^{(t)}] \quad (1)$$

Even with $k=2$ and $r=1$ or 2, the overall behaviour of cellular automata constructed in this simple way can be extremely complex.

Consider first the patterns generated by cellular automata evolving from simple 'seeds' consisting of a few non-zero sites. Some local rules ϕ give rise to simple behaviour; others produce complicated patterns. An extensive empirical study suggests that the patterns take on four qualitative forms, illustrated in Fig. 1:

(1) disappears with time;
(2) evolves to a fixed finite size;
(3) grows indefinitely at a fixed speed;
(4) grows and contracts irregularly.

Patterns of type 3 are often found to be self-similar or scale invariant. Parts of such patterns, when magnified, are indistinguishable from the whole. The patterns are characterized by a fractal dimension[5]; the value $\log_2 3 \simeq 1.59$ is the most common. Many of the self-similar patterns seen in natural systems may, in fact, be generated by cellular automaton evolution.

Figure 3 shows the evolution of cellular automata from initial states where each site is assigned each of its k possible values with an independent equal probability. Self-organization is seen: ordered structure is generated from these disordered initial states, and in some cases considerable complexity is evident.

Different initial states with a particular cellular automaton rule yield patterns that differ in detail, but are similar in form and statistical properties. Different cellular automaton rules yield very different patterns. An empirical study, nevertheless, suggests that four qualitative classes may be identified, yielding four characteristic limiting forms:

(1) spatially homogeneous state;
(2) sequence of simple stable or periodic structures;
(3) chaotic aperiodic behaviour;
(4) complicated localized structures, some propagating.

All cellular automata within each class, regardless of the details of their construction and evolution rules, exhibit qualitatively similar behaviour. Such universality should make general results on these classes applicable to a wide variety of systems modelled by cellular automata.

Applications

Current mathematical models of natural systems are usually based on differential equations which describe the smooth variation of one parameter as a function of a few others. Cellular automata provide alternative and in some respects complemen-

Fig. 1 Classes of patterns generated by the evolution of cellular automata from simple 'seeds'. Successive rows correspond to successive time steps in the cellular automaton evolution. Each site is updated at each time step according to equation (1) by cellular automaton rules that depend on the values of a neighbourhood of sites at the previous time step. Sites with values 0 and 1 are represented by white and black squares, respectively. Despite the simplicity of their construction, patterns of some complexity are seen to be generated. The rules shown exemplify the four classes of behaviour found. (The first three are $k=2$, $r=1$ rules with rule numbers[1] 128, 4 and 126, respectively; the fourth is a $k=2$, $r=2$ rule with totalistic code[2] 52.) In the third case, a self similar pattern is formed.

Fig. 2 Evolution of small initial perturbations in cellular automata, as shown by the difference (modulo two) between patterns generated from two disordered initial states differing in the value of a single site. The examples shown illustrate the four classes of behaviour found. Information on changes in the initial state almost always propagates only a finite distance in the first two classes, but may propagate an arbitrary distance in the third and fourth classes.

tary models, describing the discrete evolution of many (identical) components. Models based on cellular automata are typically most appropriate in highly nonlinear regimes of physical systems, and in chemical and biological systems where discrete thresholds occur. Cellular automata are particularly suitable as models when growth inhibition effects are important.

As one example, cellular automata provide global models for the growth of dendritic crystals (such as snowflakes)[6]. Starting from a simple seed, sites with values representing the solid phase are aggregated according to a two-dimensional rule that accounts for the inhibition of growth near newly-aggregated sites, resulting in a fractal pattern of growth. Nonlinear chemical reaction–diffusion systems give another example[7,8]: a simple cellular automaton rule with growth inhibition captures the essential features of the usual partial differential equations, and reproduces the spatial patterns seen. Turbulent fluids may also potentially be modelled as cellular automata with local interactions between discrete vortices on lattice sites.

If probabilistic noise is added to the time evolution rule (1), then cellular automata may be identified as generalized Ising models[9,10]. Phase transitions may occur if ϕ retains some deterministic components, or in more than one dimension.

Cellular automata may serve as suitable models for a wide variety of biological systems. In particular, they may suggest mechanisms for biological pattern formation. For example, the patterns of pigmentation found on many mollusc shells bear a striking resemblance to patterns generated by class 2 and 3 cellular automata (see refs 11, 12), and cellular automaton models for the growth of some pigmentation patterns have been constructed[13].

Mathematical approaches

Rather than describing specific applications of cellular automata, this article concentrates on general mathematical features of their behaviour. Two complementary approaches provide characterizations of the four classes of behaviour seen in Fig. 3.

In the first approach[2], cellular automata are viewed as discrete dynamical systems (see ref. 14), or discrete idealizations of partial differential equations. The set of possible (infinite) configurations of a cellular automaton forms a Cantor set; cellular automaton evolution may be viewed as a continuous mapping on this Cantor set. Quantities such as entropies, dimensions and Lyapunov exponents may then be considered for cellular automata.

In the second approach[3], cellular automata are instead considered as information-processing systems (see ref. 15), or parallel-processing computers of simple construction. Information represented by the initial configuration is processed by the evolution of the cellular automaton. The results of this information processing may then be characterized in terms of the types of formal languages generated. (Note that the mechanisms for information processing in natural system appear to be much closer to those in cellular automata than in conventional serial-processing computers: cellular automata may, therefore, provide efficient media for practical simulations of many natural systems.)

Entropies and dimensions

Most cellular automaton rules have the important feature of irreversibility: several different configurations may evolve to a single configuration, and, with time, a contracting subset of all possible configurations appears. Starting from all possible initial configurations, the cellular automaton evolution may generate only special 'organized' configurations, and 'self-organization' may occur.

For class 1 cellular automata, essentially all initial configurations evolve to a single final configuration, analogous to a limit point in a continuous dynamical system. Class 2 cellular automata evolve to limit sets containing essentially only periodic configurations, analogous to limit cycles. Class 3 cellular automata yield chaotic aperiodic limit sets, containing analogues of chaotic or 'strange' attractors.

Entropies and dimensions give a generalized measure of the density of the configurations generated by cellular automaton evolution. The (set) dimension or limiting (topological) entropy for a set of cellular automaton configurations is defined as (compare ref. 14)

$$d^{(x)} = \lim_{X \to \infty} \frac{1}{X} \log_k N(X) \quad (2)$$

where $N(X)$ gives the number of distinct sequences of X site values that appear. For the set of possible initial configurations, $d^{(x)} = 1$. For a limit set containing only a finite total number of configurations, $d^{(x)} = 0$. For most class 3 cellular automata, $d^{(x)}$ decreases with time, giving $0 < d^{(x)} < 1$, and suggesting that a fractal subset of all possible configurations occurs.

A dimension or limiting entropy $d^{(t)}$ corresponding to the time series of values of a single site may be defined in analogy with equation (2). (The analogue of equation (2) for a sufficiently wide patch of sites yields a topologically-invariant entropy for the cellular automaton mapping.) $d^{(t)} = 0$ for periodic sets of configurations.

$d^{(x)}$ and $d^{(t)}$ may be modified to account for the probabilities of configurations by defining

$$d^{(x)}_\mu = -\lim_{X \to \infty} \frac{1}{X} \sum_{j=1}^{k^X} p_j \log_k p_j \quad (3)$$

and its analogue, where p_j are probabilities for possible length X sequences. These measure dimensions may be used to delineate the large time behaviour of the different classes of cellular automata:

(1) $d^{(x)}_\mu = d^{(t)}_\mu = 0$
(2) $d^{(x)}_\mu > 0$, $d^{(t)}_\mu = 0$
(3) $d^{(x)}_\mu > 0$, $d^{(t)}_\mu > 0$

As discussed below, dimensions are usually undefined for class 4 cellular automata.

Information propagation

Cellular automata may also be characterized by the stability or predictability of their behaviour under small perturbations in initial configurations. Figure 2 shows differences in patterns generated by cellular automata resulting from a change in a

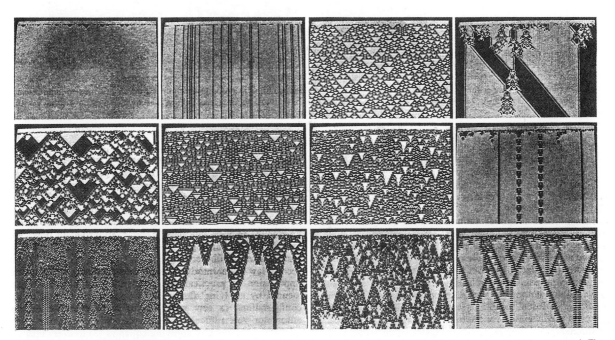

Fig. 3 Evolution of various cellular automata from disordered initial states. In many cases, ordered structure is seen to be generated. The first row of pictures show examples of the four qualitative classes of behaviour found. (The rules shown are the same as in Fig. 1.) The lower two rows show examples of cellular automata with $k = 5$ (five possible values for each site) and $r = 1$ (nearest neighbour rules). Site values 0 to 4 are represented by white, red, green, blue and yellow squares, respectively. (The rules shown have totalistic codes 10175, 566780, 570090, 580020, 583330, 672900, 5694390, 59123000.) The 'orange' discoloration is a background, not part of the pattern.

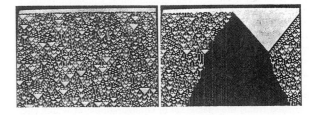

Fig. 4 Evolution of multiple phases in cellular automata. Pairs of sites are shown combined: 00 is represented by white, 01 by red, 10 by green and 11 by blue. Alternate time steps are shown. Both rules simulate an additive rule (number 90) under a blocking transformation. In the first rule (number 18), the simulation is attractive: starting from a disordered initial state, the domains grow with time. In the second rule (number 94), the simulation is repulsive: only evolution from a special initial state yields additive rule behaviour; a defect is seen to grow, and attractive simulation of the identity rule takes over.

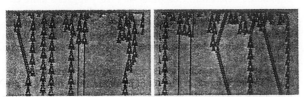

Fig. 5 Examples of the evolution of a typical class 4 cellular automaton from disordered initial states. This and other class 4 cellular automata are conjectured to be capable of arbitrary information processing, or universal computation. The rule shown has $k = 3$, $r = 1$, and takes the value of a site to be 1 if the sum of the values of the sites in its three-site neighbourhood is 2 or 6, to be 2 if the sum is 3, and to zero otherwise (totalistic code 792).

Fig. 6 Persistent structures generated in the evolution of the class 4 cellular automaton of Fig. 5. The first four structures shown have periods 1, 20, 16 and 12 respectively; the last four structures (and their reflections) propagate: the first has period 32, the next three period 3, and the last period 6. These structures are some of the elements required to support universal computation.

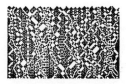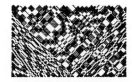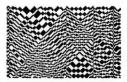

Fig. 7 Evolution of some cellular automata with reversible rules. Each configuration is a unique function of the two previous configurations. (Rule numbers 4, 22, 90 and 126 are shown.) As initial conditions, each site in two successive configurations is chosen to have value 1 with probability 0.1.

single initial site value. Such perturbations have characteristic effects on the four classes of cellular automata:

(1) no change in final state;
(2) changes only in a finite region;
(3) changes over an ever-increasing region;
(4) irregular changes.

In class 1 and 2 cellular automata, information associated with site values in the initial state propagates only a finite distance; in class 3 cellular automata, it propagates an infinite distance at a fixed speed, while in class 4 cellular automata, it propagates irregularly, but over an infinite range. The speed of information propagation is related to the Lyapunov exponent for the cellular automaton evolution, and measures the degree of sensitivity to initial conditions (see ref. 16). It leads to different degrees of predictability for the outcome of cellular automaton evolution:

(1) entirely predictable, independent of initial state;
(2) local behaviour predictable from local initial state;
(3) behaviour depends on an ever-increasing initial region;
(4) behaviour effectively unpredictable.

Information propagation is particularly simple for the special class of additive cellular automata (whose local rule function ϕ is linear modulo k), in which patterns generated from arbitrary initial states may be obtained by superposition of patterns generated by evolution of simple initial states containing a single non-zero site. A rather complete algebraic analysis of such cellular automata may be given[17]. Most cellular automata are not additive; however, with special initial configurations it is often possible for them to behave just like additive rules. Thus, for example, the evolution of an initial configuration consisting of a sequence of 00 and 01 digrams under one rule may be identical to the evolution of the corresponding 'blocked' configuration consisting of 0 and 1 under another rule. In this way, one rule may simulate another under a blocking transformation (analogous to a renormalization group transformation). Evolution from an arbitrary initial state may be attracted to (or repelled from) the special set of configurations for which such a simulation occurs. Often several phases exist, corresponding to different blocking transformations: sometimes phase boundaries move at constant speed, and one phase rapidly takes over; in other cases, phase boundaries execute random walks, annihilating in pairs, and leading to a slow increase in the average domain size, as illustrated in Fig. 4. Many rules appear to follow attractive simulation paths to additive rules, which correspond to fixed points of blocking transformations, and thus exhibit self similarity. The behaviour of many rules at large times, and on large spatial scales, is therefore determined by the behaviour of additive rules.

Thermodynamics

Decreases with time in the spatial entropies and dimensions of equations (2) and (3) signal irreversibility in cellular automaton evolution. Some cellular automaton rules are, however, reversible, so that each and every configuration has a unique predecessor in the evolution, and the spatial entropy and dimension of equations (2) and (3) remain constant with time. Figure 7 shows some examples of the evolution of such rules, constructed by adding a term $-a_i^{(t-1)}$ to equation (1) (ref. 20 and E. Fredkin, personal communication). Again, there are analogues of the four classes of behaviour seen in Fig. 3, distinguished by the range and speed of information propagation.

Conventional thermodynamics gives a general description of systems whose microscopic evolution is reversible; it may, therefore, be applied to reversible cellular automata such as those of Fig. 4. As usual, the 'fine-grained' entropy for sets (ensembles) of configurations, computed as in equation (3) with perfect knowledge of each site value, remains constant in time. The 'coarse-grained' entropy for configurations is, nevertheless, almost always non-decreasing with time, as required by the second law of thermodynamics. Coarse graining emulates the imprecision of practical measurements, and may be implemented by applying almost any contractive mapping to the configurations (a few iterations of an irreversible cellular automaton rule suffice). For example, coarse-grained entropy might be computed by applying equation (3) to every fifth site value. In an ensemble with low coarse-grained entropy, the values of every fifth site would be highly constrained, but arbitrary values for the intervening sites would be allowed. Then in the evolution of a class 3 or 4 cellular automaton the disorder of the intervening site values would 'mix' with the fifth-site values, and the coarse-grained entropy would tend towards its maximum value. Signs of self-organization in such systems must be sought in temporal correlations, often manifest in 'fluctuations' or metastable 'pockets' of order.

While all fundamental physical laws appear to be reversible, macroscopic systems often behave irreversibly, and are appropriately described by irreversible laws. Thus, for example, although the microscopic molecular dynamics of fluids is reversible, the relevant macroscopic velocity field obeys the irreversible Navier-Stokes equations. Conventional thermodynamics does not apply to such intrinsically irreversible systems: new general principles must be found. Thus, for cellular automata with irreversible evolution rules, coarse-grained entropy typically increases for a short time, but then decreases to follow the fine-grained entropy. Measures of the structure generated by self-organization in the large time limit are usually affected very little by coarse graining.

Formal language theory

Quantities such as entropy and dimension, suggested by information theory, give only rough characterizations of cellular automaton behaviour. Computation theory suggests more complete descriptions of self-organization in cellular automata (and other systems). Sets of cellular automaton configurations may be viewed as formal languages, consisting of sequences of symbols (site values) forming words according to definite grammatical rules.

The set of all possible initial configurations corresponds to a trivial formal language. The set of configurations obtained after any finite number of time steps are found to form a regular language[3]. The words in a regular language correspond to the possible paths through a finite graph representing a finite state machine. It can be shown that a unique smallest finite graph reproduces any given regular language (see ref. 15). Examples of such graphs are shown in Fig. 8. These graphs give complete specifications for sets of cellular automaton configurations (ignoring probabilities). The number of nodes Ξ in the smallest graph corresponding to a particular set of configurations may

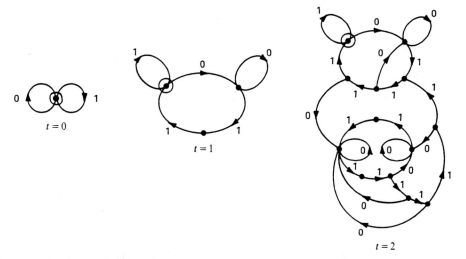

Fig. 8 Graphs representing the sets of configurations generated in the first few time steps of evolution according to a typical class 3 cellular automaton rule ($k=2$, $r=1$, rule number 126). Possible configurations correspond to possible paths through the graphs, beginning at the encircled node. At $t=0$, all possible configurations are allowed. With time, a contracting subset of configurations are generated. (After one time step, for example, no configuration containing the sequence of site value 101 can appear.) At each time step, the complete set of possible configurations forms a regular formal language: the graph gives a minimal complete specification of it. The number of nodes in the graph gives a measure of the complexity Ξ of the set, viewed as a regular language. As for other class 3 cellular automata, the complexity of the sets grows rapidly with time; for $t=3$, $\Xi = 107$, and $t=4$, $\Xi = 2,867$.

be defined as the 'regular language complexity' of the set. It specifies the size of the minimal description of the set in terms of regular languages. Larger Ξ correspond to more complicated sets. (Note that the topological entropy of a set is given by the logarithm of the algebraic integer obtained as the largest root of the characteristic polynomial for the incidence matrix of the corresponding graph. The characteristic polynomials for the graphs in Fig. 7 are $2-\lambda$ ($\lambda_{max}=2$), $1-\lambda+2\lambda^2-\lambda^3$ ($\lambda_{max} \simeq 1.755$) and $-1+\lambda-\lambda^2+2\lambda^3-4\lambda^4+\lambda^5+3\lambda^6-5\lambda^7+3\lambda^8-3\lambda^9+5\lambda^{10}-6\lambda^{11}+4\lambda^{12}-\lambda^{13}$ ($\lambda_{max} \simeq 1.732$), respectively.)

The regular language complexity Ξ for sets generated by cellular automaton evolution almost always seems to be non-decreasing with time. Increasing Ξ signals increasing self-organization. Ξ may thus represent a fundamental property of self-organizing systems, complementary to entropy. It may, in principle, be extracted from experimental data.

Cellular automata that exhibit only class 1 and 2 behaviour always appear to yields sets that correspond to regular languages in the large time limit. Class 3 and 4 behaviour typically gives rise, however, to a rapid increase of Ξ with time, presumably leading to limiting sets not described by regular languages.

Formal languages are recognized or generated by idealized computers with a 'central processing unit' containing a fixed finite number of internal states, together with a 'memory'. Four types of formal languages are conventionally identified, corresponding to four types of computer:

- Regular languages: no memory required.
- Context-free languages: memory arranged as a last-in, first-out stack.
- Context-sensitive languages: memory as large as input word required.
- Unrestricted languages: arbitrarily large memory required (general Turing machine).

Examples are known of cellular automata whose limiting sets correspond to all four types of language (L. Hurd, in preparation). Arguments can be given that the limit sets for class 3 cellular automata typically form context-sensitive languages, while those for class 4 cellular automata correspond to unrestricted languages. (Note that while a minimal specification for any regular language may always be found, there is no finite procedure to obtain a minimal form for more complicated formal languages: no generalization of the regular language complexity Ξ may thus be given.)

Computation theory

While dynamical systems theory concepts suffice to define class 1, 2 and 3 cellular automata, computation theory is apparently required for class 4 cellular automata. Examples of the evolution of a typical class 4 cellular automaton are shown in Fig. 5. Varied and complicated behaviour, involving many different time scales is evident. Persistent structures are often generated; the smallest few are illustrated in Fig. 6, and are seen to allow both storage and transmission of information. It seems that the structures supported by this and other class 4 cellular automata rule may be combined to implement arbitrary information processing operations. Class 4 cellular automata would then be capable of universal computation: with particular initial states, their evolution could implement any finite algorithm. (Universal computation has been proved for a $k=18$, $r=1$ rule[22], and for two-dimensional cellular automata such as the 'Game of Life'[22,23].) A few per cent of cellular automaton rules with $k>2$ or $r>1$ are found to exhibit class 4 behaviour: all these would then, in fact, be capable of arbitrarily complicated behaviour. This capability precludes a smooth infinite size limit for entropy or other quantities: as the size of cellular automaton considered increases, more and more complicated phenomena may appear.

Cellular automaton evolution may be viewed as a computation. Effective preidiction of the outcome of cellular automaton evolution requires a short-cut that allows a more efficient computation than the evolution itself. For class 1 and 2 cellular automata, such short cuts are clearly possible: simple computations suffice to predict their complete future. The computational capabilities of class 3 and 4 cellular automata may, however, be sufficiently great that, in general, they allow no short-cuts. The only effective way to determine their evolution from a given initial state would then be by explicit observation or simulation: no finite formulae for their general behaviour could be given. (If class 4 cellular automata are indeed capable of universal computation, then the variety of their possible behaviour would preclude general prediction, and make explicit observation or simulation necessary.) Their infinite time limiting behaviour could then not, in general, be determined by any finite computational process, and many of their limiting properties would be formally undecidable. Thus, for example, the 'halting problem' of determining whether a class 4 cellular automaton with a given finite initial configuration ever evolves to the null configuration would be undecidable. An explicit simulation could determine only whether halting occurred before some fixed time, and not whether it occurred after an arbitrarily long time.

For class 4 cellular automata, the outcome of evolution from almost all initial configurations can probably be determined only by explicit simulation, while for class 3 cellular automata this is the case for only a small fraction of initial states. Nevertheless, this possibility suggests that the occurrence of particular site value sequences in the infinite time limit is in general undecidable. The large time limit of the entropy for class 3 and 4 cellular automata would then, in general, be non-computable: bounds on it could be given, but there could be no finite procedure to compute it to arbitrary precision. (This would be the case if the limit sets for class 3 and 4 cellular automata formed at least context-sensitive languages.)

While the occurrence of a particular length n site value sequence in the infinite time limit may be undecidable, its occurrence after any finite time t can, in principle, be determined by considering all length $n_0 = n + 2rt$ initial sequences that could evolve to it. For increasing n or t this procedure would, nevertheless, involve exponentially-growing computational resources, so that it would rapidly become computationally intractable. It seems likely that the identification of possible sequences generated by class 3 and 4 cellular automata is, in general, an NP-complete problem (see ref. 15). It can, therefore, presumably not be solved in any time polynomial in n or t, and essentially requires explicit simulation of all possibilities.

Undecidability and intractability are common in problems of mathematics and computation. They may well afflict all but the simplest cellular automata. One may speculate that they are widespread in natural systems, perhaps occurring almost whenever nonlinearity is present. No simple formulae for the behaviour of many natural systems could then be given; the consequences of their evolution could be found effectively only by direct simulation or observation.

I thank O. Martin, J. Milnor, N. Packard and many others for discussions. The computer mathematics system SMP[24] was used during this work. The work was supported in part by the US Office of Naval Research (contract No. N00014-80-C-0657).

1. Wolfram, S. *Rev. Mod. Phys.* **55**, 601–644 (1983).
2. Wolfram, S. *Physica* **10D**, 1–35 (1984).
3. Wolfram, S. *Commun. Math. Phys.* (in the press).
4. Wolfram, S. *Cellular Automata* (Los Alamos Science, Autumn, 1983).
5. Mandelbrot, B. *The Fractal Geometry of Nature* (Freeman, San Francisco, 1982).
6. Packard, N. Preprint, *Cellular Automaton Models for Dendritic Growth* (Institute for Advanced Study, 1984).
7. Madore, B. & Freedman, W. *Science* **222**, 615–616 (1983).
8. Greenberg, J. M., Hassard, B. D. & Hastings, S. P. *Bull. Amer. Math. Soc.* **84**, 1296–1327 (1978).
9. Vichniac, G. *Physica* **10D**, 96–116 (1984).
10. Domany, E. & Kinzel, W. *Phys. Rev. Lett.* **53**, 311–314 (1984).
11. Waddington C. H. & Cowe, R. J. *J. theor. Biol.* **25**, 219–225 (1969).
12. Lindsay, D. T. *Veliger* **24**, 297–299 (1977).
13. Young, D. A. *A Local Activator-Inhibitor Model of Vertebrate Skin Patterns* (Lawrence Livermore National Laboratory Rep., 1983).
14. Guckenheimer, J. & Holmes, P. *Nonlinear Oscillations, Dynamical Systems, and Bifurcations of Vector Fields* (Springer, Berlin, 1983).
15. Hopcroft, J. E. & Ullman, J. D. *Introduction to Automata Theory, Languages, and Computation* (Addison-Wesley, New York, 1979).
16. Packard, N. Preprint, *Complexity of Growing Patterns in Cellular Automata* (Institute for Advanced Study, 1983).
17. Martin, O, Odlyzko, A. & Wolfram, S. *Commun. Math. Phys.* **93**, 219–258 (1984).
18. Grassberger, P. *Physica* **10D**, 52–58 (1984).
19. Lind, D. *Physica* **10D**, 36–44 (1984).
20. Margolus, N. *Physica* **10D**, 81–95 (1984).
21. Smith, A. R. *Journal of the Association for Computing Machinery* **18**, 339–353 (1971).
22. Berlekamp, E. R., Conway, J. H. & Guy, R. K. *Winning Ways for your Mathematical Plays* Vol. 2, Ch. 25 (Academic, New York, 1982).
23. Gardner, M. *Wheels, Life and other Mathematical Amusements* (Freeman, San Francisco, 1983).
24. Wolfram, S. *SMP Reference Manual* (Computer Mathematics Group, Inference Corporation, Los Angeles, 1983).

COMPUTER SIMULATIONS

Catastrophes and Self-Organized Criticality

Per Bak

Catastrophes are usually thought of as exceptional events in otherwise well-behaved dynamical systems. The traditional view of the cause of a catastrophe is that a single external, cataclysmic force acting throughout the system is responsible, e.g., the fall of a meteor or a volcanic eruption leading to the extinction of dinosaurs, an unstable fault system causing large earthquakes, and program trading leading to a collapse of the stock market. However, for specific events, such as the 1930s Depression, it is often difficult to identify exogenous forces that affect many parts of the system in a similar manner.

An alternative view, developed by Bak, Tang, and Wiesenfeld,[1,2] is that large complex systems naturally evolve to a state where events of all sizes and all durations occur. The prototypical example is a sandpile, built by slowly dropping sand on a large surface with open edges. The sandpile will grow to a statistically stationary state where the amount of grains added is balanced by the amount of sand falling off the edge. This stationary state is an attractor for the dynamics. This attractor is very different from the attractor characterizing low-dimensional chaotic systems because the number of degrees of freedom is proportional to the size of the system. Dhar has calculated analytically the number of different configurations and the entropy belonging to the attractor.[3] Michael Creutz has recently described numerical simulations of cellular automata sandpile models.[4]

The response to the dropping of a single grain of sand can be thought of as a *critical* chain reaction. At each step the probability that the activity will branch is exactly compensated by the probability that the activity will die. For sandpiles shaped so that their slope is less than the critical value, the processes will be subcritical; for steeper piles the process will be supercritical, i.e., the addition of a single grain could lead to global collapse. The critical state is self-organized in the sense that no external tuning is needed to carry the system to this state. In contrast, the concentration of fissionable material has to be carefully controlled for a nuclear chain reaction to be exactly critical.

In a self-organized critical state, the number of large events is related to the number of small events by a scaling law. The distribution function for events of size s, e.g., s grains falling, scales as $N(s) \sim s^{-b}$. If the exponent b were unity, there would be one large event of size 1000 for every 1000 events of size 1. This behavior is in contrast to the result of combining a large number of independently acting random events. In this case the distribution of the sum is Gaussian, i.e., $N(s) \sim \exp[-(s/s_0)^2]$, where there is a typical scale s_0, and the number of large events drops off exponentially. That is, in the latter case there would be no large events, whereas for a power law distribution there are events of all sizes.

Models of self-organized criticality are perfect for computation on parallel computers, because they consist of many local degrees of freedom such as the height of the sandpile at each point of a regular periodic lattice. If one was making a connection to real sandpiles then the "height" in the model would correspond to the local slope

FIG. 1. Sandpiles with grains added along the top row. The colors indicate different slopes; pink squares represent the steepest parts of the pile; black regions indicate flat regions. The simulation was done on a Connection Machine at Boston University by Eric Myers, but similar simulations can also be done on a personal computer. Such models can also be simulated very efficiently on personal computers with commercially available cellular automata model (CAM) boards.

in the actual sandpile. The height at each point is updated simultaneously and hence can be updated in parallel. Figure 1 shows a configuration of a sandpile model computed on a Connection Machine. In this model the sand is added uniformly along one closed edge. Then each point is tested to determine if its height is greater than or equal to a threshold value equal to 4. If it is, then the point

Per Bak is a senior scientist at Brookhaven National Laboratory, Department of Physics, Upton, NY 11973.

has its height reduced by 4 (for a square lattice) and the heights of its four neighbors are increased by 1. The lattice is rechecked again and again until there are no points with heights exceeding the threshold.[4]

Earthquakes. How do these considerations apply to earthquakes? Indeed, a scaling law, the Gutenberg–Richter law, has been empirically observed for earthquakes. The number of earthquakes with energy release E obeys the power law, $N(E) \sim E^{-b}$, where b is about 0.5. The magnitude of earthquakes on the Richter scale is roughly equal to the logarithm of the energy released. Since we cannot make a realistic model of California, we must resort to the study of a simplified "toy" model. Of course, we would then have to argue that the behavior that we find is robust with respect to any modification of the model.

We describe the crust of the earth as a periodic lattice with blocks on each site, connected by springs.[5] The blocks are subjected to static and dynamic friction. The blocks are pulled in a unique direction by the force exerted on them by the slow tectonic plate motion. Define a real variable $z(i,j)$ on a two-dimensional square lattice, where z represents the force on the block at position i, j, and let z_{cr} be the critical threshold value. The linear dimension of the lattice is L, and the number of sites N is given by $N = L^2$. As an example, we choose $N = 100$ and $z_{cr} = 4$. The initial state of the lattice at time $t = 0$ is found by assigning small random values to $z(i,j)$. The lattice is updated according to the following rules.

(1) Increase z everywhere by a small amount p, e.g., choose $p = 0.00001$, and increase t by unity. This step represents the increase of the force from tectonic plate motion.

(2) Check if $z(i,j)$ exceeds the critical value z_{cr} anywhere on the lattice. If not, the system is stable since there is no activity anywhere and go to step 1. If yes, go to step 3.

(3) The release of force due to slippage of a block is represented by letting $z \to z - z_{cr}$ at the appropriate position(s). The transfer of force is simulated by letting $z \to z + 1$ at the four nearest neighbors. Periodic boundary conditions are not used so a site has only three neighbors at an edge and two neighbors at the corners. Go to step 1.

In the beginning the blocks are at rest. After a while there will be small earthquakes. Later still the earthquakes become bigger, and eventually the system comes to a statistically stationary state, where the average value of the force z stops growing. We monitor the distribution of the size s, where s is the total number of sliding blocks following an initial triggering instability. Figure 2 shows the temporal activity in the critical state. Note that there are events of a great variety of sizes. Figure 3 shows the distribution of earthquake sizes on a log–log plot. The straight line behavior indicates a Gutenberg–Richter-type power law with a value of b equal to 1.4. Note that the dynamics is deterministic and that once the initial values of z are specified, the time and size of the earthquakes in the future is given. All information is contained in the initial configuration and the rules.

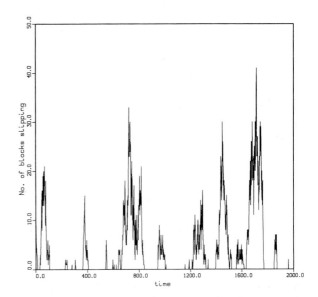

FIG. 2. Evolution of earthquakes at the self-organized critical state.

The scaling behavior for earthquake sizes indicates that there is nothing special about large events—they are simply large versions of small events. The only reason that we find earthquakes of magnitude 8 interesting is that they occur only a few times in a human lifetime and thus appear on the front page of newspapers. If a human lifetime were 10,000 years, nothing less than an earthquake of magnitude 10 would excite us.

Neither of these time scales are comparable with the geological time scale for plate tectonic motion. If we are interested in predicting the probability of very large earthquakes, where the statistics by definition are poor (large statistical fluctuations for large events are observed in the simulations), we could obtain good statistics for the more probable smaller events and then use the observed scaling behavior to find the probability for larger events.

The simulation of the Gutenberg–Richter law is the most spectacular success of the concept of self-organized criticality. However, toy models of many other natural phenomena characterized by large scale structures have been studied. We urge the reader to study one or more of these models.

The Game of Life. The "Game of Life" was conceived by the mathematician John H. Conway in an attempt to understand the emergence of organized structures in ecological systems.[7] Define an integer variable $z(i,j)$ on a two-dimensional square lattice; z can assume the values 1 (presence of life) or 0 (absence of life). For each site i,j determine n, the number of live sites of its four nearest plus its four next-nearest neighbors. Update all sites simultaneously according to the rules: (1) if $n < 2$ or $n > 3$, then $z \to 0$ (death because of loneliness or overcrowding); (2) if

13.3 Catastrophes and Self-Organized Criticality

COMPUTER SIMULATIONS

$n = 2$, then there is no change. (3) if $n = 3$, then $z \to 1$ (birth, if there were no life before). (A live site remains alive only if $n = 2$ or 3.)

Chen, Bak, and Creutz[8] have simulated this model as follows. Start from a random configuration and continue until the system comes to rest at a state with only static or simple local periodic activity (births or deaths). Then perturb the system by adding at random a live individual. This addition is analogous to adding a grain of sand to the sandpile models, or to letting $z \to z + p$ in the earthquake model. Each added individual causes a chain reaction. Update the system according to rules (1)–(3) until the system comes again to a stable configuration. The size of the response is the total number of sites that change following a single perturbation. Then add another live individual and so on. The distribution of responses follows a power law with a b-value of 1.4.

Of course, the Game of Life is not a realistic representation of anything. Nevertheless, the results have lead to speculations that fluctuations in real ecological systems are indeed critical. Kauffman has suggested that the extinction of species can be thought of as earthquakes in a coevolutionary biological system.[9]

Turbulence. Turbulence in liquids is characterized by vortices of all sizes. Mandelbrot has suggested that in turbulent systems the dissipation of energy is confined to a fractal structure with features of all length scales. This behavior can be simulated by a simple forest-fire model.[10] Distribute randomly a number of trees (green dots) and a number of fires (yellow dots) on a two-dimensional square lattice. Sites can also be empty. Update the system at each time t, as follows: (1) Grow new trees at time t randomly with a small probability p from sites that are empty at time $t-1$; (2) trees on fire at time $t-1$ die (become empty sites) and are removed at time t; (3) a tree that has a fire as a nearest neighbor at $t-1$ catches fire at time t.

Periodic boundary conditions are used. After a while the system evolves to a critical state with fire fronts of all sizes (Fig. 4). The fire indeed propagates by a self-organized critical branching process.

Economics. Conventional economic models assume the existence of a strongly stable equilibrium position for the economy, where large aggregate fluctuations can result only from exogenous shocks that affect simultaneously, in a similar manner, many different sectors of the economy. If on the other hand, the economy is a self-organized critical system, more-or-less periodic large scale fluctuations are to be expected even in the absence of any common shocks across sectors. In collaboration with economists Scheinkman and Woodford, Chen and Bak have constructed a toy model of interacting economics agents.[11] We anticipate many other applications of self-organized criticality.

Suggestions for further study

1. Simulate the two-dimensional sandpile model[4] with a threshold value $h_t = 4$. Begin with an $L_x \times L_y$ lattice with each site (i,j) assigned a random integer $h(i,j)$ in the interval 0 to 3. (In the literature the variable h is loosely referred to as the height or "local slope.") Add a grain of sand to the lattice at a random position (i,j), i.e., let $h(i,j) \to h(i,j) + 1$. If a site has h greater than 3, then this site is unstable and in the language of sandpiles such a site "topples." One updating step consists of determining all the unstable sites and then simultaneously reducing all unstable sites by 4 and increasing the value of their four neighbors by 1. This action on an unstable site is referred

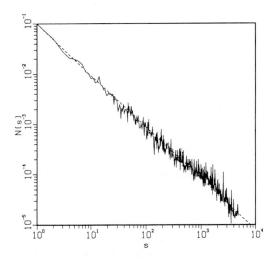
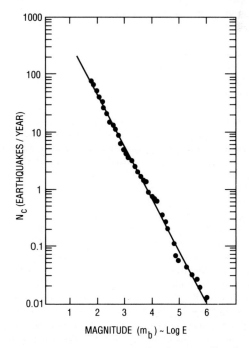

FIG. 3. Distribution of earthquake sizes from simulated earthquake catalogues. The data on real earthquakes was collected by Arch Johnston and Susan Nava.[6]

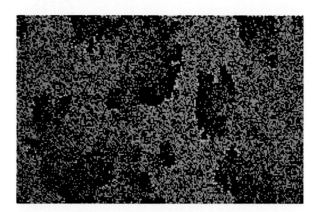

FIG. 4. Green trees grow with probability p from empty sites. Trees on fire (yellow sites) form fractal frontlike structures.

to as a "toppling." Continue this step until no further changes are needed. Determine s, the total number of topples. Add another grain at random and repeat the updating procedure. Use open boundary conditions so that a toppling on an edge loses one grain of sand and a toppling at a corner loses two. To visualize this process represent the heights of the sandpile in different colors. When the system has relaxed, $h(i,j) < 4$ for all sites and there are only four colors for $h = 0$ to 3. For intermediate states h can be as large as 7, since a site can have $h = 3$ and have all four neighbors over the threshold. Compute the distribution $N(s)$ that the addition of a grain at random causes s topples.

2. A non-trivial one-dimensional sandpile model has been formulated by Kadanoff et al.[12] When a site topples, its height is reduced by two, and the height of its right nearest and next-nearest neighbors increase by one unit each. The boundary conditions are such that grains can flow out of the system from the right side only, i.e., there is a closed boundary at the left edge and an open boundary at the right edge. Simulate this model as in Problem 1 with $h_t = 2$ and use different colors to represent different heights.

3. Write a program to simulate the forest-fire model. You will need to specify the four parameters: L, the linear dimension of the lattice; p_t, the initial probability of a site having a tree; p_f, the initial probability of a site having a tree on fire; and p, the probability that an empty site will grow a tree. Use periodic boundary conditions. Some questions to explore include: (a) For what values of p_t, p_f, and p will the forest maintain fires for all time? Note that as long as $p > 0$, new trees will always grow. (b) What is the distribution of the number of sites on fire? Does it depend on the probability parameters? (c) What is the distribution of the number of trees? Is the distribution simply related to the answer to (b)? Be sure to average over many updates of the lattice.

4. Write a program to implement the Game of Life on a square lattice and the perturbation procedure discussed in the text. Use open (absorbing) boundary conditions. Start with a random distribution of live sites and evolve the system until it reaches a stable configuration. Compute the size distribution $N(s)$, where s is defined as the total number of births and deaths that follow a single perturbation until the lattice settles to a stable state or periodic sequence of states. One site might be changed several times after a perturbation; each change counts as part of the response. The difficult part of the program is determining whether a configuration is part of a periodic cycle. For small periods the lattice can be stored for a few times and then compared to previous configurations to check whether the state has repeated itself. Note that it is very difficult to check for all types of periodic states but that if the system is started from a random distribution of live sites, cyclic structures with long periods are very rare.

Acknowledgments

Supported by the Division of Materials Science, U.S. Department of Energy under contract DE-AC02-76CH00016. The author would like to thank Harvey Gould and Jan Tobochnik for helpful comments. ∎

References

1. P. Bak, C. Tang, and K. Wiesenfeld, Phys. Rev. A **38**, 364 (1987).
2. For a popular account of self-organized criticality see P. Bak and K. Chen, Sci. Am. **246**(1), 46 (1991).
3. D. Dhar, Phys. Rev. Lett. **64**, 1613 (1990).
4. M. Creutz, Comput. Phys. **5**, 198 (1991).
5. P. Bak and C. Tang, J. Geophys. Res. **94**, 15635 (1989); P. Bak and K. Chen, "Fractal Dynamics of Earthquakes" in *Fractals and their Application to Geology*, edited by P. LaPointe and C. Barton (Geological Society of America, Denver, in press).
6. For an excellent study of earthquake statistics for the New Madrid zone, see A. Johnson and S. J. Nava, J. Geophys. Res. B **90**, 6737 (1985).
7. See M. Gardner in Sci. Am. **223**(4), 120 (1970); **223**(5), 118 (1970); **223**(6), 114 (1970); **224**(1), 108 (1971); **224**(2), 112 (1971); **224**(3), 108 (1971); **224**(4), 116 (1971); **224**(5), 120 (1971); **226**(1), 107 (1971); **233**(6), 119 (1975).
8. P. Bak, K. Chen, and M. Creutz, Nature **342**, 780 (1989).
9. S. A. Kauffman and S. Johnson, "Coevolution to the Edge of Chaos: Coupled Fitness Landscapes, Poised States, and Coevolutionary Avalanches," J. Theor. Biol. (in press).
10. P. Bak, K. Chen, and C. Tang, Phys. Lett. A **147**, 297 (1990).
11. P. Bak, K. Chen, J. Scheinkman, and M. Woodford, Santa Fe Institute Working Papers (1991).
12. L. P. Kadanoff, S. R. Nagel, L. Wu, and S. Zhou, Phys. Rev. A **39**, 6524 (1990).

From the editors: Future columns are planned on quasicrystals, classical spin models, and quantum Monte Carlo. Please address comments and requests to hgould @clarku or jant @kzoo.edu.

COMPUTER SIMULATIONS

ACTIVE-WALKER MODELS: GROWTH AND FORM IN NONEQUILIBRIUM SYSTEMS

Lui Lam and Rocco Pochy

Department Editors: Harvey Gould
hgould@vax.clark.edu

Jan Tobochnik
jant@kzoo.edu

Almost all interesting growth phenomena in nature occur in open systems under nonequilibrium conditions. Life is a case in point—the human body changes continuously with the influx of air and food and never reaches an equilibrium state. The continuous expansion or shrinkage of a river is another example. On smaller scales, numerous examples are found in various physical, chemical, and biological systems and include the evolution of a cell, the growth of snowflakes, the formation of electrodeposit patterns, and the aggregation of colloids, soot, and molecules in thin films.[1-5]

Consider the phenomenon of pattern formation. One can hardly fail to notice the striking similarity between the ramified patterns formed by rivers, trees, leaf veins, and by lightning. These branching patterns are different from the compact patterns observed in clouds and algae colonies. How does nature generate these patterns? Is there a simple principle or universal mechanism behind these pattern-forming phenomena?

Among the many growth processes underlying pattern formation, there are some exceptional, abnormal cases. We have in mind two types of abnormal growth: *transformational growth*, the occurrence of a (usually abrupt) qualitative change in behavior during a growth process, and *irreproducible growth*, the occurrence of distinctively different and unexpected patterns obtained from presumably identical samples. An example of transformational growth is the abrupt change in morphology observed in electrodeposit patterns.[6] The evolution of cancerous cells[7] is a well-known example belonging to both categories; the transformation of the AIDS virus[8] might be another example.

The cause for both types of growth is usually attributed to an extrinsic mechanism. For transformational growth, it is assumed that an external control parameter changes during the growth process, e.g., a mother's intake of drugs can cause brain damage to the growing fetus. We might blame irreproducible growth on unnoticed differences in the preparation of the samples, the fluctuation of temperature, or perhaps the carelessness of the student who performs the experiment. Although extrinsic mechanisms are responsible in many cases, is it possible that some growth patterns are intrinsically abnormal?

The active-walker models[9] discussed here serve as a starting point for addressing these important questions. The fact that many of the growth patterns produced by the active-walker models agree well with experiments suggests that there is a reasonable basis for these models in spite of their simplicity.

The basic idea of an active-walker model (AWM) is simple. A walker walks on a landscape. At each step the walker changes the landscape according to a *landscaping*

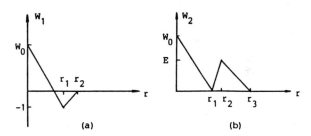

Fig. 1: Two types of isotropic landscaping functions, W_1 and W_2.

rule; the walker's next step is determined by the landscape near it according to a *stepping rule*. The landscape is represented by a single-valued scalar field, a potential or height function $V(i,n)$ defined at every site i at time n, where $n = 0,1,2,...$. We let $V(i,n) = V_0(i,n) + V_1(i,n)$, where V_0 is the external background. If the landscape changes are due only to the walkers, then the external background $V_0(i,n) = V_0(i)$ is independent of time. At $n = 0$, a walker is placed on the initial landscape $V_0(i,0)$. V_1 is updated by

$$V_1(i,n+1) = V_1(i,n) + W(\mathbf{r}_i - \mathbf{R}(n)), \quad (1)$$

where $V_1(i,0) = 0$, W is the *landscaping function*, $\mathbf{r}_i$ is the position of site i, and $\mathbf{R}(n)$ is the position of the walker at time n. Two types of isotropic forms of W are illustrated in Fig. 1.

After the landscape is updated, the walker takes a step according to the rule specified by $P_{ij}(n)$, the

---

Lui Lam is a professor of physics at San Jose State University, San Jose, CA 95192-0106, email: luilam@sjsuvm1.bitnet. Rocco Pochy is a physics graduate student at San Jose State University and software engineer at Lighthouse Associates, Milpitas, CA.

probability for the walker to step from its present site i to a site j. The set of *available sites* for the walker could be chosen to be the nearest-neighbor sites of $\mathbf{R}(n)$. If self-crossing of the walker's track is forbidden, the available sites are restricted to the sites not yet visited by the walker. We assume self-crossing of the walker's track is forbidden unless otherwise specified. (A slight modification of this procedure consists of first allowing the walker to walk and then updating the landscape. If self-crossing is forbidden, the two procedures are not equivalent.)

An example of an AWM is the formation of a river, with the walker representing the flow of water from the melting of ice on a mountain. As the water flows, it erodes the landscape; the subsequent flow of the water is influenced by the eroded landscape. In this case, we would use an anisotropic W function that is more negative in the forward direction. The function V_0 can be used to mimic the initial landscape, such as an inclined plane or a valley.

Three forms of P_{ij} have been studied:[9–12]

- The deterministic active walk (DAW). Among the available sites with potential $V(j,n)$ less than or equal to that at $\mathbf{R}(n)$, the site with the lowest potential is chosen as the next position of the walker.

- The probabilistic active walk (PAW). This walk is specified by

$$P_{ij}(n) \propto \begin{cases} [V(i,n) - V(j,n)]^\eta, & \text{if } V(i,n) \geqslant V(j,n), \\ 0, & \text{otherwise.} \end{cases} \quad (2)$$

The parameter η is to be specified.

- The Boltzmann active walk (BAW). This walk is given by

$$P_{ij}(n) \propto \exp\{[V(i,n) - V(j,n)]/T\}, \quad (3)$$

where the parameter T may be interpreted as the "temperature."

If all the available sites have the same value of V as $\mathbf{R}(n)$, an available site is chosen at random. In the DAW and PAW models, the walk terminates if the potential at $\mathbf{R}(n)$ is a local minimum. In contrast, the walker can go uphill and downhill in the BAW model, and if self-crossing of the walker's track is allowed, the walker can continue indefinitely.

The track of the walker forms a filamentary pattern.[9,10] An example of the PAW model is shown in Fig. 2. For this model the exponent v defined by the asymptotic n dependence of the ensemble averaged end-to-end distance $R_e \sim n^v$ differs from a random walk ($v = 0.5$) and a self-avoiding walk ($v = 0.75$).[10] The exponent v is not universal, but depends on the parameters of W.[9,10] In this and the following examples, unless otherwise specified, all the results are for a square lattice.

Ants with a scent

A landscape function such as W_1 (see Fig. 1), might represent the distribution of scent left by an ant that continuously emits a scent as it walks. The track of the walker represents the path of a single ant. An alternative interpretation is important for understanding the behavior of an ant swarm, a subject of serious study in the context of complex, biological systems.[13] If the scent is carried instantaneously by the ant, then the walker's track corresponds to a file of ants coming out of an anthill. Roughly speaking, the strength of the scent is proportional to $-W_1$. The dip in W_1 creates a trough on either side of the track so that if the track turns around, it tends to

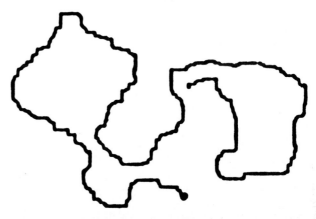

Fig. 2: An example of a filamentary pattern formed by the track of a single walker generated by the PAW model with $\eta = 1$. Self-crossing of the walker's track is not allowed. The pattern is for $V_0 = 0$ and $W = W_1$, with $W_0 = 15$, $r_1 = 10$, and $r_2 = 20$.

grow parallel to the previous steps. This effect mimics the behavior of ants that chat with their friends while keeping the file intact. If the ants wish to avoid each other, we could choose $W = W_2$, for which the peak at r_2 creates a barrier. The background landscape V_0 can be used to mimic many situations. For a flat surface, $V_0 = 0$, the ants might have a leisurely walk. If V_0 represents an inclined plane, the ants might be fetching food on the other side. A V_0 with random or regular hills could represent the presence of obstacles such as grass.

Let us return to the single-ant interpretation discussed above. One extension of the model is to allow the scent template to be a time-dependent function, since we would expect that the ant's scent spreads out in time. Let us assume that W maintains its form for a finite amount of time and introduce two characteristic times, the time lag m and the template lifetime τ, both of which are non-negative integers. The time lag m represents the time it takes for the scent to be effective. We assume that W is zero (the scent is absent) for $n \leqslant m$ and $n > m + \tau$ and generalize Eq. (1) by

$$V_1(i, n+1) = V_1(i,n) + W(\mathbf{r}_i - \mathbf{R}(n-m)) \\ - W(\mathbf{r}_i - \mathbf{R}(n-m-\tau)). \quad (4)$$

Each W term is zero if the argument of $\mathbf{R}$ is negative. Equation (4) reduces to Eq. (1) when $m = 0$ and $\tau = \infty$.

Multiwalkers

The AWM can be generalized to include many walkers. For simplicity we assume that each walker has the same W function. When multiwalkers coexist, the walkers influ-

COMPUTER SIMULATIONS

ence each other through the shared landscape. The evolution of their tracks depends on the order in which the walkers are updated.

The track pattern becomes particularly interesting when *branching* is introduced and multiwalkers are generated as part of the dynamics. Suppose that the walker is at site i and moves to one of the available sites j. Then choose one of the remaining available sites k and determine the difference $[V(\mathbf{R}(n),n) - V(k,n)]$. If this difference is greater than $\gamma[V(\mathbf{R}(n),n) - V(j,n)]$, then this site k is occupied by a new walker and branching occurs. The parameter γ is called the branching factor. Sequential and random methods of updating the walkers (with and without branching) are introduced in Ref. 10. In the former, the walkers move one step according to their seniority based on age (old walker moves first); in the latter they are chosen at random. In both versions, one cycle is completed after each walker is chosen.

The filamentary patterns formed by the AWM are similar to the patterns found in many biological, chemical, and physical systems, including spirals,[9] retinal neurons,[11] dielectric breakdown of liquids in a thin cell,[9-11] and the dense radial morphology in electrodeposits[10] (see Figs. 3 and 4). In these systems, the filamentary patterns are frozen in space once they are formed. For dielectric breakdown,[9] the walker's track represents the places where "burning," a chemical reaction between the electrons and ions liberated by the dielectric breakdown process and the conducting materials on the inner surfaces of the cell, has occurred. The quantity $-V(i,n)$ represents the readiness of the surface reaction at site i to occur once the material is visited by the walker; the readiness results from the combined effects of the amount of unburnt chemicals and heat. Similar interpretations can be made

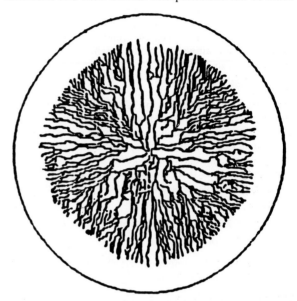

Fig. 4: A multiwalker track pattern from the probabilistic active walker model. The parameters are $\eta = 1$, $m = 0$, $\tau = \infty$, $\gamma = 0.99999$, and $W = W_1$, with $W_0 = 5$, $r_1 = 12$, and $r_2 = 15$. The height of the coneshaped V_0 is -500 at (0,0), the center of the 100×100 lattice, and is zero when it intersects the boundary of the lattice. The circle has diameter 100. At $n = 0$, four walkers are placed at $(0, \pm 2)$ and $(\pm 2, 0)$. The seniority of the four initial walkers is in the clockwise direction starting from the west. The multiwalkers are updated sequentially. This pattern resembles the experimental electrodeposit pattern of Zn aggregates (Ref. 16). Dense radial morphology has been observed in various systems (Ref. 17).

for the surface reaction of CO on Pt,[14] where the diffusion of CO replaces the motion of the electrons and ions, and for the formation of retinal neurons (see Fig. 1 of Ref. 11). These processes can be described[18] by two coupled partial-differential equations such as in diffusion-reaction processes found in combustion and thus can be studied theoretically.

If the word is readiness is replaced by excitability, the system can be interpreted as an excitable medium.[19] The motion of the walker favoring lower-potential sites (as in the DAW and PAW models) can be interpreted as the evolution of a species, if the potential $-V$ is understood as the fitness landscape; multiwalkers correspond to the case of adaptive coevolution.[20] Equivalently, the landscape may be interpreted as the set of allowable configurations in various optimization problems considered in computer science and in a spin glass.

A moving worm

Consider a single walker without branching and suppose

Fig. 3: A multiwalker (without branching) track spiral pattern from the PAW model. The parameters are $\eta = 1$, $m = 1$, $\tau = \infty$, $V_0 = 0$, and $W = W_1$, with $W_0 = 5$, $r_1 = 4.4$, and $r_2 = 11$. The ten walkers are placed randomly at $n = 0$, updated sequentially and terminate naturally. The cross shows the center of the lattice and has a width of r_2. Both left- and right-handed spirals may appear (not shown here). This pattern resembles the experimental growth of a spiral due to the oxidation of CO on a Pt(110) surface (Ref. 14). Similar spiral patterns are found in many other systems (Ref. 15).

that at time n, only the latest N sites in the walker's track are retained as visited sites. As depicted in Fig. 5, the time evolution of the track can be interpreted as a moving worm of length N. The different parts of the worm can influence each other and the environment, tunable by the W function. The influence of obstacles can be described by a nonzero V_0. If $\tau = N$, the worm moves in a tube of the same length in a way similar to that found in polymer reptation.[21,22] The length of the tube can be longer than the polymer by varying m and τ. The application of the AWM for studying polymer dynamics remains to be explored. This type of multiwalker model might also be used to simulate a can of worms.

Abnormal growth

The patterns generated by the AWM shown in Figs. 2–5 are filamentary. We now describe a way of generating compact patterns. Begin with the central site occupied by a seed (an active walker). The walker changes the local landscape according to Eq. (1) and is frozen. One of the perimeter sites is then chosen and occupied by a new walker. The probability that perimeter site j is chosen is given by

$$P_j \propto \begin{cases} [V(i,n) - V(j,n)]^\eta, & \text{if } V(i,n) > V(j,n), \\ 0, & \text{otherwise.} \end{cases} \quad (5)$$

If there is more than one occupied site adjacent to j, then i is the position of the nearest-neighbor-occupied site that maximizes the difference $V(i,n) - V(j,n)$. The newly added walker changes the local landscape as before and the process is repeated until the aggregate of occupied sites touches the boundary of the lattice. We call this model the boundary PAW. In this description of the model, the active walker does not walk, but we can equally interpret the model as follows. Almost every walker at the edge of the aggregate has a chance of being chosen, but only one is allowed to walk one step with the PAW stepping rule. See Ref. 23 for a comparison between this model and the dielectric breakdown model.[24]

The consequences of the boundary PAW are dramatic. By varying the parameters of W, we can obtain patterns ranging from compact to filamentary, with the Eden-like pattern and semicompact patterns in between (see Fig. 6). In some cases, there is a morphological change in the growth pattern (see Fig. 7), and we say that there is

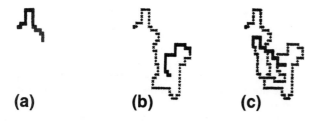

Fig. 5: A moving worm generated by the PAW. The parameters are $\eta = 1$, $m = 0$, $\tau = \infty$, $V_0 = 0$, $W = W_1$, with $W_0 = 15$, $r_1 = 4$, and $r_2 = 20$, and $N = 25$. The worm is in black and the previous track is in gray. The worm is not allowed to cross itself, but is allowed to cross its previous tracks. (a) Time $n = 25$; the worm just climbs out of the hole at the center; (b) $n = 100$; (c) $n = 150$.

intrinsic transformational growth. Some of these patterns closely resemble those observed in electrodeposits [see Fig. 9(b) of Ref. 6]. For $\eta < 5$ and $\eta > 6$, we always obtain a statistically similar pattern for different runs. However in the range $5 \leq \eta \leq 6$, the outcome of different runs may be one of two different types, i.e., the growth pattern is intrinsically irreproducible (see Fig. 6). (To our knowledge, patterns from any other model are always statistically reproducible.)

This behavior is represented schematically in Fig. 8. Within the sensitive zone ($\eta \sim 5.5$), the growth pattern depends sensitively on the random number sequence generated in each run, i.e., the order in which the perimeter sites are occupied. Of course, this kind of behavior is possible because there is randomness built into the model. But stochasticity always is present in an open system, due to interactions with the environment. If we interpret the needles in Fig. 6 with $\eta > 6$ as a normal, benign growth and the more compact ones with $\eta < 5$ as a dangerous, cancerous growth, then we see that a normal cell can become cancerous if η is shifted to low values and can be intrinsically cancerous if η is in the sensitive zone.

Rough surfaces

The landscape[10–12] in the AWM forms a rough surface[25] resulting from the landscaping action of the walker. Examples of one-dimensional landscapes are shown in Figs. 9(a) and 9(b). Due to space limitations, we only give some highlights here.

● Grooved surfaces, similar to those observed in Si films grown by MBE,[26] are formed using the PAW[10] and BAW[27] models (see Fig. 9). These grooved surfaces result from the trapping of the walker that keeps digging on the landscape when $W_0 < 2$.

● For these grooved states, when $W_0 < 2$ the surface width σ does not appear to saturate. The width σ is defined as the rms height averaged over many runs,[25] and $\sigma \sim n^\beta$. The exponent $\beta = 1$.[10,11]

● For the BAW, the average height remains constant or changes with constant speed according to whether $W_0 = 2$ or $W_0 \neq 2$, respectively. The roughness exponents α and β can be tuned by W_0 and T; they behave differently from the exponents found in other models.[25] (Here α is defined by $\sigma \sim L^\alpha$, where L is the lattice size.) For example, for $W_0 = 2$, $\sigma(L,n) \sim T^\gamma g(n/T^{\gamma/\beta(T)})$, independent of L, where γ is a constant.[12] This scaling expression has the same form as found for a flux line in the superconducting sandpile,[28] a system similar to the sandpile model of self-organized criticality (SOC).[29] This coincidence between the BAW and the flux-line motion may be more than accidental; both are thermally activated with a nonconstant activation energy.[18] (See Ref. 9 for an interpretation of the SOC sandpile as an AWM.)

● The AWM is capable of generating two-dimensional rough surfaces,[11] which resemble sand-blasted brass surfaces produced in the laboratory [see Fig. 7.2(a) of Ref. 30].

● We also note that for the BAW, a first-order transition between grooved surfaces and fractal surfaces

13.4 Active Walker Models: Growth and Form in Nonequilibrium Systems

COMPUTER SIMULATIONS

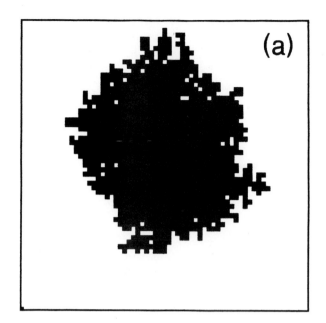
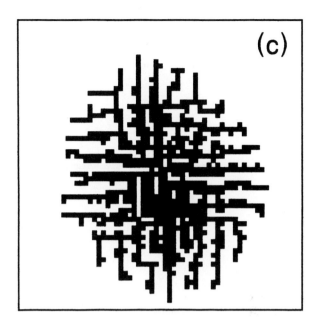
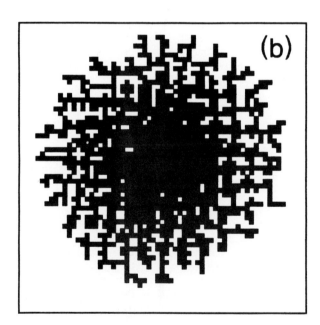
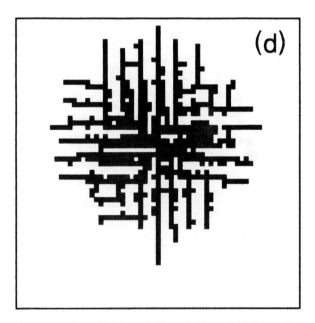

Fig. 6: Patterns generated by the boundary PAW model. $W = W_1$, with $W_0 = 20$, $r_1 = 4$, and $r_2 = 20$. (a) $\eta = 0$; (b) $\eta = 2$; (c) $\eta = 4$; (d), (e) $\eta = 5$; (f), (g) $\eta = 6$; (h) $\eta = 7$–9. The square represents the boundary of the 64×64 lattice.

(as W_0 is varied) is found[11] and a reentrant soliton exists at $T = 0$.[12]

We hope that our discussion of AWMs convinces you that they are capable of generating, in a simple and unified way, many types of growth patterns that are observed in nonequilibrium systems. AWMs also provide an intrinsic mechanism for explaining both irreproducible and transformational growth. Much work remains to be done including investigating different landscaping functions and stepping rules to determine what other patterns can be found, studying the models analytically, and relating the functions and parameters in the active walker models to the relevant physical quantities and parameters of real systems.

Chaos[31] and SOC[29] are two important paradigms that have been discovered recently. These paradigms have provided unexpected, intrinsic mechanisms for explaining phenomena that were thought to have extrinsic origins. The existence of chaos reminds us that some apparently random behavior can be explained by deterministic

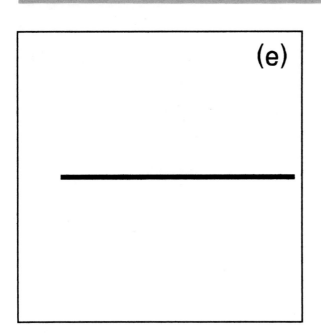
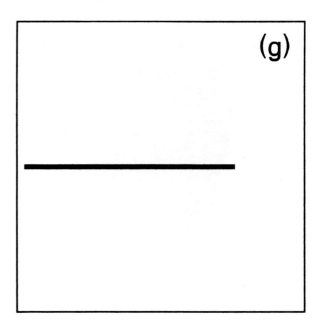
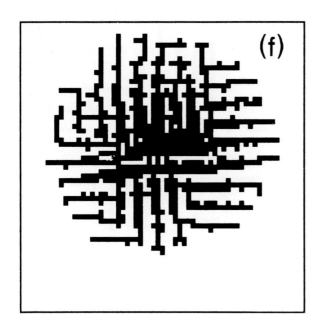
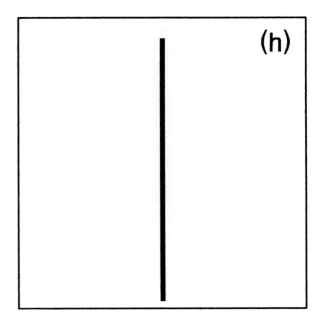

nonlinear equations; the signature is strange attractors and sensitive dependence on initial conditions. The existence of SOC suggests that some complex, dissipative systems can adjust by themselves towards a critical state; the signature is power laws in space and time. The intrinsic abnormal growth exemplified by AWMs points to a new paradigm—the sensitive dependence of growth and form on noise; the signature is sensitive zones and perhaps something else still to be discovered.

Suggestions for further study

Active-walker models have been proposed only recently, and hence almost anything you study will be new. The following suggestions may be considered as possible research projects. Please communicate your results to us.

1. Write a program for the probabilistic active walk with a single walker and reproduce the results shown in Fig. 2. Vary the parameters and the W function to see what patterns can be produced. From the statistics of the tracks calculate, e.g., R_e as a function of n, and N_n, the number of filaments with length greater than or equal to n. (See Refs. 9 and 10 for details.)

2. Start from some experimental patterns, e.g., the set of electrode posit patterns in Ref. 32, and construct a model that produces similar patterns. You might wish to begin with the model summarized in the caption of Fig.

COMPUTER SIMULATIONS

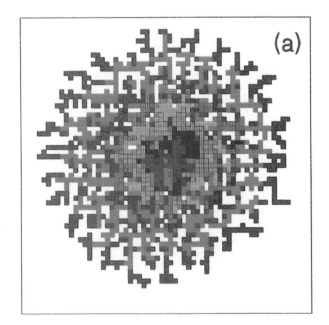
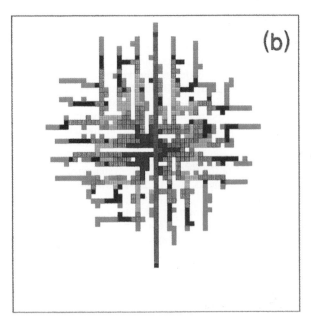

Fig. 7: Time evolution of transformational growth from the boundary PAW model. Color code: red, blue,..., green, black, as time increases. (a) Same as Fig. 6(b); the pattern grows from a compact core to filamentary at the rim. (b) Same as Fig. 6(d); the pattern grows from a needle to a radially filamentary pattern.

10(d) of Ref. 10, which is similar to that in Fig. 4 except that the V_0 cone has a height of 600 at the center, $\gamma = 0.8$, and $W = W_2$ ($W_0 = E = 5$, $r_1 = r_2 = 10$, and $r_3 = 15$). See if you can tune only one parameter to generate the entire set of observed patterns. If this approach does not work, try a slightly different W function.

3. The moving worm is fun to watch on the computer screen. Write a program for the probabilistic active-

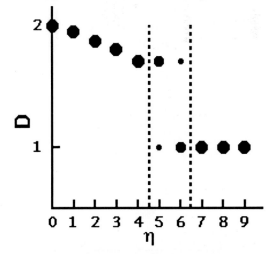

Fig. 8: Schematic representation of the outcome from 20 runs corresponding to the case of Fig. 6. The size of the dot represents the percentage of the outcome. The largest dot corresponds to 100%, the medium-size dot about 70%, and the smallest dot about 30%. The sensitive zone is between the two broken lines. D is the box dimension and is taken to be the best linear fit in the log–log plot of the number of covering boxes versus the box size, even in the presence of a crossover due to morphological change. Although the box dimension is not a good quantity to characterize a pattern with morphological change, it is nevertheless used here, in lieu of a better choice, to produce a concise and quantitative summary of the results.

multiwalker model using the parameters given in Fig. 5 and see how a can of worms will coexist. In two dimensions, some worms will go into a temporary sleep or die, but this state will be rare in three dimensions. You can confine the worms by letting the worms rebound from the boundary of the lattice. Read Refs. 21 and 22 to learn what quantities you need to calculate to compare your results with experiments in polymer dynamics and the predictions of de Gennes's reptation model.[21]

4. Plot the landscape found in the above problems and study its time evolution. Calculate the surface width σ as a function of the time n and the system size L and estimate the exponents α and β if power law behavior exists.[25]

5. Write a program for the boundary PAW model for two or three dimensions. Repeat the results shown in Fig. 6, and then use your imagination to invent your own model by changing the site-selection rule.

Acknowledgments

This work is supported by the NSF-REU program and a Syntex Corporation grant from the Research Corporation. We are grateful to the many undergraduate students, Rolf Freimuth, Daniela Kayser, Jeff Fredrick, Lance Aberle, and Mike Veinott, who contributed to the development of the AWMs. In particular, Daniela Kayser, and Mike Veinott helped to produce some of the diagrams.

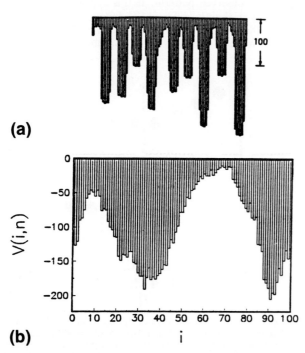

Fig. 9: Grooved surfaces. In (a) and (b), $m = 0$, $\tau = \infty$, $V_0 = 0$, and $W = W_1$, with $W_0 = 1$, $r_1 = 1$, and $r_2 = 2$. Self-crossing is allowed. The initial one-dimensional surface is a flat line of length 100; periodic boundary conditions are used. (a) Simulation from the PAW model (Ref. 10) with $\eta = 1$. Multiwalkers are dropped sequentially; a new one is added randomly after the previous one terminates naturally and is removed. (b) A pattern from the BAW model (Ref. 27) with $T = 1000$. A single walker walks forever in this case.

References

1. *Modeling Complex Phenomena*, edited by L. Lam and V. Naroditsky (Springer, New York, 1992).
2. *Fractals and Disorder*, edited by A. Bunde (North-Holland, Amsterdam, 1992), or Physica A **191**, Nos. 1–4 (1992).
3. *Pattern Formation in Complex Dissipative Systems*, edited by S. Kai (World Scientific, River Edge, 1992).
4. *Nonlinear Structures in Physical Systems*, edited by L. Lam and H. C. Morris (Springer, New York, 1990).
5. *Random Fluctuations and Pattern Growth*, edited by H. E. Stanley and N. Ostrowsky (Kluwer, Boston, 1988).
6. L. Lam, R. D. Pochy, and V. M. Castillo, in Ref. 4.
7. J. D. Watson, N. H. Hopkins, J. W. Roberts, J. A. Steitz, and A. M. Weiner, *Molecular Biology of the Gene* (Benjamin/Cummings, Menlo Park, 1987).
8. P. W. Ewald, Sci. Am. **266**, No. 4, 86 (1993).
9. R. D. Freimuth and L. Lam, in Ref. 1.
10. L. Lam, R. D. Freimuth, M. K. Pon, D. R. Kayser, J. T. Fredrick, and R. D. Pochy, in Ref. 3. In the caption of Fig. 8, 1.82 should be replaced by 1.61.
11. D. R. Kayser, L. K. Aberle, R. D. Pochy, and L. Lam, Physica A **191**, 17 (1992).
12. R. D. Pochy, D. R. Kayser, L. K. Aberle, and L. Lam, Physica D **66**, 166 (1993).
13. E. O. Wilson, *The Insect Societies* (Harvard University, Cambridge, 1971); M. M. Millonas, J. Theor. Biol. **159**, 529 (1992).
14. S. Jakubith, H. H. Rotermund, W. Engel, A. von Oertzen, and G. Ertl, Phys. Rev. Lett. **65**, 3013 (1990).
15. *Spiral Symmetry*, edited by I. Hargittai and C. A. Pickover (World Scientific, River Edge, 1992); T. A. Cook, *The Curves of Life* (Dover, New York, 1979).
16. Y. Sawada, A. Dougherty, and J. P. Gollub, Phys. Rev. Lett. **56**, 1260 (1986).
17. E. Ben-Jacob and P. Garik, Nature **343**, 523 (1990).
18. L. Lam, in *Introduction to Nonlinear Physics*, edited by L. Lam (Springer, New York, 1993).
19. A. T. Winfree, Chaos **1**, 303 (1991).
20. S. Kauffman, *The Origins of Order: Self-Organization and Selection in Evolution* (Oxford University, New York, 1993).
21. P. G. de Gennes, J. Chem. Phys. **55**, 572 (1971).
22. M. Doi and S. F. Edwards, *The Theory of Polymer Dynamics* (Clarendon, Oxford, 1988).
23. L. Lam, M. C. Veinott, and R. D. Pochy, in *Proceedings of the NATO Advanced Research Workshop*, Spatiotemporal Patterns in Nonequilibrium Complex Systems, Santa Fe, 13–17 April 1993, edited by P. E. Cladis and P. Palffy-Muhoray (Addison–Wesley, Redwood City, 1993).
24. L. Niemeyer, L. Pietronero, and H. J. Wiesmann, Phys. Rev. Lett. **52**, 1033 (1984).
25. *Dynamics of Fractal Surfaces*, edited by F. Family and T. Vicsek (World Scientific, River Edge, 1991).
26. D. J. Eaglesham, H.-L. Gossmann, and M. Cerullo, Phys. Rev. Lett. **65**, 1227 (1990).
27. D. R. Kayser, R. D. Pochy, and L. Lam, Am. Phys. Soc. Bull. **38**, 700 (1993).
28. C. Tang, Physica A **194**, 315 (1993).
29. P. Bak, C. Tang, and K. Wiesenfeld, Phys. Rev. Lett. **59**, 381 (1987); P. Bak, Comput. Phys. **5**, 430 (1991).
30. *Fractals and Disordered Systems*, edited by A. Bunde and S. Havlin (Springer, New York, 1992).
31. S. Eubank and D. Farmer, in *Introduction to Nonlinear Physics*, edited by L. Lam (Springer, New York, 1993).
32. W. Y. Tam and J. J. Chae, Phys. Rev. A **43**, 4528 (1991).

Active Walks and Path Dependent Phenomena in Social Systems

Lui Lam
Nonlinear Physics Group, San Jose State University, San Jose, CA 95192-0106, USA

Chang-Qing Shu
BBN Systems & Technologies, 70 Fawcett Street, Cambridge, MA 02138, USA

Sabine Bödefeld
*Physics Building, Swiss Federal Institute of Technology (ETH),
Hönggerberg/HIL, CH-8093, Zürich, Switzerland*

Active walk is a paradigm for pattern formation and self-organization in complex systems. An active walker changes the landscape as it walks, and is influenced by the changed landscape in choosing its next step. Active walk models have been applied successfully to various biological and physical systems, including the formation of filamentary patterns in retinal neurons and surface reaction patterns in thin cells of liquids, anomalous ion transport in glasses, and food collection by ant swarms. In this paper, the basic ideas and important applications of active walks are summarized, with new applications in urban growth and path dependent phenomena suggested. In particular, the use of active walk in describing increasing returns in economics is demonstrated.

1. Introduction

In the long pursuit of an understanding of the *universe* in which the humankind exists and of the human society itself [1], the emphasis has always been on the *universal* aspect of behaviors. As pointed out by Schrödinger [2], this fact is evident from the observation that *university* is the very name given to our institutions of highest learning. In this regard, one may note that the highest degree of learning conferred by a university is the Doctor of Philosophy (Ph.D.), independent of the specialties studied (see Section 1.3.6 of [3]), while *philosophy* is a Greek word meaning the "love of wisdom."

It is interesting to recall that Aristotle (384-322 BC), a Greek scholar, did not focus his attention on just one or two branches of knowledge but actually studied and contributed significantly to biology, psychology, physics, and literary theory, as well as invented formal logic and pioneered zoology. The fragmentation of learning and in the pursuit of knowledge into different disciplines (such as physics, chemistry, economics, etc.) is a rather recent phenomenon, occurring only in the last few centuries.

Recent attempts to recoup the unified approach to knowledge, with a view to providing a common description of both natural and social systems, include the movement of cybernetics or general system theory [4] from the 1940s to 1960s, Prigogine's dissipative structure [5] and Haken's synergetics [6] in the 1970s (see [7] for a review of both), and the studies under the banner of complex systems since the 1980s [8,9].

The two signatures of a successful universal law applicable to all complex systems, in our opinion, should be: (i) The law should be simple enough to be stated in one or two reasonably short sentences. (ii) The law should conform to our daily experience. The former follows from the wide applicability of the law. The latter is due to the fact that each one of us humans is a complex system by herself or himself. (As an example, these two requirements are easily satisfied by the second law of thermodynamics which states that "heat cannot flow on its own from cold to hot bodies.")

In principle, such a universal law for complex systems, if exists, could be (i) on the phenomenological level, extracted from the observation of the behavior of many complex systems. Or, the law could be (ii) on the organizational level, obtained by looking into the mechanism with which the complex systems self-organize themselves. The theory of self-organized criticality [10] and the principle of active walks [11] are two important results that came out by following the second approach. They seem to apply to many complex systems. Moreover, they do possess the two signatures required for any universal law of complex systems. In the rest of this paper, only the principle of active walks will be discussed.

2. Active Walks

The use of a walker in modeling physical and other phenomena runs a long history. The most well known is a random walker, which has been used in mimicking the motion of a completely drunk person, the Brownian motion of a particle suspended in a liquid, or the fluctuations in a financial market. A random walker does not change anything in its environment and is what we call a passive walker. In contrast, an active walker is one who changes the landscape during its motion and is influenced by the changed landscape in choosing its next step [12,13].

The description of an active walk (AW) thus involves two interacting components, viz., the location of the walker $\mathbf{R}(t)$ at time t and the deformable landscape $V(\mathbf{r},t)$, a scalar potential, where $\mathbf{r}$ is the spatial coordinate. (The landscape can be a vector potential in other cases [14].) The dynamics of an active walk are determined by three constituent rules: (i) the landscaping rule, which specifies how the walker changes the landscape as it walks; (ii) the stepping rule, which tells how the walker chooses its next step; and (iii) the landscape's self-evolving rule, which specifies any change of the landscape due to factors unrelated to the walker, such as diffusion and external influences. The details of these three constituent rules should depend on the system under study

The track of the active walker forms a filamentary pattern. The landscape becomes a rough surface with scaling properties. Several active walkers may coexist; they interact indirectly with each other through the shared landscape. Branching, the process giving rise to the birth of new walkers, can be incorporated. (See [14,15] for reviews.)

The modification of the landscape due to the active walker at every time step could have a finite lifetime τ. For τ equals to one time step, the landscape modification is carried along by the walker (such is the case of the gravitational potential carried by a mass when the mass is considered as an active walker). For τ greater than one time step, the future steps of the walker are influenced by the walker's past locations; the system becomes a path (or history) dependent system. Many natural systems and almost all social systems (e.g., a person's career path [16]) are path dependent. It is then not surprising that all the previous applications of AW [15] and the cases discussed in this paper, are path dependent systems.

Examples of AW systems with physical landscapes include a woman walking on a sand dune, percolation in soft materials, chemotaxis of ants [13,17,18], and movement of bacteria [19,20] or fish. In particular, we will like to mention three successfully worked-out examples: the filamentary patterns in dielectric-breakdown induced surface reactions and retinal neurons (see Fig. 5.4 of [21], and [22]), the mixed alkali effect in glasses [14,23], and food collection by ants [18]. All these cases show good agreement between simulations and experiments.

In other cases, the landscapes are purely mathematical artifacts. For example, urban growth can be modeled by the aggregation of active walkers. Initially, a "value" can be assigned to every piece of

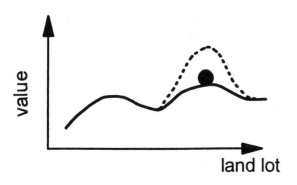

Fig. 1. The value landscape in an urban growth model. Once a house, represented by the black dot, is added to a land lot, the value nearby is altered, as indicated by the broken line.

vacant land lot according to its location (Fig. 1). For example, a lot on the flatland will have a higher value than one located on the hills; a river nearby can increase the value of the lot. The value could be taken to be the probability that the lot will be developed, by having a house or factory built on it. Once this happens, the value of the lands nearby will be increased and a new house will be added somewhere. The process is repeated. In this scheme, the house acts like an active walker, except that it does not actually walk. Such a model is more realistic and flexible than the correlated percolation model used by Makse et al [24]. (See also [25].)

Another example is the fitness landscape employed in evolution biology [26]. Every species is in coevolution with other species. The presence of species A (Fig. 2a) affects the fitness landscape of species B (Fig. 2b), which in turn changes the fitness landscape of A (Fig. 2c); the changed landscape of A then determines how A will move. Now if we want to describe this evolution process by a simple model involving A alone, we will go directly from Fig. 2a to Fig. 2c with Fig. 2b hidden. It then seems that A deforms its own fitness landscape at every step of its movement, i.e., A acts like an active walker. The third example concerns the phenomenon of increasing returns in economics [27] and is discussed in Section 3.

It turns out that chance (or noise or contingency) plays a more important role in AW models than in other probabilistic models (such as the diffusion-limited aggregation model). Repeated runs of the same computer algorithm for a probabilistic AW growth model, corresponding to the same set of model parameters but different sequences of random numbers, may give same or different morphological growth patterns, depending on where the system sits in the parameter space [13,28]. If our real world is believed to be such an AW system, then when "the life's tape is replayed," we may or may not recover a similar (but never identical) history of life (see Chapter 7 of [21]). This observation differs from that of Gould [29] who argues that contingency is so crucial that history would always be very different if life's tape was ever replayed. (See also [10].)

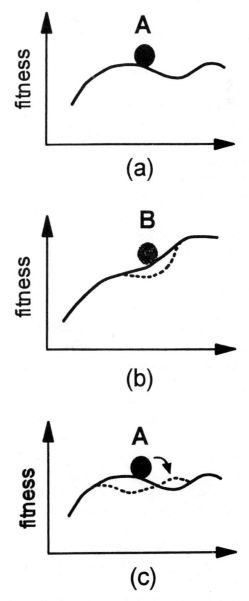

Fig. 2. Sketch of coevolution of two species A and B (see text).

3. Increasing Returns in Economics

Increasing returns refer to self-reinforcing mechanisms in the economy, whereby chance events can tilt the competitive balance of technological products [30]. This phenomenon can be modeled [31] by the motion of an active walker jumping among the sites on a fitness landscape of the product types (Fig. 3). The visit of the walker at a site represents that the product there is chosen by the customer, and so the local fitness is increased, as represented by the broken line in Fig. 3.

Specifically, consider the simple case of two products A and B (e.g., the VHS and Beta versions of videotapes) so that there are only two sites on the landscape. Let the fitness of A be $u(t)$, and that of B be $v(t)$, where t is (discrete) time. At each time step, the probability that site A (B) will be occupied by the walker, the black dot in Fig. 3, is proportional to $f(u)$ [$f(v)$]. After one of the sites is chosen, the fitness of the chosen site is increased by an amount a and the other site has its fitness decreased by b. The process is repeated.

Numerical results of a special case is shown in Fig. 4, where $a = 0.1$, $b = 0$, $u(0) = v(0) = 0$; $f(u) = \exp(\beta u)$; $d = (u - v)/\max\{u,v\}$, where u and v are the time averages of $u(t)$ and $v(t)$, respectively, and $t_m = 10^4$, where t_m is maximum of t used in the calculation. Here β can be interpreted as an inverse temperature. For $\beta = 0$, the "noise" level is high, the consumer is completely ignorant of the fitness of the products and chooses the product randomly. Each product has 50 % share of the market on the average, with fluctuations, and hence $d = 0$. The opposite case of β being infinity corresponds to a perfectly

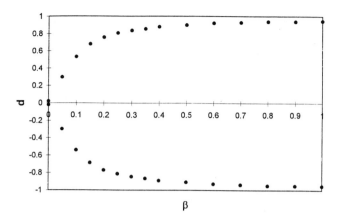

Fig. 4. Variation of d, normalized time-averaged fitness difference between two products, versus β, the inverse temperature, in an AW model of increasing returns. For each β, the results from several runs of the algorithm are shown.

informed and rational consumer who always chooses the product with the higher fitness. Depending on which product is first chosen by accident, this product will go on be bought by every subsequent consumer, resulting in a complete domination of the market and hence $d = 1$ or -1 (the former corresponds to product A wins out, the latter product B). For β between zero and infinity, we have the usual case of a customer who is only partially rational or has only partial information about the products or both. In the initial stage, the two products dominate the market in turns but only slightly, until one of them wins out clearly over the other. Which product wins out cannot be predicted and is determined by chance. (These results are comparable to that in [32] concerning information contagion, even though the approach there is different from ours.)

Further results from our model are presented in Figs. 5-10, which have the same parameters as in Fig. 4, unless otherwise specified. Figure 5 shows the effect of t_m. For larger t_m, d tends to -1 faster as β increases. This means that for the same value of β, longer computing time will result in smaller value of d. This can be understood in that once one product gained sufficient advantage over the other, it will more likely to be chosen in following steps and hence d tends to -1 (here we show only the case of v winning).

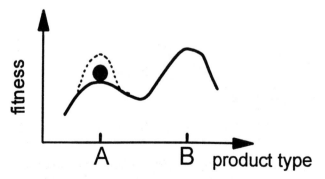

Fig. 3. The AW description of competition between different products (see text).

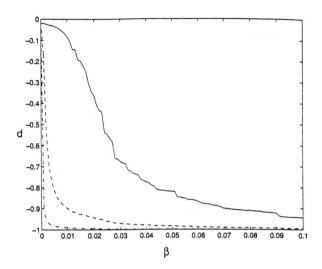

Fig. 5. Dependence of d on β for different maximum times t_m. $t_m = 10^4$ (solid line), 10^5 (dash) and 10^6 (dash-dot), respectively.

and both products are chosen with equal probability. For β = 0.1 (Fig. 10), the customer has some information about the products. The initial advantage of u cannot be overcome by v and u clearly wins.

Analytic solutions for the model studied in this section are found by Shu and Lam, and will be presented elsewhere [33]. In this paper, we are content to show that important problems in social systems, such as increasing returns in economics, can indeed be modeled by AW. The simple model proposed here can obviously be generalized (e.g., to more than two products) and made more realistic.

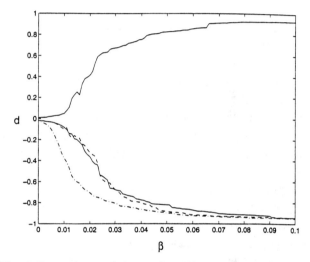

Fig. 6. Dependence of d on β for different random number seeds N_s. $N_s = 10$ (upper solid line), 1 (lower solid), 3 (dash) and 6 (dash-dot), respectively. Here, $t_m = 10^4$.

In Fig. 6, we show the dependence of d on the random number seed N_s (which dictates the sequence of random numbers used in the computer run). For different N_s, the curves are different for low β, i.e., for processes that are completely or partially random. The higher β, the less random is the process and consequently, the less the influence of the random number seed. The choice of N_s may also influence which of the products wins, as exemplified by the upper curve in Fig. 6, where product A wins over product B.

In Figs. 7 and 8 we show plots of the time dependence of the fitness functions u(t) and v(t). Figure 7 corresponds to β = 0. The choice of the customer is completely random and none of the products win; each product has a market share of 50%. In Fig. 8 we show the case of β = 0.1. The customer has some information about the products. After an initial phase where the two products are chosen with equal probability, one product (here v) starts to gain some advantage. The partially informed customer will now start to prefer that product, and so v wins.

Up to now, the two products have equal start chances: u(0) = v(0) = 0. In Figs. 9 and 10 we show the case that u has an initial advantage. For β = 0 (Fig. 9), the choice of the customer is random. After some time, the initial advantage of u is washed out

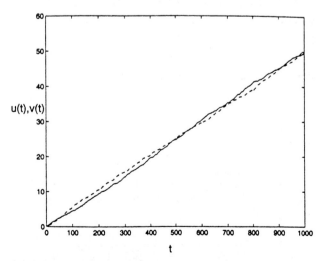

Fig. 7. The fitness u(t) and v(t) as a function of time t. $t_m = 10^3$; β = 0. u(0) = v(0) = 0. Here and in all following figures, u(t) is plotted with a solid line, and v(t) with a dash line.

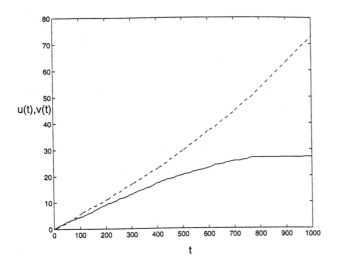

Fig. 8. Same as Fig. 7, except $\beta = 0.1$.

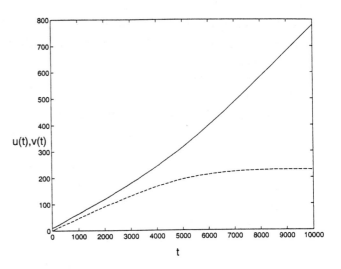

Fig. 10. Same as Fig. 9, except $\beta = 0.1$.

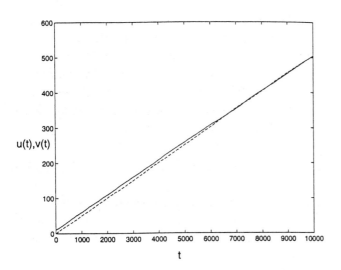

Fig. 9. The fitness u(t) and v(t) as a function of time t. $t_m = 10^4$; u(0) = 10 and v(0) = 0. $\beta = 0$.

4. Discussions

Active walk, as a field of research, is relatively new. For example, there is practically no systematic study of the statistical properties of even a single walker, in contrast to the case in passive walks such as random walk or Levy walk. For a large number of active walkers, in the continuum limit and in special cases, AW can become a reaction-diffusion system.

As shown in this paper, many non-physical systems with abstract landscapes can be handled by the AW model. Active walk could be understood as a general principle of organization in complex systems [11], in the sense that self-similarity, leading to fractals, is another such universal principle. In essence, AW is a description of nature using a potential function, the landscape; it is particularly effective in handling path dependent phenomena. It remains to be understood why nature seems to prefer a potential theory, not just for elementary particles but even for complex systems, and whether some basic symmetry principles are associated with the potential in these cases.

References

1. C. Van Doren, *A History of Knowledge: Past, Present, and Future* (Random House, New York, 1991).
2. E. Schrödinger, *What is Life?* (Cambridge University Press, Cambridge, 1967), Preface.
3. *Introduction to Nonlinear Physics*, edited by L. Lam (Springer, New York, 1997).
4. L. von Bertalanffy, *General System Theory: Foundations, Development, Applications* (George Braziller, New York, 1968).

5. I. Prigogine, *From Being to Becoming: Time and Complexity in the Physical Sciences* (Freeman, New York, 1980).
6. H. Haken, *Synergetics: An Introduction* (Springer, New York, 1983).
7. M. Bushev, *Synergetics: Chaos, Order, Self-Organization* (World Scientific, River Edge, 1994).
8. *Complexity: Metaphors, Models and Reality*, edited by G.A. Cowan, D. Pines and D. Meltzer (Addison-Wesley, Menlo Park, 1994).
9. *Modeling Complex Phenomena*, edited by L. Lam and V. Naroditsky (Springer, New York, 1992).
10. P. Bak, *How Nature Works: The Science of Self-Organized Criticality* (Copernicus, New York, 1996).
11. L Lam, in *Lectures on Thermodynamics and Statistical Physics*, edited by M. Costas, R. Rodriquez and A.L. Benavides (World Scientific, River Edge, 1994).
12. R. Freimuth and L. Lam, in [9].
13. L. Lam and R. Pochy, *Comput. Phys.* 7 (1993) 534.
14. L. Lam, *Chaos Solitons Fractals* 6 (1995) 267.
15. L. Lam, in [3], Chapter 15.
16. G. Sonnert and G. Holton, *Who Succeeds in Science?: The Gender Dimension* (Rutgers University Press, New Brunswick, 1995).
17. E.M. Rauch, M. Millonas and D.R. Chialvo, *Phys. Lett. A* 207 (1995) 185.
18. F. Schweitzer, K. Lao and F. Family, *BioSystems* 41 (1997) 153.
19. D.A. Kessler and H. Levine, *Phys. Rev. E* 48 (1993) 4801.
20. E. Ben-Jacob, O. Schochet, A. Tenenbaum, I. Cohen, A. Czirok and T. Vicsek, *Nature* 368 (1994) 46.
21. L. Lam, *Nonlinear Physics for Beginners* (World Scientific, River Edge, 1997).
22. R.P. Pan, C.R. Sheu and L. Lam, *Chaos Solitons Fractals* 6 (1995) 495; Ching-Yen Cheng, "Analysis and Simulation of Dielectric Breakdown Patterns in Thin Layers of Media," M.S. Thesis, National Chiao Tung University, Hsinchu, 1996.
23. P. Maass, A. Bunde and M.D. Ingram, *Phys. Rev. Lett.* 68 (1992) 3064.
24. H.A. Makse, S. Havlin and H.E. Stanley, *Nature* 377 (1995) 608.
25. F. Schweitzer and J. Steinbrink, in *Self-Organization of Complex Structures: From Individual to Collective Dynamics*, edited by F. Schweitzer (Gordon and Breach, London, 1997); and references therein.
26. S. Kauffman, *The Origins of Order: Self-Organization and Selection in Evolution* (Oxford University, New York, 1993).
27. W.B. Arthur, *Sci. Am.* (Feb. 1990) 92.
28. L. Lam, M.C. Veinott and R.D. Pochy, in *Spatial-Temporal Patterns in Nonequilibrium Complex Systems*, edited by P.E. Cladis and P. Palffy-Muhoray (Addision-Wesley, Menlo Park, 1995).
29. S.J. Gould, *Wonderful Life: The Burgess Shale and the Nature of History* (Norton, New York, 1989).
30. W.B. Arthur, in *The Economy as an Evolving, Complex System*, edited by P. Anderson, K. Arrow and D. Pines (Addison-Wesley, Menlo Park, 1988).
31. L. Lam and C.Q. Shu, "Active Walks and Increasing Returns in Economics" (preprint, 1996).
32. W.B. Arthur and D. Lane, *Structural Change and Economic Dynamics* 4 (1993) 81.
33. Li Shu and Lui Lam, to be published.

PART III

PROJECTS

The projects presented in Part III are the works of students from San Jose State University. Here are the stories behind their creations.

In my teaching of the freshman physics course on mechanics, after finishing the chapter on oscillations I would usually show the class the one-hour videotape of the Nova program on Chaos, which was followed by one lecture using the logistic map as an illustrative example. The idea was to emphasize to the students that not every oscillation in the world was small and simple harmonic and, of course, to introduce them to something new and exciting that was not yet in their textbooks. (That video program was splendidly made, in color, and guaranteed to excite anybody from all walks of life. The program included many demonstrations from the experts in chaos. For example, near the beginning, the chaos game was demonstrated by Michael Barnsley.) On one of these occasions, the day after I showed the video, a student in my freshman class, Prasanna Pendse, walked into my office, inserted his floppy disk into the obsolete Apple IIe computer belonging to my officemate, and—voila!—the chaos game just jumped out of the screen. Prasanna's effort, as inspired by the Nova progam, was recorded in Section 14.1.1.

Another pleasant surprise occurred one day, a few years ago. A sophomore, Rolf Freimuth, whom I had never met before, walked in and said he wanted to do some nonlinear physics with me. As was routine, I asked why and what he had been up to. As it turned out, Rolf had seen the Nova program on TV and read James Gleick's book on Chaos. He had written a program describing the chaotic waterwheel and its butterfly strange attractor (see Section 14.2.2). He wanted to learn more about chaos. Rolf subsequently joined my nonlinear physics group in research and took up my two courses in nonlinear physics. By the time he received his B.S. degree, he had written six papers with me.

The rest of the projects collected in Part III have their origins in the two courses, PHYS 255N—Nonlinear Physics and PHYS 255S—Nonlinear Systems, that

I initiated and taught. These two courses were offered in alternate years in the fall term. The students were graduate students plus a few senior undergraduates. Occasionally, guest speakers came to the class to speak about their own research. There were homework assignments, a midterm exam and a final exam. The final consisted of a term project by the student, the topic of which was determined during the first month of the course. The student could and usually did choose the topic by himself or herself, subject to the approval of the instructor. Joint projects by two, and in very exceptional cases up to four, students were allowed. The project could be computational, theoretical or experimental; it could be a repeat of some published work, a review of literature on some special topic or a completely original undertaking. Original work was not required but *highly* encouraged. A written progress report was submitted by the student around the middle of the term, which sometimes was counted as the midterm exam in lieu of a written test. At the end of the term, a final report was demanded, and a one-day workshop *open to the public* was organized in which every student had to orally present his or her work.

One such project, "Instabilities of Finite Water Columns" (Section 16.1), won the Allied-Signal Award of the Society of Physics Students. This project was initiated by Mark Fallis, a senior at the time, who discovered the undulatory instability of a water column when washing his hands some time back.

A word should be said about the backgrounds of our graduate students. Since our university is located in the heart of the Silicon Valley, many of these students hold full-time jobs in the neighborhood industries and have special skills or access to equipment which may not be readily available to students in other cities. Nonlinear physics offers these students the chance to combine their specialties, whatever that may be, with something exciting in the making. The students themselves become, in the process of learning, contributors to the development of nonlinear physics. Rocco Pochy and Victor Castillo, who wrote a number of the projects included here in Part III, are two such students.

Apart from Sections 14.1.1 and 14.2.2, the projects presented in Part III are selected from the homework and term projects of the two courses described above. They serve the purposes of (i) showing what the student projects can be in a course on nonlinear physics and (ii) supplementing the materials in Parts I and II, in particular, by providing computer programs for calculating some of the quantities or simulating some of the models described before. The reader can easily repeat or improve on the projects, or use them as the basis in formulating his or her own projects.

In Part III the computer languages adopted in the various listed programs range from Basic to C; personal computers from Apple II to PC are used; printers from dot matrix to ink jet to laser are used in the drawings. These choices made by the contributors are deliberately kept here, to convey the message that one does not need to be rich or have expensive equipments to do nonlinear science.

Part III Projects

Not included in Part III are some more original works which have been published as research papers, as listed in Appendix A2.

14 Computational

14.1 Fractals

14.1.1 The Chaos Game and Sierpinski Gasket
Prasanna U. Pendse

The Sierpinski gasket (SG) is defined as a recursive process which begins by cutting away the middle part of an equilateral triangle, followed by an infinte number of cutting and removal of smaller triangles (see Chapter 2). To carry out such a process is simple in principle but difficult in practice.

There is another way of generating the SG by using random numbers. Pick any three points on the paper and label the first point 1,2; the second point 3,4; and the third point 5,6. Now pick any point on the paper as the initial point. Next, roll a die which has six faces representing the numbers 1 to 6. According to the number on the die, construct a point that is the midpoint of the line segment joining the initial point and one of three points according to the number of the die. Now treat this midpoint as the new initial point and repeat the above process many number of times; the resulting figure will be the SG. This process of constructing the SG is called the chaos game [1].

To carry out the chaos game by hand is a long and tedious process. Programming the computer to do the job is much easier. The following is a program in BASIC that will run on an Apple II computer.

```
10      DIM X(6), Y(6)           ) Set up a one-dimensional arry for the
                                 ) X and Y data points
20      XX=150: YY=150           ) Randomly selected initial point
30      HGR                      ) High resolution graphics command
40      HCOLOR=3                 ) Set white color
50      FOR K = 1 TO 6           ) Set up a loop to read in the data values
60      READ X(K), Y(K)          ) Read in the data values
```

```
70      NEXT K
80      FOR I = 1 TO 50000          ) Repeat process 50,000 times
90      R = INT(RND(1)*6) + 1       ) Generate a random number from 1 to 6
100     XX = (XX+X(R))/2            ) Find x coordinate of midpoint
110     YY = (YY+Y(R))/2            ) Find y coordinate of midpoint
120     HPLOT XX,YY                 ) Plot the midpoint
130     NEXT I
140     DATA 140,5,140,5,40,140,    ) Data
             40,140,240,140,240,140
```

If QuickBASIC is used, simply replace line 30 by "30 SCREEN 9", discard line 40, and replace line 120 by "120 PSET(X,Y)". Unfortunately, the computer does not allow us to print out the picture. But one can see how it is formed step by step, and may use a (Polaroid) camera to take a picture of the screen. Although it may seem that generating a number of random points may fill up the whole triangle, it does not. It forms a definite shape; after the first few points are excluded, the shape is that of the SG.

The sides of the triangle may in fact be of any length. Upon experimentation with 4, 5 or 6 points, it can be concluded that the method works only for 3 points.

Reference

1. M. F. Barnsley, "Fractal Modelling of Real World Images," in *The Science of Fractal Images*, edited by H.-O. Peitgen and D. Saupe (Springer, New York, 1988).

14.1.2 Iteration Maps and the Sierpinski Fractals

Rolf D. Freimuth

Figures like the Sierpinski gasket, carpet, and tetrahedron can be constructed by a computer using an iteration loop which generates a new point from an old one repeatedly. After many iterations, a definite pattern will evolve, but even more astounding is that such interesting fractal patterns can be generated by a single map.

The Map

The iteration map in a two-dimensional space is defined by

$$x_{n+1} = ax_n + X_i, \quad y_{n+1} = ay_n + Y_i$$

which, after one iteration, transforms the point (x_n, y_n) to (x_{n+1}, y_{n+1}). Here a is a constant, and the points $P_i \equiv (X_i, Y_i)$, $i = 1, \ldots, m$, are given constants. Before each new point (x_{n+1}, y_{n+1}) is calculated, one of the P_i points is chosen randomly by the computer. The randomly selected P_i is then used in the above equations to calculate the new point from (x_n, y_n). The first point (x_1, y_1) is selected arbitrarily. The resulting pattern, after a large number of iterations (excluding the first few points), depends on a and the set of P_i.

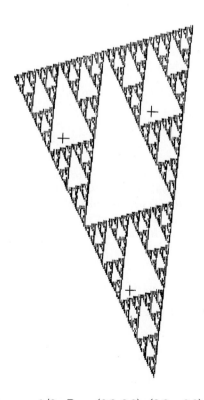

Fig. 1. The Sierpinski gasket. $a = 1/2$; $P_i = (0.3, 0.9), (0.2, -0.9), (-0.4, 0.6)$. Here and in the following figures, "+" denotes locations of the P_i points.

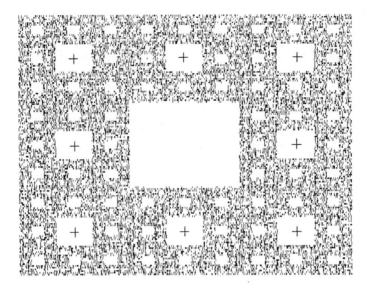

Fig. 2. The Sierpinski carpet. $a = 1/3$; $P_i = (1,1), (-1,1), (-1,-1), (1,-1), (0,1), (-1,0), (0,-1), (1,0)$.

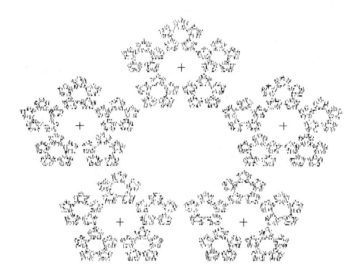

Fig. 3. The pentagonal fractal. $m = 5$, $a = 1/2.7$. For the values $m = 5$ and $a > 1/2.7$ (approximately), the five clusters of points will overlap. When the clusters overlap they do not interfere with each other, but they can make this ordered pattern look like a big mess of dots. For smaller values of a, the clusters will become distinct and more distant from each other, and the overall figure will become smaller. Increasing R can help if one wants to try small values of a.

When $a = 1/2$, $m = 3$, and the three points P_i are not collinear, the Sierpinski gasket is obtained (Fig. 1). Note that, in general, the m points P_i are not necessarily part of, within, or even near the figure.

Another fractal that may be generated is the Sierpinski carpet (Fig. 2). For this case, let $a = 1/3$, $m = 8$. The eight fixed points P_i are the four vertices and the four midpoints of the four sides of an *arbitrary* square.

14.1.2 Iteration Maps and the Sierpinski Fractals

As well as producing the well-known figures above, the map can also generate a wide variety of other self-similar patterns. For example, the P_i points can be arranged symmetrically in a circle of radius R. For values of $m > 3$ and $a < 1/2$ some interesting patterns can develop. An example of a fractal generated by this method is given in Fig. 3.

Though the previous examples are all symmetric (except the Sierpinski gasket), their self-similarity does not depend on their symmetry. Figure 4 shows a slightly modified Sierpinski carpet which is not symmetrical but is self-similar.

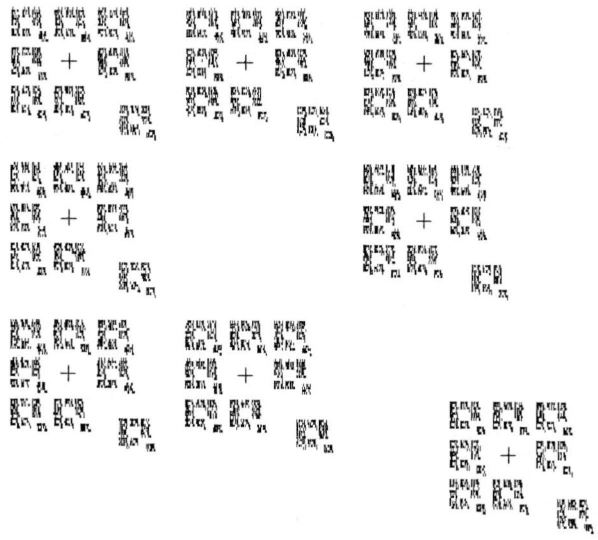

Fig. 4. This figure is generated by the program in Appendix B, after the following modifications have been typed:

130 a = 1/4
170 X(4) = 1.5 : Y(4) = −1.5 : X(8) = 1 : Y(8) = 0

The smaller value for a causes the eight clusters to separate so that they do not overlap, and the redefinition of X(4), Y(4) has an effect not only on the lower right cluster but on the whole figure.

The Sierpinski Tetrahedron

The two-dimensional map can be generalized to three dimensions,

$$x_{n+1} = ax_n + X_i, \quad y_{n+1} = ay_n + Y_i, \quad z_{n+1} = az_n + Z_i$$

where $i = 1, \ldots, m$. The Sierpinski tetrahedron (the fractal skewed web) is generated using four points in three dimensions (Fig. 5). The values of the set (X_i, Y_i, Z_i) can be any four noncoplanar points, but for simplicity we have let them be the vertices of a regular tetrahedron (having four equilateral triangular faces), defining three of them to be in the xy plane ($z = 0$), and the fourth to be above them at a height h ($x = 0 = y$). See Appendix D.

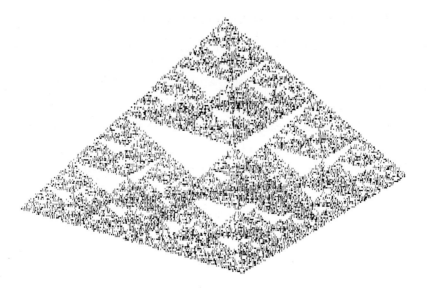

Fig. 5. The Sierpinski tetrahedron. $a = 1/2$, $m = 4$.

Discussions

By using different factors for each coordinate in the maps, e.g., $x_{n+1} = ax_n + X_i$, $y_{n+1} = by_n + Y_i$, $a \neq b$, self-affine (instead of self-similar) fractals are generated. Note that each figure shown here is made up of m smaller copies of the entire figure, each of which is scaled down by a factor of a. The maps presented here are special cases of some more general *iterated function systems* [1] in which rotating as well as rescaling and translating the "subpatterns" is involved.

The programs given in the Appendices are written in QUICK BASIC.

Appendix A: Program for the Sierpinski Gasket

```
100 CLS : SCREEN 2                          ' Setup the
110 WINDOW (-2, -2)-(2, 2)                  '   graphics screen
120 DIM X(3), Y(3)                          ' Dimension the arrays
130 A = 1 / 2                               ' Define a
140 X(1) =  .3 :Y(1) =  .9                  ' Define constants for
150 X(2) =  .2 :Y(2) = -.9                  '   the Sierpinski
160 X(3) = -.4: Y(3) =  .6                  '   gasket
170 X = 0: Y = 0                            ' Arbitrary initial pt
180 FOR L = 1 TO 25000                      '
190     R = INT(RND * 3 + 1)                ' Choose a random #
200     X = A * X + X(R): Y = A * Y + Y(R)  ' Calculate the new pt
210     PSET (X, Y)                         ' Plot the new pt
220 NEXT L                                  '
230 END
```

14.1.2 Iteration Maps and the Sierpinski Fractals

Appendix B: Program for the Sierpinski Carpet

```
100 CLS : SCREEN 2                                        ' Setup the
110 WINDOW (-2, -2)-(2, 2)                                '   graphics screen
120 DIM X(8), Y(8)                                        ' Dimension the arrays
130 A = 1 / 3                                             ' Define a
140 X(1) =  1:   Y(1) =  1:   X(5) =  0:   Y(5) =  1 '    Define constants
150 X(2) = -1:   Y(2) =  1:   X(6) = -1:   Y(6) =  0 '      for the
160 X(3) = -1:   Y(3) = -1:   X(7) =  0:   Y(7) = -1 '      Sierpinski
170 X(4) =  1:   Y(4) = -1:   X(8) =  1:   Y(8) =  0 '      carpet
180 X = 0: Y = 0                                          ' Arbitrary initial pt
190 FOR L = 1 TO 25000                                    '
200    R = INT(RND * 8 + 1)                               ' Choose a random #
210    X = A * X + X(R): Y = A * Y + Y(R)                 ' Calculate the new pt
220    PSET (X, Y)                                        ' Plot the new pt
230 NEXT L                                                '
240 END
```

Appendix C: Program for the Pentagonal Fractal

```
100 CLS : SCREEN 2                       ' Setup the
110 WINDOW (-2, -2)-(2, 2)               '   graphics screen
120 DIM X(5), Y(5)                       ' Dimension the arrays
130 A = 1 / 2.7                          ' Define A
140    X(1) =  .951: Y(1) =  .309        ' These points
150    X(2) =  .588: Y(2) = -.809        '   are the
160    X(3) = -.588: Y(3) = -.809        '   vertecies of
170    X(4) = -.951: Y(4) =  .309        '   symmetric
180    X(5) = 0: Y(5) = 1                '   pentagon

190 X = 0: Y = 0                         ' Arbitrary initial pt
200 FOR L = 1 TO 25000                   '
210    I = INT(RND * 5 + 1)              ' Choose random #
220    X = A * X + X(I): Y = A * Y + Y(I)' Calculate the new pt
230    PSET (X, Y)                       ' Plot the pt
240 NEXT L
250 END
```

Appendix D: Program for the Sierpinski Tetrahedron

```
100 CLS : SCREEN 2                       ' Setup the graphics
110 WINDOW (-2,-2)-(2,2)                 '   screen
120 PI=3.14159                           ' Define pi
130 ROT= 5*PI/180                        ' Base rotation angle
140 DIM X(4),Y(4),Z(4)                   ' Dimension arrays
150 A=1/2                                ' Define A
160 FOR I=1 TO 3                         ' Calculate the
170    X(I)= COS(I*PI*2/3 + ROT)         '   three
180    Y(I)= SIN(I*PI*2/3 + ROT)         '   base
190    Z(I)= 0                           '   points
200 NEXT I                               '
210 X(4)= 0 : Y(4)= 0 : Z(4)= SQR(2 )    ' Define the "peak" pt
220 X=0:Y=0:Z=0                          ' Arbitrary initial pt
230 FOR L=1 TO 2500
```

```
240      I=INT(RND*4+1)                            ' Choose random #
250      X=A*X+X(I) : Y=A*Y+Y(I) : Z=A*Z+Z(I)      ' Calculate the new pt
260      PSET ( Y-.3*C ,Z-.3*X-1   )               ' Plot the new pt
270 NEXT L
280 END
```

Reference

1. Michael Barnsley, *Fractals Everywhere* (Academic, New York, 1988), Sec. 3.8.

14.1.3 Calculating the Box Dimension

Victor M. Castillo

To calculate the fractal dimension of an irregular shape, we can use a box counting method (see Chapter 2). The program in Appendix C is used to find the box dimensions of two typical fractals. (The programs listed below are written in Microsoft QuickBASIC version 4.5.)

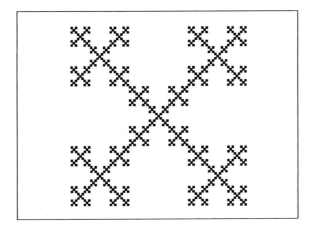

Fig. 1. The "snowflake" prefractal ($n = 4$).

In the first example, the deterministic "snowflake" prefractal with $n = 4$ (shown as Fig. 1 here) is used as the fractal object. It is generated by inserting the program of Appendix A into that in Appendix C, and stored in a 256×256 array. The array is then divided into a grid of non-overlapping blocks of size $\varepsilon \times \varepsilon$. The program then scans to count the number of blocks that cover the pattern. This number is stored along with the block size ε. The blocks are then subdivided into $\varepsilon/2 \times \varepsilon/2$ blocks and the scan is repeated. This continues down to the pixel (or cell) level. Next a linear least-squares routine is used to determine the slope of the logarithm of the count vs the logarithm of the block size. The fractal dimension is obtained as the negative of the slope. For the snowflake example, the dimension is found to be 1.445 which is less than 2% off the expected value of 1.465 ($\approx \log 5/\log 3$).

For the second example, one may use the random walks generated by the program in Appendix B as the fractal object. More generally, one can place any set of fractal points in the space provided for the Pattern Generating Program in Appendix C, and try to find its box dimension.

Appendix A: Pattern Generating Program: "Snowflake" Fractal

```
'*****************************************************************
'  Pattern Generating Program: "Snowflake" fractal
'*****************************************************************

pmax = 4                              'number of fractal iterations
t = (3 ^ (pmax + 1) - 1) / 2
w = 2 ^ 8
LET x0 = 1:  y0 = 1:  x1 = w:  y1 = w:  c = 10
SCREEN 9: WINDOW (x0 - 1, y0 - 1)-(x1 + 1, y1 + 1): CLS

t = 0
PSET (t + w / 2, t + w / 2), c
FOR p = 0 TO pmax
    s = 3 ^ p
    FOR i = -t TO t
        FOR j = -t TO t
            IF POINT(i + w / 2, j + w / 2) = 10 THEN
                PSET (i + s + w / 2, j + s + w / 2), c
                PSET (i + s + w / 2, j - s + w / 2), c
                PSET (i - s + w / 2, j + s + w / 2), c
                PSET (i - s + w / 2, j - s + w / 2), c
            END IF
        NEXT
    NEXT
    t = (3 * s - 1) / 2
NEXT
```

Appendix B: Pattern Generating Program: Random Walks

```
'**************************************************
'  Pattern Generating Program: Random Walks
'**************************************************

LINE (x0 - 1, y0 - 1)-(x1 + 1, y1 + 1), , B     'draws bounding box
j0 = (y1 - y0) / 2
FOR i = x0 + 1 TO x1
    j1 = j0 + 20 * RND - 10
    LINE (i - 1, j0)-(i, j1), c
    j0 = j1
NEXT
```

Appendix C: The Box Counting Program

This program uses the graphics memory to store the fractal pattern for the analysis. $PSET(x,y), c$ and $POINT(x,y)$ commands are used to store and read to pattern. Simple modifications can be done in order to analyze a pattern in a memory array.

A possible improvement to this routine is having a collection of random x and y offsets for the grids. This must still enforce non-overlapping blocks and must require wrap-around searching. The average values would then be used for the least-squares analysis.

14.1.3 Calculating the Box Dimension

```
DEFINT I-J, M, P, T, W-Y
DIM logC(20),logN(20)


'*****************************************************************
'Place "Pattern Generating Program" in the space below.
'*****************************************************************




'*****************************************************************
' Now the image is scanned to determine the fractal dimension
'*****************************************************************

m = INT(LOG(x1 - x0) / LOG(2))
FOR boxdiv = 2 TO m
      count = 0
      LET B = (x1 - x0 + 1) / 2 ^ boxdiv
      FOR i0 = x0 TO x1 STEP B
            FOR j0 = y0 TO y1 STEP B
                  LET flag = 0
                  FOR j = j0 TO j0 + (B - 1)
                        FOR i = i0 TO i0 + (B - 1)
                              IF POINT(i, j) = c THEN
                                    LET count = count + 1
                                    i = i0 + B: j = j0 + B
                              END IF
                        NEXT
                  NEXT
            NEXT
      NEXT
      SOUND 1000, .5
      logC(boxdiv) = LOG(count)
      logL(boxdiv) = LOG(B)
NEXT boxdiv

'*****************************************************************
'Linear least squares method is used to find the slope.
' (Only the straight-line part of the log-log curve should be
'  used in finding the slope. Generally, this part should be found
'  by examining the log-log curve before using the program below.)
'*****************************************************************

FOR boxdiv = 2 TO m
      sumy = sumy + logC(boxdiv)
      sumx = sumx + logL(boxdiv)
      sumxy = sumxy + logL(boxdiv) * logC(boxdiv)
NEXT
barx = sumx / (m - 1)
bary = sumy / (m - 1)
FOR boxdiv = 2 TO m
      u = u + (logL(boxdiv) - barx) ^ 2
NEXT
a = (sumxy - (m - 1) * barx * bary) / u
PRINT "Box Dimension = "; -a

END
```

14.1.4 Diffusion-Limited Aggregates in Radial Geometry
Rocco D. Pochy

The program below uses the random walk model to create a two-dimensional, diffusion-limited aggregation (see Section 9.1). The process is based on the idea that a particle will wander in space until it comes in contact with a cluster. There it will stick and become part of the cluster and a new particle will wander in.

This implementation is designed to display the growth of this DLA fractal pattern. The system requirement are an IBM-PC compatible with at least 640K of memory and a VGA graphics display. This code was written to take advantage of the highest graphics resolution available on a generic VGA display. Changes to the size of the field array and graphics display initialization will allow porting to any other system. The computer language is C: Microsoft 6.00a.

A typical DLA cluster generated by this program is shown in Fig. 1.

Fig. 1. A computer generated diffusion-limited aggregate (DLA).

14.1.4 Diffusion-Limited Aggregates in Radial Geometry

```
#include <stdio.h>
#include <stdlib.h>
#include <math.h>
#include <conio.h>
#include <graph.h>

/*---[ Define Global Constants ]-----------------------------------*/
#define  TRUE         1
#define  FALSE        0
#define  PI           3.141592
#define  SET          1
#define  EMPTY        0
#define  RELEASE      5                      /* Release distance of  */
                                             /* new particle.        */
#define  NMAX         480                    /* Size of Field Array  */
#define  RANDOM       rand()/(RAND_MAX+1.0)  /* Random Number (0..1] */
#define  COIN_FLIP    rand()>(RAND_MAX/2)    /* Random Value (0|1)   */

/*---[ Define Global Variables ]-----------------------------------*/
static   char
         huge   field[NMAX][NMAX];           /* Growth Field Array   */
static   int    offset = NMAX/2;             /* Offset to center of  */
                                             /* growth.              */

/*---[ Define External Functions ]---------------------------------*/
extern   void   move(int*,int*,int*);
extern   void   pick(int,int*,int*);
extern   void   init_display(void);
extern   void   close_display(void);
extern   void   display(void);

/*================================================================*/
main()
{
   int      x,y;              /* Position in Field Array              */
   int      r;                /* Radius from center of Field Array    */
   int      radius= 1;        /* Radius is circle enclosing growth    */
   int      rmax  = offset-1; /* Maximum radius for growth.           */
   unsigned seed;             /* Random Number Generation Seed        */
   long     stuck = 0;        /* # of particles in Growth Pattern     */

/*---[ Initialation ]---------------------------------------------*/

   /*
    * Initialize Field Array to EMPTY and set a seed in the center of
    * the array.
    */
   for (x=0;x<NMAX;x++)
      for (y=0;y<NMAX;y++)
         field[x][y] = EMPTY;

   field[offset][offset] = SET;
   printf("Enter Random Seed:");  /* Get Seed for Random Number Gen. */
   scanf("%d",&seed);
   srand(seed);
   init_display();                /* Initialize Graphics Display     */

/*---[ Main Loop ]------------------------------------------------*/

   /*
    * Loop until growth pattern is greater than rmax or keyboard hit.
    */
   while (radius<rmax)
   {
      r = radius;
```

```c
            move(&r,&x,&y);              /* Create particle & let it walk.*/
            stuck++;                     /* Count the # of particles stuck*/
            field[x][y] = SET;           /* Log where the particle stuck  */

            if (r>radius)                /* Define the maximum radius of  */
                radius = r;              /* growth.                       */

            if (stuck%1000==0)           /* Update display every thousand */
                display();               /* particles.                    */

            if (kbhit()) break;          /* Exit Loop if key is hit.      */
        }

        display();                       /* Display Final Configuration   */
        printf("# of particles:%ld",stuck);
        while (!kbhit());                /* Exit program when key is hit. */
        close_display();                 /* Close Graphics Display        */
        return(0);
}
/*================================================================*/


void
pick(int r, int* x, int* y)
/*
 * Pick a random position on the grid that is r distant from
 * the center.
 */
{
    double theta = 2.0 * PI * RANDOM;
    *x = (int)((double)(r)*sin(theta)) + offset;
    *y = (int)((double)(r)*cos(theta)) + offset;
}

int
test(int x, int y)
/*
 * Test if particle is next to a cluster site. If so, return a
 * ono-zero value.
 */
{
    int count = 0;
    if (field[x+1][y]==SET) count++;
    if (field[x-1][y]==SET) count++;
    if (field[x][y+1]==SET) count++;
    if (field[x][y-1]==SET) count++;
    return(count);
}

void
move(int* r,int* x,int* y)
/*
 * Create a wandering particle and let it random walk until it
 * finds the cluster.
 */
{
    int *temp;
    int done = FALSE;
    int rel  = *r+RELEASE;          /* Radius of Release of new particle*/
    int step = 1;                   /* Step Size of particle movement   */
    long r2;                        /* Square of distance of particle   */
                                    /* from the center.                 */
    long rel2  = (long)(rel)*rel;   /* Square of Release radius.        */
    long rmax2 = 4 * rel2;          /* Square of Escape Radius.         */
```

14.1.4 Diffusion-Limited Aggregates in Radial Geometry

```
        long dx, dy;             /* X and Y position of particle with*/
                                 /* respect to the center.           */
        pick(rel,x,y);           /* Release a new particle into the  */
                                 /* system.                          */
        /*
         * Loop until wandering particle stick to cluster. If particle
         * wanders more than 2*rmax, then particle is considered "lost"
         * and a new particle is created.
         */
        while (!done)
        {
            if (COIN_FLIP)       /* Choose walk direction (x or y)   */
                temp = x;
            else
                temp = y;

            if (COIN_FLIP)       /* Choose step interval             */
                *temp += step;
            else
                *temp -= step;

            dx = *x-offset;
            dy = *y-offset;
            r2 = dx*dx + dy*dy;  /* Calculate distance from center   */

            if (r2 > rmax2)      /* If distant too great, pick a new */
                pick(rel,x,y);   /* particle.                        */
            if (r2 > rel2+100)   /* To optimize perfomance, step size*/
                step = 2;        /* is increased the further away from*/
            if (r2 > rel2+400)   /* the center the particle drifts.  */
                step = 4;        /* If r>R +10 step size = 2         */
            if (r2 > rel2+1600)  /*      +20           = 4           */
                step = 8;        /*      +40           = 8           */
            if (r2 > rel2+6400)  /*      +80           = 16          */
                step = 16;       /* R = Radius of Release Zone       */
            if (r2 < rel2)       /* If particle is within range, then*/
                if (test(*x,*y)) /* test if particle is adjacent to  */
                    done = TRUE; /* "stuck" cluster particles.       */
        }
        *r = (int)(sqrt((double)(r2)));/* Return the distance at which */
                                 /* particle stuck from the center. */
}
void
init_display(void)
/*
 * Set Graphics Display to VGA 640x480
 */
{
    _setvideomode(_VRES16COLOR);
}

void
close_display(void)
/*
 * Reset Graphics Display to default display Mode
 */
{
    _setvideomode(_DEFAULTMODE);
}

void
display(void)
/*
```

```
 * Display Field Array.  Cells that contain cluster particle are
 * drawn in white.
 */
{
    short    i,j;

    for (i=0;i<NMAX;i++)
       for (j=0;j<NMAX;j++)
          if (field[i][j]!=EMPTY)
             _setpixel(i,j);
}
```

14.1.5 The Dielectric Breakdown Model with Noise Reduction

Rocco D. Pochy

The program below is designed to simulate the growth patterns found in the dielectric breakdown model with the option of noise reduction (see Section 12.2 and the comments in the program below). The computer language used is Microsoft C v.5.00. Typical fractal patterns produced by the program are shown in Fig. 1 (without noise reduction) and Fig. 2 (with noise reduction).

```
/*   The model used defines a potential field on a lattice, exis-  */
/*   ting between two coaxial electrodes. The center electrode     */
/*   assumes a fixed min. potential (MIN_VALUE) and the outer,     */
/*   radial electrode with the max. potential (MAX_VALUE). The     */
/*   potential field between the electrodes is calculated by       */
/*   Laplace's equation solved numerically via the relaxation      */
/*   method. The allowed error in the solution is define by an     */
/*   epsilon (epsilon), if found within the maximum number of      */
/*   interations (ITER).                                           */
/*                                                                 */
/*          ..........                                             */
/*          ....o.......        o=site (immediate neighbor)        */
/*          ...oxo.o....        x=node (aggregation point)         */
/*          ..ooxooxo...        .=mesh (lattice points)            */
/*          .oxxxxxxoo..                                           */
/*          ..ooxooxxxo.                                           */
/*          ....o..ooxo.                                           */
/*          .........o..                                           */
/*          ..........                                             */
/*                                                                 */
/*   Once the potential field of the system is defined, one of     */
/*   sites surrounding the center electrode is chosen via a        */
/*   probability assigned to each of the sites based on the        */
/*   potential field strength. To simulate the rapid tip growth    */
/*   associated with some growth patterns, a parameter "eta" is    */
/*   assigned:                                                     */
/*                                                                 */
/*      probability(site) = (field[site])^eta / SUM(field^eta)     */
/*                                                                 */
/*   By varying eta, the growth pattern changes from "blobs" to    */
/*   "stringy."                                                    */
/*                                                                 */
/*   There is a COUNT array to track the number of times a site    */
/*   has been visited during an iteration. If the site has been    */
/*   visited min_count, then "growth" will occur at that site and  */
/*   the COUNT array will be reset to zero.                        */
/*                                                                 */
/*   Growth occurs at nearest neighbor and next-nearest neighbor   */
/*   sites.                                                        */
/*                                                                 */
/*   After run is completed, data array is stored to disk for      */
/*   later analysis.                                               */
/*   Files generated have the extensions:                          */
/*          HDR = header information                               */
/*          DAT = Data                                             */
/*                                                                 */
/*----------------------------------------------------------------*/
```

```c
/*--[INCLUDE FILES]----------------------------------------------*/
#include <stdio.h>
#include <stdlib.h>
#include <float.h>
#include <math.h>
#include <graph.h>

/*--[FORWARD DEFINITIONS OF RELAX FUNCTIONS]---------------------*/
extern void Init_Mesh(void);            /* Initialize Mesh points    */
extern void Relax_Mesh(int,float);      /* Solve Mesh via Relaxation */
extern int
      Update_Mesh(float,float,int);     /* Choose growth site        */
extern void Display_Mesh(void);         /* Graphic Display of Mesh   */
extern void Display_Node(int,int,int);  /* Display points on Screen  */
extern void Set_Display(int);           /* Set Video Graphics Display*/
extern void Reset_Display(void);        /* Restore to "normal" display*/
extern void Pause_X(void);              /* Pause unitl 'X' is pressed */
extern void Save_Array(char *);         /* Save Field Array to file  */
extern void
      Save_Header(char *,float,float);  /* Save Header Info to file  */
extern void
      Save_Data(char *,float,float);    /* Save Data (array & header) */
extern void Print_Array(void);          /* Print Contents of Array to */
                                        /* hardcopy                   */
/*--[DEFINITIONS]------------------------------------------------*/
#define TRUE          1
#define FALSE         0

#define BOUND         1              /* Site bounded to aggregate */
#define UNBOUND       0              /* Site not bounded, possible*/
                                  /* growth site              */
#define BOUNDARY      2              /* Site is part of the out-  */
                                  /* side boundary            */
#define UNAVAILABLE   3              /* Site not bounded, but un- */
                                  /* available as growth site */

#define NMAX          256            /* Max. Size of System Field */
#define NODE          7000           /* Max. # of neighbor Sites  */

#define MAX_VALUE     10000          /* Max. Potential Value      */
#define MIN_VALUE     0              /* Min. Potential Value      */
#define EPSILON       10             /* Max. Allowed Error        */
#define ITER          400            /* Max. # of Iterations to   */
                                  /* find solution            */

#define OUT_COUNT     5              /* First Series Ouput        */
#define OUT_INT1      100            /* Graphic Output Every      */
                                  /* interval up to OUT_COUNT */
#define OUT_INT2      1000           /* Graphic Output Interval   */
                                  /* for greater than OUT_COUNT*/

#define NCOLOR        6              /* Number of Colors define   */
                                  /* contour plotting         */
#define BLACK         0              /* Max. Potential Field Color*/
#define WHITE         15             /* Min. Potential Field Color*/

/*--[Global Variables]-------------------------------------------*/
static unsigned
      int   huge   field[NMAX][NMAX]; /* Integer Field Potential  */
                                  /* Array                    */
static unsigned
      char  huge   tag[NMAX][NMAX];   /* Status Array of sites in */
                                  /* Potential Field          */
static unsigned
      char  huge   count[NMAX][NMAX]; /* Count Array of site vists */
                                  /* of Potential Field sites  */
```

14.1.5 The Dielectric Breakdown Model with Noise Reduction

```c
    static int         f_size;             /* Size of System Field      */
                                /* where f_size<NMAX         */

    static float huge  node[NODE];         /* Probablity value of site  */
    static int         xtemp[NODE];        /* x location of site        */
    static int         ytemp[NODE];        /* y location of site        */
    static int         minval[NODE];       /* Minimum value of bounded  */
                                /* neighbor                  */
    static float       r_value=0;          /* Sum of Raduii^2 measured  */
                                /* for Radius of Gyration    */

    int           video;                   /* Video Display Type Flag   */
                                /* for graphics functions    */
    int           graph_flag;              /* Graphics Output Flag      */
    int           txt_loc;                 /* X location of text        */
                                /* position on screen        */
/*--[Error Handling]------------------------------------------*/

/*------------------------------------------------------------*/
void main(void)
{
/*------------------------------------------------------------*/
    int       done    = FALSE;      /* End of Program Flag        */
    int       i       = 0;          /* Number of Sites occupied   */
    int       seed;                 /* Seed for Random Generator  */
    int       gr_count=0;           /* Graphics Output Counter    */
    int       vtype;                /* Video Type Flag            */
    float     eta;                  /* Probability Parameter      */
    float     epsilon;              /* Smallest Non-zero value    */
    int       min_count;            /* Minimum Number of vists to */
                            /* "bound" a site             */
    char      filename[13];         /* File Name where to store   */
                            /* data array and infomation  */
    char      *fmode="wb";          /* Set Binary File Write mode */
    FILE      *fp;                  /* File Pointer               */
    int       fsize;                /* File Size                  */
    int       counter=OUT_INT1;     /* Graphics Output Count      */
                            /* Interval                   */
/*------------------------------------------------------------*/

    printf("Enter Size [max=%d]:",NMAX-1);
    scanf("%d",&f_size);

    printf("Enter Random Seed (1-65000):");
    scanf("%d",&seed);
    srand(seed);

    printf("Enter Eta Value (0.0-10.0) :");
    scanf("%f",&eta);

    printf("Enter Epsilon Value (1.0-0.001):");
    scanf("%f",&epsilon);

    printf("Enter Maximum Visit Count (1-255):");
    scanf("%d",&min_count);

    printf("Graphics Display Output (1=Yes or 0=No)?:");
    scanf("%d",&graph_flag);

    if (graph_flag==TRUE)
    {
        printf("Enter Display Type (1=BW, 2=COLOR, 3=HiRes):");
        scanf("%d",&vtype);
        if (vtype==3)
        txt_loc = 65;
        else
        txt_loc = 27;
    }
```

```c
printf("Enter Filename (max. 8 characters):");
scanf("%s",filename);

printf("Initializing Potential Field \n");

Set_Display(vtype);
Init_Mesh();

Display_Mesh();

done = FALSE;

if (graph_flag==TRUE)
{
    _settextposition(1,txt_loc);
    printf("eta : %3.3f",eta);

    _settextposition(2,txt_loc);
    printf("eps : %3.3f",epsilon);

    _settextposition(3,txt_loc);
    printf("count:%d",min_count);
}

while (done==FALSE)
{
    /* Solve Laplacian Every 4th Iteration to */
    /* enhance performance                    */
    if (i%4==0)
    {
     Relax_Mesh(ITER,epsilon);
    }


    /* Update Display of Graphics Display Devices */
    if ((graph_flag==TRUE)&&(i%counter==0))
    {
     Display_Mesh();

     ++gr_count;
     if (gr_count>OUT_COUNT) counter=OUT_INT2;
    }

    done = Update_Mesh(eta,epsilon,min_count);

    ++i;

    if (graph_flag==TRUE)
    {
     _settextposition(20,txt_loc);
     printf("Points: %d ",i);
    }
}
/* Final Display of Plot */
if (graph_flag==TRUE)
{
    Display_Mesh();

    _settextposition(21,txt_loc);
    printf("Size : %d",f_size);

    /* Clear Area to Print File Name of Data Stored */
    _settextposition(23,txt_loc);
    printf("File :          ");

    _settextposition(24,txt_loc);
    printf("                ");
```

14.1.5 The Dielectric Breakdown Model with Noise Reduction

```
      _settextposition(24,txt_loc);
      printf("%s",filename);

      Pause_X();
      Reset_Display();
   }

   /* Save Data to Disk */
   Save_Data(filename,epsilon,eta);
}

void Init_Mesh(void)
/*------------------------------------------------------------------*/
/* Description:                                                     */
/*     This routine initializes the values at each site on the      */
/*     mesh. Boundary conditions are set, all other sites are set   */
/*     to zero. Count Array is set to zero.                         */
/*                                                                  */
/*------------------------------------------------------------------*/
{
/*------------------------------------------------------------------*/
   register int    i,j;               /* Site indices              */
   register int    itemp,jtemp;       /* Relative site position    */
   register int    offset=f_size/2;   /* Location of center of mesh */
   float           r = f_size/2-1;    /* Radius of active mesh     */
   float           radius;            /* Distance from center      */
/*------------------------------------------------------------------*/

   for (i=0;i<=f_size;i++)
   {
      itemp = i - offset;
      itemp = itemp * itemp;

      for (j=0;j<=f_size;j++)
      {
         /* Clear count array */
         count[i][j] = 0;

         jtemp = j - offset;
         jtemp = jtemp * jtemp;

         radius = sqrt((double)(itemp) + jtemp);
         /* Define circular boundary for radial electrode of radius */
         /*   offset (center of mesh)                               */
         if (radius>=r)
         {
            field[i][j] = MAX_VALUE;
            tag[i][j]   = BOUNDARY;
         }
         else
         {
            /* If center of mesh, initialize center electrode       */
            /* else set site to zero                                */
            if (radius==0.0)
            {
               field[i][j] = MIN_VALUE;
               tag[i][j]   = BOUND;
            }
            else
            {
               field[i][j] = MAX_VALUE;
               tag[i][j]   = UNBOUND;
            }
         }
      }
   }
}
```

```
void Relax_Mesh(int iter, float eps)
/*-----------------------------------------------------------------*/
/* Description:                                                    */
/*      This routines calculates the field potential in the mesh   */
/*      model using Laplace's equation via the Relaxtion Method.   */
/*-----------------------------------------------------------------*/
/* Parameters:                                                     */
/* int         iter            /* Maximum # of interactions        */
                               /* to solve field potential  */
/* float       eps             /* Maximum allow percentage         */
                               /* value of solution to the  */
                               /* potential field (barring  */
                               /* # of iterations).         */
/*-----------------------------------------------------------------*/
{
/*-----------------------------------------------------------------*/
   register int   i,j;           /* Site indices                   */
   register int   l=0;           /* Current # of iterations        */
   int            average;       /* Calculated value of a site     */
   int            residual;      /* Calculated error of a site     */
   int            norm;          /* Current error of the mesh      */
   int            max_err;       /* Maximum Error Value allowed    */
/*:----------------------------------------------------------------*/

   max_err = MAX_VALUE * eps;
   do
   {
      norm = 0;
      for (i=1;i<f_size;i++)
      {
       for (j=1;j<f_size;j++)
       {
          if (tag[i][j]==UNBOUND)
          {
             average = (field[i][j-1] + field[i][j+1] +
                    field[i-1][j] + field[i+1][j])/4;
             residual= abs(average - field[i][j]);
             if (residual>norm) norm = residual;
             field[i][j]  = average;
          }
       }
      }
      l++;
   } while ((norm>max_err)&&(l<iter));

   if (graph_flag==TRUE)
   {
      _settextposition(24,txt_loc);
      printf("Iter: %3d ",l);
   }
}


int Update_Mesh(float exp,float eps,int min_count)
/*-----------------------------------------------------------------*/
/* Description:                                                    */

/*      This routine finds the neighboring sites of the growth     */
/*      pattern and assigns them a growth probability based on     */
/*      their potential field strenght. The equation to find the   */
/*      probability is:                                            */
/*                                                                 */
/*         probability(site) = (field[site])^exp / SUM(field^exp)  */
/*                                                                 */
/*      A site becomes BOUND if the site is visted min_count times.*/
/*                                                                 */
/*      Note: All values less than MAX_VALUE*eps are treated as zero.*/
/*            This affects the probabiTity distribution.           */
```

14.1.5 The Dielectric Breakdown Model with Noise Reduction

```
/*                                                                      */
/*--------------------------------------------------------------------- */
/*                                                                      */
/* Parameter:                                                           */
/* float          exp                /* Exponent modifer for proba-*/
/*                                   /* bility growth factor       */
/* float          eps                /* Maximum percentage error   */
/*                                   /* allowed in calculation     */
/* int            min_count          /* Minimum # of Vists to "set"*/
/*                                   /* a site                     */
/*--------------------------------------------------------------------- */
{
/*--------------------------------------------------------------------- */
    register int i,j,k;              /* Indices                        */
    register int m = 0;              /* # of neighboring sites         */
    register int neighbor;           /* #of neighbors around a site    */
    register int itemp,jtemp;        /* Position wrt Center of Mesh    */
    int          offset=f_size/2;    /* Offset of Center of Mesh       */
    int          max_err=MAX_VALUE*eps; /* Max. error value            */
    int          found;              /* flag if site found             */
    int          min;                /* Minimum value of bound site    */
    float        interval;           /* prob. for a given site         */
    float        norm;               /* Normalization factor           */
    float        x;                  /* random number [0..1]           */
/*--------------------------------------------------------------------- */
    for (i=1;i<f_size;i++)
    {
       for (j=1;j<f_size;j++)
       {
          /* Test location is occupied */
          if (tag[i][j]==UNBOUND)
          {
             /* Test if next to aggregation growth     */
             /* Count number of neighboring occupied cells */
             neighbor = 0;
             min = MAX_VALUE;

             /* Nearest Neighbors */
             if (tag[i+1][j]==BOUND)
             {
                if (min>field[i+1][j]) min = field[i+1][j];
                neighbor++;
             }

             if (tag[i][j+1]==BOUND)
             {
                if (min>field[i][j+1]) min = field[i][j+1];
                neighbor++;
             }

             if (tag[i-1][j]==BOUND)
             {
                if (min>field[i-1][j]) min = field[i-1][j];
                neighbor++;
             }

             if (tag[i][j-1]==BOUND)
             {
                if (min>field[i][j-1]) min = field[i][j-1];
                neighbor++;
             }

             /* Neighbor is present */
             if (neighbor>0)
             {
                /* Test for potential error */
                if (m<NODE)
                {
```

```c
            /* Get value of node */
            int value = field[i][j];

            if (value<max_err)
            {
                tag[i][j] = UNAVAILABLE;
            }
            else
            {
                /* Save Value and Location */
                node[m]  = ((float)(value))/MAX_VALUE;
                xtemp[m] = i;
                ytemp[m] = j;
                minval[m]= min;
                m++;
            }
          }
          else
          {
            if (graph_flag==TRUE)
            {
                _settextposition(16,txt_loc);
                printf("overflow");
            }
          }
       }
     }
   }
}

/* Over-shot by one (sites = 0 .. m-1) */
m = m - 1;

/* Define Normalization Constant */
norm = 0.0;
for (k=0;k<=m;k++)
{
   /* Modify node[k] by algorithm to scale the growth */
   /*  probability based on potential and eta.        */
   node[k] = pow(node[k],exp);

   /* Create normalization factor for growth probability */
   norm = norm + node[k];
}

/* Iterate until a site has been found to have been visted */
/* min_count times (associated with "sticking" probability)*/
do
{
   /* Define Random Number [0,1] and initialize index k */
   k = 0;
   x = rand()/32768.0;
   interval = 0.0;
   do
   {
    interval = interval + node[k]/norm;
    if (interval>=x)
    {
        i = xtemp[k];
        j = ytemp[k];
        count[i][j]++;
        found = TRUE;
    }
    else
    {
        found = FALSE;
        if (k>m)
```

14.1.5 The Dielectric Breakdown Model with Noise Reduction

```
                {
                  /* Diagnostic Ouput */
                  if (graph_flag==TRUE)
                  {
                    _settextposition(20,20);
                    printf("x=%5.4f interval=%5.4f",x,interval);
                    Pause_X();
                  }
                  found=TRUE;
                }
                k++;
              }
          } while (found==FALSE);
      } while (count[i][j]<min_count);

      /* Set k-node to aggregation */
      i = xtemp[k];
      j = ytemp[k];
      tag[i][j] = BOUND;
      field[i][j] = MIN_VALUE + minval[k];

      /* Add new radius to sum */
      itemp = offset - i;
      jtemp = offset - j;
      r_value = r_value + ((float)(itemp*itemp) + (float)(jtemp*jtemp));

      /* Clear Count Array */
      for (itemp=0; itemp<f_size; ++itemp)
         for (jtemp=0; jtemp<f_size; ++jtemp)
              count[itemp][jtemp] = 0;

      /* Test if edge of field reached (system shorted) */
      if ((tag[i+1][j]==BOUNDARY)||
          (tag[i][j+1]==BOUNDARY)||
          (tag[i-1][j]==BOUNDARY)||
          (tag[i][j-1]==BOUNDARY))
         return (TRUE);
      else
         return (FALSE);
}


void Display_Mesh(void)
/*----------------------------------------------------------------*/
/* Description:                                                   */
/*     This routine displays the mesh model graphically           */
/*----------------------------------------------------------------*/
{
/*----------------------------------------------------------------*/
      register int    i,j;         /* Indices                     */
      int             value;       /* State of site               */
                                   /* BOUND = Crystal site        */
/*----------------------------------------------------------------*/

      for (i=1;i<f_size;i++)
      {
         for (j=1;j<f_size;j++)
         {
           value = tag[i][j];
           Display_Node(value,i,j);
         }
      }
}


void Display_Node(int value,int i,int j)
/*----------------------------------------------------------------*/
/* Description:                                                   */
```

```c
/*      This routine draws a color point on the graphics screen based*/
/*      on the value of the location (value) and the graphics display*/
/*      mode.                                                        */
/*-------------------------------------------------------------------*/
/* Parameters:                                                       */
/* int            value              /* Value of a site              */
/* int            i,j                /* Location of the site         */
/*-------------------------------------------------------------------*/
{
/*-------------------------------------------------------------------*/
    int            color;            /* color of the site            */
/*-------------------------------------------------------------------*/

    switch (video)
    {
    case 1 : if (value==BOUND)
                _setcolor(WHITE);
             else
                _setcolor(BLACK);
             break;
    /* This requires modification for it to work properly in color */
    case 2 : color = (NCOLOR * (1 - ((float)(value))/MAX_VALUE))+1;
             if (value==MAX_VALUE)
             {
                _setcolor(BLACK);
             }
             else
             {
                if (value==BOUND)
                {
                  _setcolor(WHITE);
                }
                else
                {
                  _setcolor(color);
                }
             }
             break;
    case 3 : if (value==BOUND)
                _setcolor(WHITE);
             else
                _setcolor(BLACK);
             break;
    }

    _setpixel(i,j);
}


void Set_Display(int type)
/*-------------------------------------------------------------------*/
/*-------------------------------------------------------------------*/
{

/*-------------------------------------------------------------------*/
    video = type;                    /* Set Video Type Value */
/*-------------------------------------------------------------------*/

    switch (video)
    {
       case 1 : _setvideomode(_MRESNOCOLOR);
                break;
       case 2 : _setvideomode(_MRES16COLOR);
                break;
       case 3 : _setvideomode(_ERESCOLOR);
                break;
    }
}
```

14.1.5 The Dielectric Breakdown Model with Noise Reduction

```c
void Reset_Display(void)
/*------------------------------------------------------------*/
/*------------------------------------------------------------*/
{
    _setvideomode(_DEFAULTMODE);
}

void Pause_X(void)
/*------------------------------------------------------------*/
/*------------------------------------------------------------*/
{
    char   x='.';
    _settextposition(5,txt_loc);
    printf("*\a");
    while (x!='X')
    {
        scanf("%c",&x);
    }
}

void Save_Array(char *filename)
/*------------------------------------------------------------*/
/*------------------------------------------------------------*/
{
    char       out_file[16];
    FILE       *output;
    long       fil_size;

    strcpy(out_file,filename);
    strcat(out_file,".DAT");

    /* Create File and Store Data */
    output = fopen(out_file,"wb");
    fil_size = fwrite((char *) field, sizeof(int), NMAX*NMAX, output);

    fclose(output);
}

void Save_Header(char *filename, float epsilon, float eta)
/*------------------------------------------------------------*/
/*------------------------------------------------------------*/
{
    char       out_file[16];
    FILE       *output;

    strcpy(out_file,filename);
    strcat(out_file,".HDR");

    output = fopen(out_file,"w");

    fprintf(output,"%s\n",out_file);
    fprintf(output,"%d %g %g\n",f_size,epsilon,eta);

    fclose(output);
}

void Save_Data(char *filename,float epsilon,float eta)
/*------------------------------------------------------------*/
/*------------------------------------------------------------*/
{
    Save_Header(filename,epsilon,eta);
    Save_Array(filename);
}
```

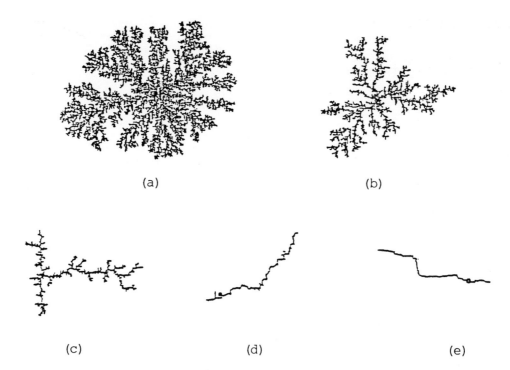

Fig. 1. Fractals generated by the dielectric breakdown model (DBM), without noise reduction. In the DBM, the aggregate (growing from a seed at the center of the lattice) assumes potential $\phi = 0$; an outer circle, with the seed at the center, assumes potential $\phi = 1$. The potential distribution between the aggregate and the circle is given by the Laplace equation $\nabla^2 = 0$. The probability of a perimeter site adjacent to the aggregate being chosen is proportional to $(\nabla \phi)^\eta$. This process of choosing the perimeter sites is repeated until one of them is chosen s times. Then a new particle is added to this site. In (a)–(e) here, $s = 1$. The values of η are 0.5, 1, 2, 4 and 8 in (a)–(e), respectively.

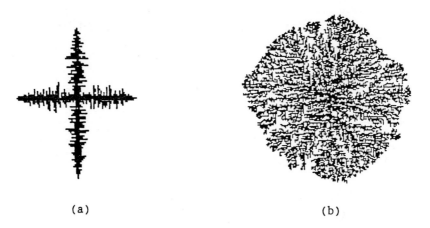

Fig. 2. Fractals generated by the dielectric breakdown model, with noise reduction. (a) $\eta = 1$, $s = 50$. (b) $\eta = 0.1$, $s = 2$.

14.2 Chaos

14.2.1 The Tent Map

Rocco D. Pochy, Yuk S. Yung and *William A. Baldwin*

The bifurcation diagram of the tent map is studied analytically and numerically. It is shown that period doubling does not occur in this case. Beyond a critical value of the control parameter, the system becomes immediately chaotic. A computer program to generate the bifurcation diagram is given.

Absence of Periodicity

The tent map, also called the Lozi map [1], is defined by the difference equation

$$x_{n+1} = a\left(1 - 2\left|x_n - \frac{1}{2}\right|\right) \equiv f(x_n) \qquad (1)$$

where $n = 0, 1, 2, \ldots$; a is a constant satifying $0 < a \leq 1$, and $0 \leq x_n \leq 1$. Using the program in Appendix A, the bifurcation diagram of the tent map is obtained numerically and shown in Fig. 1.

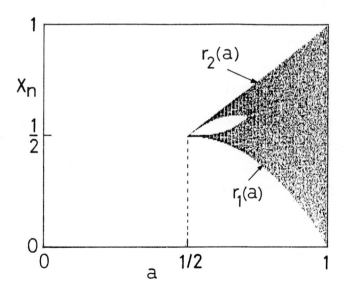

Fig. 1. The bifurcation diagram of the tent map obtained from the program in Appendix A.

As shown in Fig. 1, there is no periodicity beyond $a = 0.5$. This is due to the fact that the slope at any point in the f^m map is given by $|df^m/dx| = (2a)^m$, which is always greater than one when $a > 0.5$. Consequently, no fixed points of f^m can be stable and hence no periodicity of any order can exist if $a > 0.5$. (In Fig. 1, the apparent periodicity of two just beyond $a = 0.5$ is an illusion due to the finite resolution of the graphical display.)

Upper and Lower Bounds

For $0.5 \leq a \leq 1$, let the the upper and lower bounds in the bifurcation diagram be denoted by the two functions $r_2(a)$ and $r_1(a)$, respectively. As suggested by Fig. 1, we assume

$$0 \leq r_1 < 1/2, \quad \text{and} \quad 1/2 \leq r_2 \leq 1 \tag{2}$$

For $x \geq 1/2$, i.e., $r_2 \geq x \geq 1/2$, one has $f(x) = a[1 - 2(x - 1/2)] = 2a(1 - x)$. Hence $r_1 \leq 2a(1 - x) \leq r_2$, which implies $r_2 = 2a(1 - x_{\min}) = 2a(1 - 1/2) = a$, and $r_1 = 2a(1 - x_{\max}) = 2a(1 - r_2) = 2a(1 - a)$. In short,

$$r_2 = a \tag{3}$$

$$r_1 = 2a(1 - a) \tag{4}$$

For $x \leq 1/2$, i.e., $1/2 \geq x > r_1$, one has $f(x) = a[1 - 2(1/2 - x)] = 2ax$. Hence $r_1 \leq 2ax \leq r_2$. But $x_{\max} = 1/2$ implies that $(2ax)_{\max} = a = r_2$, and $x_{\min} = r_1$ implies that $(2ax)_{\min} = 2ar_1 \geq r_1$, since $1 \geq a \geq 0.5$. Consequently, the r_2 and r_1 of Eqs. (3) and (4) indeed satisfy the inequality $r_1 \leq 2ax \leq r_2$ in the region of $x \leq 1/2$, and are the desired upper and lower bounds. As a consistency check, note that r_2 and r_1 do satisfy the initial assumption of Eq. (2), and agree beautifully with the numerical results in Fig. 1.

Appendix A: Program for the Bifurcation Diagram of the Tent Map

This program is written in Microsoft QuickBASIC 4.0 and runs on an IBM PC/AT computer. The results are output to an Epson RX-80 dot matrix printer.

```
PRINT "Tent Map Program"
INPUT "Enter Display Type (0=CGA,1=EGA,3=Herc):", vt

If vt = 0 THEN
    video = 2
ELSEIF vt = 1 THEN
    video = 9
ELSE
    video = 3
END IF

SCREEN video

start:
CLS
'*
'* Window of Graphics Display
'*
```

14.2.1 The Tent Map

```
WINDOW (0, 0)-(1, 1)
LINE (0, 0)-(1,1), 15, B
S1# = .5
S2# = .5

FOR ii = 1 TO 640
'** Interval of Interest
 A# = ii / 640#

'** Plot "trajectory" of series in A(n) vs A(n+1) space
    x1# = S1#
    x2# = S2#

'** Skip first 200 iterations
    FOR i = 1 TO 200
      x2# = A# * (1! - 2! * ABS(.5 - x1#))
      x1# = x2#
    NEXT

'** Plot "points" of series for given A
    FOR i = 1 to 200
      x2# = A# * (1! - 2! * ABS(.5 - x1#))
      PSET (A#, x1#), 14
      x1# = x2#
    NEXT i

NEXT ii
WHILE INKEY$ = "": WEND
END
```

Reference

1. R. Lozi, "Un Attracteur Etrange (?) du Type Attracteur de Hénon," J. Phys. (Paris) **39**, Coll. **C5**, 9 (1978).

14.2.2 The Waterwheel

Rolf D. Freimuth

Chaos theory has recently received a great deal of attention from scientists and popularists. The scientific explanations of chaos are usually incomprehensible to those of us without an extensive background in higher mathematics, and the popularists' accounts are aggravating because of their lack of specifics. For example, one segment of a recent Nova TV program on Chaos talked about Edward Lorenz and his model of a chaotic system. Although the equations that mimic the behavior of the system were given, there was no mention of where the equations came from, or how they were derived. Here, I will explain how I managed to work out an explanation of Lorenz equations with the help of a computer model of a similar but simpler system, the waterwheel.

The Lorenz Equations

The Lorenz equations were described in the Nova program as a model of heat driven convection current, in which heat is applied at the bottom of a container of liquid. The application of heat drove the liquid around in a circular convection rolls which will stabilize if the temperature of the applied heat source is not too high. If the source temperature is too high, the roll will not stabilize and the fluid flow velocity will fluctuate and even reverse with seeming randomness.

The Lorenz equations involve three quantities that represent the state of the roll at any given time, and along with three constants, provide a scheme for predicting how the system will change depending on its current state. Specifically, the equations are given by

$$dX/dt = -\sigma X + \sigma Y$$

$$dY/dt = -XZ + rX - Y \qquad (1)$$

$$dZ/dt = XY - bK$$

The X coordinate represents the speed of the fluid flow in the rolls, Y characterizes the temperature difference between ascending and descending fluid elements, and Z corresponds to the deviations of the vertical temperature profile from its equilibrium value. Of these three quantities, the only one that I understand is X; the other two are meaningless to me (but very interesting) without more information. The constants σ, r and b were not explained at all in the Nova program, but they probably have to do with the amount of heat applied, etc.

The Lorenz equations can be graphed in three-dimensional space provided an initial set of values for X, Y and Z is supplied. If X, Y and Z are considered to be space coordinates then their time derivatives represent velocity at each point in that space. The given initial point will move according to that velocity and be brought to a nearby point which may have a different velocity (with different direction or magnitude), and from there the point will move again and so on. The path of the point is called a "trajectory" or a "flowline." When a flowline is calculated by numerical integration, the resulting trajectory is not regular or periodic, but it never strays far from a specific region. The trajectory resembles an owl's mask or a butterfly's wings if the values of the constants are such that the equations describe a chaotic motion. For another set of the constants the trajectory may spiral in to a single point at which dX/dt, dY/dt and dZ/dt equal zero. This stable state represents a steady motion (constant velocity) of the convection rolls.

I do not know anything about fluid dynamics and would have little hope of finding out any more about the Lorenz equations if it was not for another system that is described by the same equations. That system, described in the first chapter of *Chaos: Making a New Science* by James Gleick [1], is the Lorenzian waterwheel.

The Waterwheel

The waterwheel was designed with the equations in mind and is closely analogous to the convection roll. The waterwheel is similar to a Ferris wheel but has buckets instead of chairs. Water flows down onto the wheel from the top middle into the buckets which have small holes in their bottoms (Fig. 1). The wheel is free to turn and does so if it is out of balance due to differing amounts of water in the buckets. If the flow of water into the buckets is slow, the turning of the wheel will eventually become stable; but if the flow is large, the wheel's velocity will fluctuate and reverse just like that in the convection roll.

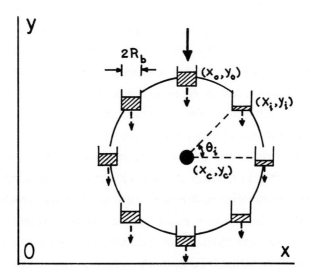

Fig. 1. Sketch of the vertical waterwheel. Each bucket is idealized as a mass point in the treatment of its dynamics. In the computer model, as water filling is concerned each bucket has a width $2R_b$; $(x_c, y_c) = (0, 0)$ and $(x_0, y_0) = (0, R)$ where R is the radius of the wheel.

It is important to realize that the wheel must be analogous to the convection roll if it is to be described by the same equations. The roll is driven by heat which causes the water at the bottom of the vessel to become less dense. Being less dense, the water is compelled to rise. On the other hand, the added water from the top of the wheel is the driving force. The waterwheel is essentially the convection roll turned upside down. In both systems, some "stuff" is added at one end of a kind of "disk." The "stuff" is then forced to move to the other end by an outside force which, in the case of the wheel, is gravity.

Computer Simulations

The behavior of the wheel can be predicted in a simple manner by Newton's laws. Unfortunately, Newton's laws (without integration) do not give directly the position or velocity of the wheel as a function of time, even though they can be used to calculate the torque on the wheel as a function of the weight and position of each bucket.

As a first step, it would be interesting to see if a computer model of the wheel could be made to exhibit chaotic motion. (It is not clear at this point, to me at least, exactly what is meant by "chaotic motion" so we will have to loosely define it as "a motion that does not seem to be regular or periodic." An example of motion that is not irregular or nonperiodic would be the motion of a simple pendulum which wings back and forth taking equal amounts of time for each swing.)

The general method of setting up my computer model of the wheel is similar to the method of numerical integration used in the calculation of the butterfly curve. For small periods of time, the wheel can be treated as if it is moving with constant angular acceleration. At the beginning of each time interval Δt, after calculating the change in mass of each bucket the instantaneous torque τ on the wheel is calculated, and from that the angular acceleration α. The change of angular velocity during each time interval is simply given by $\alpha \Delta t$. In the model, constant filling and draining rates are assumed such that the total water mass in the buckets is conserved. Each bucket is assumed to have a radius R_b as filling of water is concerned, but is treated as a mass point in its dynamics. (See Appendix A.)

My model of the waterwheel, after some difficulties, did exhibit chaotic behavior. The first of these difficulties was the problem of friction. In the first version of the program, without putting in friction, the wheel would continuously accelerate until the wheel was rotating through nearly an entire revolution through each iteration in time. For a reasonably accurate model, the maximum angular displacement of the wheel should be no more than about five degrees per iteration, which is about seventy iteration per revolution. To keep the wheel from running out of control, I decided to introduce a frictional torque τ_f but it was unclear at first how τ_f should vary with angular velocity ω.

14.2.2 The Waterwheel

A real model would have frictional torques due to bearing friction at the hub, and due to the viscosity of air. A very simple relation would be to use a constant τ_f, but that does not seem realistic because of the viscosity of air. The faster you try to move something through the air, the more resistance you get. There should be some dependence on velocity. So the first relation I tried was a direct proportionality between frictional torque and velocity,

$$\tau_f = (\text{const})\omega \qquad (2)$$

After a few unsuccessful runs of the program I decided to try a proportionality with the square of velocity,

$$\tau_f = (\text{const})\omega^2 \qquad (3)$$

The velocity squared term did produce a chaotic motion in which the angular velocity of the wheel would fluctuate and reverse (Fig. 2(a)).

However, there are two things wrong with the velocity squared relation. The first is that it is not realistic. At very high speeds, an airplane may experience resistance that goes nearly as the square of velocity, but the wheel does not seem to turn at velocities high enough for this to happen. Second, if we look at the Lorenz equations, there is no X^2 term anywhere in the equations. If the wheel is truly analogous to the Lorenz equations, the frictional torque should be linearly proportional to the velocity.

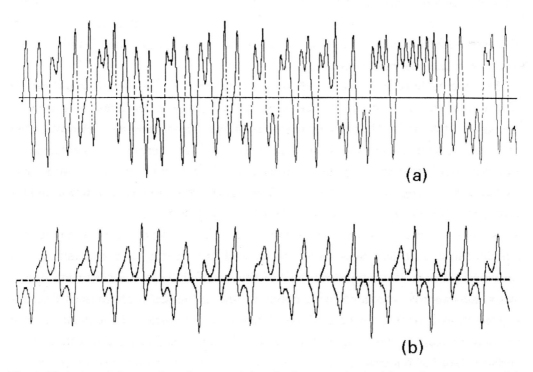

Fig. 2. Variation of the angular velocity ω of the wheel vs time obtained from the computer model. The points above (below) the center lines represent clockwise (anticlockwise) rotation of the wheel. (a) Results from an early version using Eq. (3). (b) Results from the final version listed in Appendix A with the use of Eq. (2).

It turned out that my original linear friction equation did not produce chaotic motion because of a mistake in the sign of the frictional torque. The fact that squaring a number makes it positive was responsible for the success of the velocity squared relation. If I had used the linear relation with the absolute value of the velocity, it would have worked. Though the absolute value would have made the model go chaotic, it is possible to make a linear relation without the absolute value. This was just sloppy programing on my part.

From the final program listed in Appendix A in which Eq. (2) is used, the chaotic motion of the waterwheel is demonstrated in Fig. 2(b) and Fig. 3.

Though a real waterwheel has a finite number of buckets, to be truly analogous to the convection roll the wheel must have an extremely large number of buckets. The reason for this is that the fluid in the convection roll can absorb heat continuously. So to make the wheel have the same kind of continuity, it should have a large number of, or infinitely many buckets. As it turned out, in the computer model, as few as eight buckets will be enough to produce chaotic motion.

Dynamics of the Waterwheel

Now that it has been shown that a computer model of the wheel can exhibit chaotic behavior, it is reasonable to assume that all relevant considerations of the dynamics of the wheel have been included in the model. Even though the model is highly idealized, the assumptions do not leave out anything critical. From the results depicted in Figs. 2 and 3, ω may be identified with X. The constants σ, r and b, and the variables Y and Z are still unknown, but we now have enough information to try to figure out what they are.

The first to look at is the dX/dt term in the first of the Lorenz equations. If $X = \omega$, then $d\omega/dt = \alpha$, the angular acceleration. Angular acceleration is equal to the torque divided by the rotational inertia of the wheel, $\alpha = \tau/I$, provided that the total rotational inertia I is constant in time. [More generally, $d(I\omega)/dt = \tau$.] It is reasonable to guess that in the equation $dX/dt = \sigma X - \sigma Y$, the term $-\sigma Y$ (constant times angular velocity) is the frictional torque, because it always opposes the wheel's current angular velocity. The σY term probably determines the angular acceleration of the wheel due to the imbalance in the amounts of water in the buckets (located on the two sides of the vertical line passing through the center of the wheel). But in the computer model of the wheel above, the acceleration due to water imbalance is calculated by summing up the individual contribution of torque applied by each bucket. In light of this, it would be interesting to see if we can find a way of calculating the torque on the wheel that does not depend on a knowledge of the number of buckets.

To proceed, I choose a Cartesian coordinate system with the positive y axis to be the "up" direction. The hub is located at (x_c, y_c) and the water pours down onto

14.2.2 The Waterwheel

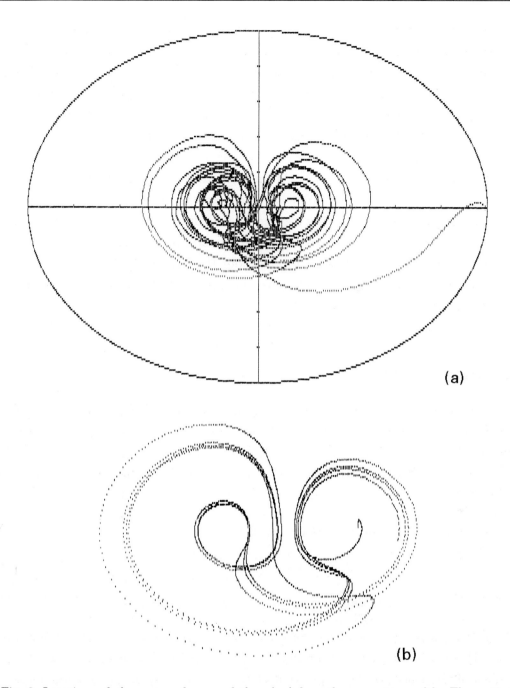

Fig. 3. Locations of the center of mass of the wheel from the computer model. The vertical (horizontal) coordinate is y_{cm} (x_{xm}). (a) In this plot, the outer ellipse is not part of the location curve. (b) Another run with shorter run time and different parameters. The innermost paths are moving upwards because the dominant influence on the change of the center of mass is the filling of the buckets. Near the outside the motion is downwards and predominantly a rotational motion. The bend at the bottom is caused by a tendency of the wheel to act like a pendulum after swinging down around the outside, but the motion is very quickly dampended. A low frictional constant would produce a large number of oscillations or bends at the bottom of the plot.

the wheel at the point (x_0, y_0), i.e., the water need not be pouring from the top if one so chooses. Each bucket is represented as a point on the rim of the wheel. The rotational inertia of the wheel is assumed to come completely from the water in the buckets (i.e., the wheel frame and the buckets are massless). The center of mass of all the water, or the whole wheel, is located at (x_{cm}, y_{cm}) which varies with time t.

From the calculations of torque and the center of mass, it is possible to show that for a wheel with any number of buckets, the angular acceleration does has exactly the form $-\sigma Y$ (constant times the unknown quantity Y) in the dX/dt equation! (See Appendix B.) It is reasonable at this point to identify Y with x_{cm}, the x component of the wheel's center of mass.

We must now ask the question: What changes the x component of the center of mass? That is, we need to know what changes the quantity Y. The draining and filling water change the mass of the wheel at different places, so they will also change the center of mass. The wheel is also in rotation, so the point at which the center of mass is located will also rotate about the hub of the wheel. But, when a point (x, y) is rotated in two dimensions, the amount that x changes is dependent on the quantity y (and vice versa). So if we are to calculate the time derivative of the x_{cm}, we will also need to know y_{cm}. The quantity y_{cm} also changes with time; so we might guess that y_{cm} is equal to Z. Each of the dY/dt and dZ/dt equations has a binary term of velocity multiplied by the other variable (i.e., $-XZ$ in dY/dt, and XY in dZ/dt), which is what we might expect for a rotational motion.

If we calculate the time change of x_{cm} and y_{cm} due to rotation and the change of mass arising from filling and draining, we get equations that look very much like the Lorenz equations (see Appendix C)! In short, we have now identified X, Y and Z as ω, x_{cm} and y_{cm}, respectively. The equations of motion of the waterwheel are given by

$$d\omega/dt = -k\omega - (g/R^2)x_{cm} + (g/R^2)x_c \qquad (4)$$

$$dx_{cm}/dt = -\omega y_{cm} + y_c\omega - sx_{cm} + s(x_0 + x_c) \qquad (5)$$

$$dy_{cm}/dt = \omega x_{cm} - x_c\omega - sy_{cm} + s(y_0 + yc) \qquad (6)$$

Note that these equations have constants not found in the Lorenz equations. In Eq. (4), $-k\omega$ is the frictional torque, g the acceleration due to gravity, and R the radius of the wheel. In the derivation of these equations, the total water mass is assumed to be constant. In Eqs. (5) and (6) the filling rate is assumed constant, but the draining rate of each bucket is proportional to the mass there, i.e., $dm_i/dt = -sm_i$. [In contrast, in the computer model $dm_i/dt = $ const. If constant draining rate is used in the mathematical model here and *if* one is willing to assume that $x_{cm}^2 + y_{cm}^2 \approx $ const, then one can still obtain a set of equations similar in structure to those in Eqs. (5) and (6). But this assumption is not very physical as can be seen from Fig. 3.]

As seen from the numerical solutions of the wheel equations plotted in Fig. 4, the general characteristics of the system is *very* similar to those of the Lorenz attractor (see p. 28 of Ref. [1]).

14.2.2 The Waterwheel

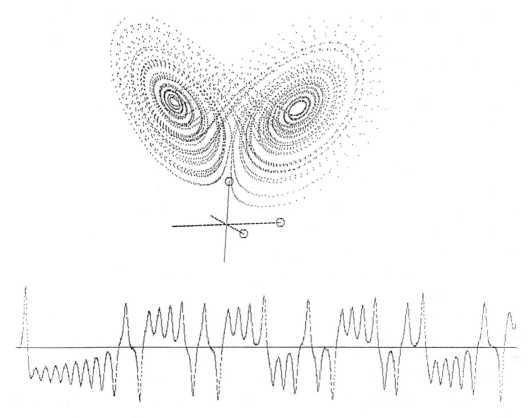

Fig. 4. Numerical solutions of equations of motion of the waterwheel. The upper plot is of the quantities ω, x_{cm} and y_{cm}. The lower plot is of ω vs time, in which points above (below) the line represent clockwise (anticlockwise) rotation of the wheel.

The main idea of the process used to derive the wheel equations can conceivably be applied to many other dynamical systems. In this process the system is analyzed in order to determine what quantities are necessary to uniquely determine the state of the system at any given time. The rate of change of each quantity as a function of some or all of the other quantities is then determined. Though this process is applicable to the study of other phenomena, the study of chaos seems to be primarily concerned with the finding of order in situations that once were thought to be random or inconceivably complicated in their behavior.

Editor's note: Rolf Freimuth did this work when he was a sophomore, just finishing an introductory course in calculus physics. He was not aware of and did not have the benefit of the works of Refs. [2] and [3], in which a continuum version of the waterwheel is presented.

The set of equations, Eqs. (4)–(6), is in fact closer to that of Eq. (1) than it looks. To see this, in Eqs. (4)–(6), let $x_c = y_c = x_0 = 0$, $y_0 = R$, $\omega = X$, $x_{cm} = -Y$, $y_{cm} = -Z + R$, $s = 1$, and $R = (g/k)^{1/2}$. One then obtains Eq. (1) with $\sigma = k$, $r = R$, and $b = 1$. Consequently, the Lorenz equations is a special case of Eqs. (4)–(6).

Appendix A: Computer Program for Simulating the Motion of the Waterwheel

In this simulation of the motion of the waterwheel, everything is treated in a two-dimensional space. The wheel has radius R and is centered at (0,0). There are N buckets equally spaced on the rim of the wheel. The ith bucket is located at $(x_i, y_i) = (R\cos\theta_i, R\sin\theta_i)$, where θ_i is the polar angle which is a function of time t with $i = 1, 2, \ldots, N$; $d\theta_i/dt = \omega$. Each bucket has a width of $2R_b$. Water is injected from the top at $(0, R)$ with a constant rate f. Consequently, water is added to the ith bucket if both conditions, $|x_i| < R_b$ and $y_i > 0$, are satisfied. The constant draining rate of each bucket δ is given by $\delta = f/N$. The water mass in the ith bucket is denoted by m_i.

The initial conditions written into the program below are: $\theta_1 = 0$, $m_1 = 1$, and $m_i = 0$ for $i \geq 2$ (which can be easily changed). The iteration time interval is Δt. At the end of each time interval, $m_i \rightarrow m_i - \delta(\Delta t)$ due to draining alone. The total torque $\tau = \tau_f + \tau_g$, where the frictional torque is given by $\tau_f = -\mu\omega$ and the gravitational torque $\tau_g = \Sigma_i m_i g(R\cos\theta_i)$. The wheel frame and the empty buckets contribute WR^2 to the total rotational inertia I, which is given by $I = (W + m)R^2$ where $m = \Sigma_i m_i$. The angular acceleration $\alpha = \tau/I$, and $\omega \rightarrow \omega + \alpha(\Delta t)$, $\theta_i \rightarrow \theta_i + \alpha(\Delta t)^2/2$ at the end of the iteration.

Finally, the center of mass is given by $x_{cm} = \Sigma_i m_i x_i/m$, and $y_{cm} = \Sigma_i m_i y_i/m$.

```
100 'Computer model of the Lorenzian waterwheel (in quickBASIC)
110 '
120 'Define parameters (in SI units)
130    N = 8         ' number of buckets
140    R = .5        ' wheel radius (m)
150    W = 5         ' WR² is rotational inertia due to wheel frame/buckets (kg)
160    BR = .19      ' nonzero bucket radius Rb (m)
170    G = 9.8       ' gravitational acceleration (m/s²)
180    FC = 2        ' frictional constant μ
190    FR = .25      ' rate of flow onto the wheel f (kg/s)
200    DR = .25/N    ' rate of flow out of each bucket δ (kg/s)
210    P = 0         ' initial angular position θ₁ (rad)
220    V = .2        ' initial angular velocity ω (rad/s)
230    DT = .1       ' time interval of each iteration Δt (s)
240    PI = 3.141592654#
250    DIM P(N),C(N),S(N),M(N) : M(1)=1
260 '
270    CLS:    SCREEN 2 : WINDOW (0,-6)-(300,6)
280    LINE (0,0)-(300,0)
300 'Assign initital angular positions to the buckets
310    For Z=1 TO N : P(Z)=(Z-1)*2*PI/N : NEXT Z
320 '
330 'BEGIN LOOP
340    M = 0 : TG=0 : CMX=0 : CMY=0
350 '
360 'Compute sine and cosine for each bucket
370    FOR Z=1 TO N : C(Z)=COS(P+P(Z)) : S(Z)=SIN(P+P(Z)) : NEXT Z
380 '
```

```
390 'Determine which if any buckets are being filled and fill them
400   FOR Z = 1 TO N : IF ABS(R*C(Z))<BR AND S(Z)>0 THEN M(Z)=M(Z)+FR*DT
410   NEXT Z
420 '
430 'Drain all buckets and compute total mass
440   FOR Z = 1 TO N : M(Z)=M(Z)-DR*DT : IF M(Z)<0 THEN M(Z) = 0
450   M=M+M(Z) : NEXT Z
460 '
470 'Compute the torque on the wheel due to gravity
480   FOR Z = 1 TO N : TG=TG-M(Z)*G*R*C(Z) : NEXT Z
490 '
500 'Add gravitational torque and frictional torque to obtain total torque
510 'and compute the angular acceleration, new position and angular velocity
520   T=TG-V*FC : A=T/((W+M)*R^2) : P=P+V*DT+DT^2*A/2 : V=V+A*DT
530 '
540 'Compute the center of mass
550   FOR Z=1 TO B : CMX=CMZ+M(Z)*R*C(Z) : CMY=CMY+M(Z)*R*S(Z) : NEXT Z
560   CMX=CMX/M : CMY=CMY/M
570   PSET (CMX,CMY) : IF INKEY$=" " THEN CLS
580 GOTO 330      'Repeat loop
```

Appendix B: Derivation of Eq. (4)

The total torque has two parts, given by $\tau = \tau_f + \tau_g$. Adopting the assumption of Eq. (2) for the frictional torque, one has $\tau_f = -k\omega$ where k is a constant. Referring to Fig. 1, the gravitational torque $\tau_g = \Sigma_i m_i (-g) R \cos\theta_i$ (with the torque vector $\vec{\tau}_g = \tau_g \hat{z}$). Since

$$x_{cm} = x_c + (R/m)\Sigma_i m_i \cos\theta_i \tag{7}$$

thus $\tau_g = -gm(x_{cm} - x_c)$.

The rotational inertia $I = mR^2$. With the assumption of $m = $ const, one obtains $I(d\omega/dt) = \tau$ from the Newton's second law. Putting everything together Eq. (4) is obtained.

Appendix C: Derivation of Eqs. (5) and (6)

From Eq. (7), one has

$$dx_{cm}/dt = m^{-1}\Sigma_i (dm_i/dt)(R\cos\theta_i) + (R/m)\Sigma_i m_i(-\sin\theta_i)(d\theta_i/dt) \tag{8}$$

We adopt three assumptions here. (i) The draining rate of each bucket is proportional to its mass, i.e., $dm_i/dt = -sm_i$ due to draining alone. (ii) There are enough number of buckets so that there is always a bucket at the location (x_0, y_0) to receive the injected water at any time. (iii) The total water mass m is conserved. These three assumptions imply that water must be injected with a fixed rate given by sm. It follows that the first term on the right hand side of Eq. (8) becomes $m^{-1}[smx_0 + \Sigma_i(-sm_i)(R\cos\theta_i)] = s[x_0 - (x_{cm} - x_c)]$.

Since $d\theta_i/dt = \omega$ and

$$y_{cm} = y_c + (R/m)\Sigma_i m_i \sin\theta_i \tag{9}$$

the second term on the right hand side of Eq. (8) becomes $-\omega(y_{cm} - y_c)$. Equation (8) thus reduces to Eq. (5).

Similarly, Eq. (9) implies

$$dy_{cm}/dt = m^{-1}\Sigma_i(dm_i/dt)(R\sin\theta_i) + (R/m)\Sigma_i m_i(\cos\theta_i)(d\theta_i/dt) \qquad (10)$$

On the right hand side of Eq. (10), the first term equals $m^{-1}[smy_0 + \Sigma_i(-sm_i)(R\sin\theta_i)] = s[y_0 - (y_{cm} - y_c)]$, and the second term equals $\omega(x_{cm} - x_c)$. Consequently, Eq. (10) reduces to Eq. (6).

Reference

1. J. Gleick, *Chaos: Making a New Science* (Viking, New York, 1987), p. 27.
2. C. Sparrow, *The Lorenz Equations: Bifurcations, Chaos, and Strange Attractors* (Springer, New York, 1982), Appendix B.
3. M. Kolář and G. Gumbs, Phys. Rev. A **45**, 626 (1992).

14.3 Pattern Formation

14.3.1 Biased Random Walks

Mark A. Guzman and *Rocco D. Pochy*

The program below generates patterns from the three-directions biased random walk (BRW) model [1], which was introduced to simulate the tree patterns observed in electrodeposits in a linear cell [2,3]. In this model the random walker is allowed to walk sideward or downward, but not upward. There is only one parameter R with $R = (p_R + p_L)/p_D$, where p_R, p_L and p_D are the probabilities of stepping to the right, left and down directions, respectively; $p_R = p_L$ is always assumed.

In this BRW model, a random walker is released at random from a top horizontal line, which is frozen when it touches the bottom line. Another walker is then released from the top line and sticks when it reaches the bottome line or the perimeter sites of the growing aggregate. The process is repeated. Periodic side boundary condition is assumed. Typical patterns obtained from this program are presented in Fig. 1.

```
/*------------------------------------------------------------*/
/* BIASWALK.C                                                 */
/* Language: Microsoft C v7.0                                 */
/*                                                            */
/* The R parameter defines the ratio between                  */
/* the side to side motion vs the downward motion.            */
/*    R = 0, only downward motion, ballistic behavior.        */
/*    R = 10000, moves downward with the probability of       */
/*        1 out of 10001.                                     */
/*------------------------------------------------------------*/

#include <stdio.h>
#include <stdlib.h>
#include <conio.h>
#include <graph.h>

/*---[ Define Constants ]-------------------------------------*/
#define XMAX    128
#define YMAX    128

#define TRUE    1
#define FALSE   0

#define OCCUPIED 1
#define EMPTY    0

/*---[ Define Global Variables ]------------------------------*/
static char field[XMAX][YMAX];

/*---[ Define Functions ]-------------------------------------*/
void Initialize();
void Random_walk();
void Display_field();
int  Wrap();
double random();
```

```c
/*================================================================*/
main()
{
    int R, i;

    printf("Enter Bias parameter R [0-10000]:");
    scanf("%d",&R);
    R = R + 1;                          /* converted ratio */

    printf("Enter Random Seed (integer):");
    scanf("%d",&i);
    srand(i);

    Initialize();

    Random_walk(R);

    printf("\007");
    while (!kbhit());
}

void Initialize()
{
    int i,j;

    _setvideomode(_MRESNOCOLOR);

    /*---[ Clear Field ]------------------*/
    for (i=0;i<XMAX;i++)
        for (j=0;j<YMAX;j++)
            field[i][j] = (char) 0;

    /*---[ Initialize "Bottom" ]----------*/
    j = YMAX - 1;
    for (i=0;i<XMAX;i++)
        field[i][j] = OCCUPIED;
}

void Random_walk(int R)
{
    int i,j,n;
    int jmax = YMAX-1;
    int stuck=FALSE;
    int done =FALSE;
    int count;
    char text[40];

    while (!done)
    {
        /*---[ Get Starting Location ]--------*/
        i = XMAX*random();
        j = jmax-1;

        stuck = FALSE;
        count = 0;

        /*---[ Walking "Down" to the Surface ]*/
        while (!stuck)
        {
            i = Wrap(i);

            count = field[i][j+1] + field[Wrap(i-1)][j] +
                    field[Wrap(i+1)][j] + field[i][j-1];

            if (count!=0)               /* Is it next growth */
            {
```

14.3.1 Biased Random Walks

```
                stuck = TRUE;
                field[i][j] = OCCUPIED;
                if (j<=jmax) jmax--;        /* Decrement Release Boundary */
            }
            else
            {

                if (random() > 1.0/(float)R)
                   if (random()<0.5)        /* Move left or Right*/
                      i++;
                   else
                      i--;
                else
                   j++;                     /* Move downward */


            }
            if ((j<1) || (j<(jmax-3))) stuck=TRUE;   /* Lost particle */

        }
        if (jmax==1)
          done = TRUE;
        Display_field();
    }
    _settextposition(23,0);
    sprintf(text,"R = %d",R);
    _outtext(text);
}
void Display_field()
{
    short i,j;

    for (i=0;i<XMAX;i++)
       for (j=0;j<YMAX;j++)
         if (field[i][j]==OCCUPIED)
            _setpixel(i,j);

}

int Wrap(int n)
{
    if (n>=XMAX) return (n-XMAX);
    if (n<0)     return (n+XMAX);
    return(n);
}

double random()
{
    return(rand()/32767.0);
}
```

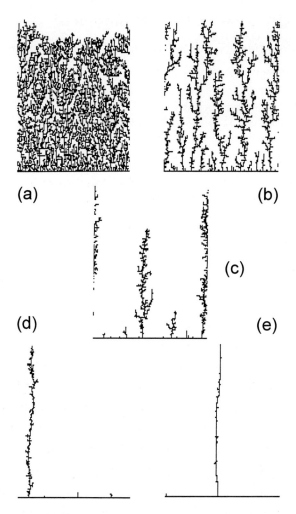

Fig. 1. Patterns generated by the bias random walk (BRW) model. $R = 0, 10, 100, 1000$ and 10000, respectively, in (a)–(e). The case $R = 0$ is called the ballistic deposition model.

References

1. L. Lam, R. D. Pochy and V. M. Castillo, "Pattern Formation in Electrodeposits," in *Nonlinear Structures in Physical Systems*, edited by L. Lam and H. C. Morris (Springer, New York, 1990).
2. M. A. Guzman, R. D. Freimuth, P. U. Pendse, M. C. Veinott and L. Lam, "Experiments on Electrodeposit Patterns," in *Nonlinear Structures in Physical Systems*, edited by L. Lam and H. C. Morris (Springer, New York, 1990).
3. R. D. Pochy, A. Garcia, R. D. Freimuth, V. M. Castillo and L. Lam, Physica D **51**, 539 (1991).

14.3.2 Surface Tension and the Evolution of Deformed Water Drops

Victor M. Castillo

A program is developed for the PC that simulates the effect of surface tension on fluid patterns. An algorithm similar to this was originally used by Liang [1] to model viscous fingers. Here the program is used to model the evolution of an irregular shape to a stable, round form similar to the way a deformed water droplet redistributes its mass as a result of surface tension.

Surface Tension

Surface tension is the result of the affinity that a fluid has for itself. Each discrete particle of a fluid is subjected to an attractive force from every other particle in that fluid. This phenomenon is responsible for keeping a liquid in a condensed phase and has a great influence on the shape that it takes. A liquid is not normally seen to have sharp edges because this would create points at the tip that have strong net forces opposing that geometry. It can be seen that the net force at a given point on the interface is inversely proportional to the local radius of curvature. If the point is near or on a sharp tip, the local radius of curvature is small, and the net force inward is large. If the point lies on a straight or slightly rounded edge, the curvature is large and the net force is small. If the point lies near the bottom of a crevice, the local radius is small, but negative; thus, the point is subject to a large net force away from the center of mass.

A lattice gas model could be used to calculate the effects of the short range interactions of all the particles. However, to make this model workable on a PC, the local radius-of-curvature approximation is used instead. Details of this program is described below. The evolution of an irregular shape into a rounded shape obtained from the program is depicted in Fig. 1; it works as expected. Of course, other initial shapes can be used.

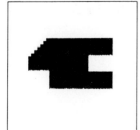   

Fig. 1. Evolution of an irregular shape into a rounded shape. The time from left to right is 0, 200, 800 and 1600 iterations, respectively.

The Program

This program was written in Microsoft QuickBASIC 4.5 for use on an IBM PC with graphics capability. First, an arbitrary, irregular shape is defined as a closed curve on a two-dimensional square lattice. The program then enters a loop of choosing two random points that exist on the boundary, determining their local "radius of curvature" by counting the number of occupied neighboring sites, and deciding which location will receive one particle and which will lose one particle.

Cells of the lattice are chosen at random. The cell is determined to exist on the boundary if it satisfies two conditions: (i) The cell must be occupied. (ii) The cell must have less than 8 occupied neighbors in the 3 × 3 local neighborhood (the so-called Moore neighborhood in the language of cellular automata, which consists of 4 nearest neighbors and 4 next-nearest neighbors in the diagonal directions); otherwise it must exist within the interior. The local "radius of curvature" is then determined by counting the number of occupied neighbors that a cell has. If the cell has a large number of occupied neighbors, then it must lie in a crevice with a negative radius of curvature. If the cell has a small number of occupied neighbors, then the cell must lie on a peak with a positive local radius of curvature. Since the local radius of curvature is related inversely to the number of neighbors, the program takes points with few occupied neighbors and puts them into the neighborhood of cells with more occupied neighbors. After a number of iterations, the pattern starts to lose the sharp corners and cavities.

```
'*********************************************************************
' BUBBLE.BAS
'*********************************************************************
RANDOMIZE TIMER
DIM a%(50, 50)
wx0 = 0: wy0 = 0: wx1 = 40: wy1 = 40
SCREEN 9: WINDOW (wx0 - 10, wy0)-(wx1 + 10, wy1)
mcn = (wx1 - wx0) * (wy1 - wy0)      'monte carlo number
iter = 10 * mcn
D = 3: surten = 9

'*********************************************************************
'Loads Data Array with initial pattern
'*********************************************************************
FOR i = 13 TO 33
    FOR j = 13 TO 18
        LET a%(i, j) = 1
    NEXT j
NEXT i
FOR i = 6 TO 25
    FOR j = 19 TO 23
        LET a%(i, j) = 1
    NEXT j
NEXT i
C = 0
FOR j = 24 TO 29
    C = C + 1
    FOR i = 6 + C TO 33
        LET a%(i, j) = 1
```

14.3.2 Surface Tension and the Evolution of Deformed Water Drops

```
            NEXT i
     NEXT j
     FOR i = wx0 TO wx1                           'Displays array data
         FOR j = wy0 TO wy1
             IF a%(i, j) THEN CIRCLE (i, j), .1, 10
         NEXT
     NEXT

     '***********************************************************************
     'Main loop
     '***********************************************************************
     FOR s = 1 TO iter
         flag = 0                                 'look for boundary point #1
         DO
             DO
                 x1 = INT(RND * wx1) + wx0
                 y1 = INT(RND * wy1) + wy0
             LOOP UNTIL a%(x1, y1)
             cnt1 = 0
             FOR i = x1 - 1 TO x1 + 1
                 FOR j = y1 - 1 TO y1 + 1
                     cnt1 = cnt1 + a%(i, j)
                 NEXT
             NEXT
             IF cnt1 < 9 THEN flag = 1
         LOOP UNTIL flag
         flag = 0                                 'look for boundary point #2
         DO
             DO
                 x2 = INT(RND * wx1) + wx0
                 y2 = INT(RND * wy1) + wy0
             LOOP UNTIL a%(x2, y2) = 1
             cnt2 = 0
             FOR i = x2 - 1 TO x2 + 1
                 FOR j = y2 - 1 TO y2 + 1
                     cnt2 = cnt2 + a%(i, j)
                 NEXT
             NEXT
             IF cnt2 < 9 THEN flag = 1
             IF x1 = x2 AND y1 = y2 THEN flag = 0
             IF cnt1 = cnt2 THEN flag = 0
         LOOP UNTIL flag
         cnt1 = 0                                 'calculate local radius of curvature
         FOR i = x1 - D TO x1 + D
             FOR j = y1 - D TO y1 + D
                 cnt1 = cnt1 + a%(i, j)
             NEXT
         NEXT
         cnt2 = 0
         FOR i = x2 - D TO x2 + D
             FOR j = y2 - D TO y2 + D
                 cnt2 = cnt2 + a%(i, j)
             NEXT
         NEXT
         IF cnt1 > cnt2 THEN
             recx = x1: recy = y1
             givx = x2: givy = y2
         ELSE
             recx = x2: recy = y2
             givx = x1: givy = y1
         END IF
         IF ABS(cnt1 - cnt2) > surten THEN
             a%(givx, givy) = 0
             CIRCLE (givx, givy), .1, 0
             DO
                 i = INT(3 * RND - 1) + recx
                 j = INT(3 * RND - 1) + recy
             LOOP UNTIL a%(i, j) = 0
```

```
            a%(i, j) = 1
            CIRCLE (i, j), .1, 10
        END IF
NEXT s
```

References

1. S. Liang, "Random-Walk Simulations of Flow in Hele-Shaw Cells," Phys. Rev. A **33**, 2663 (1986).

14.3.3 Ising-like Model of Ferrofluid Patterns
Victor M. Castillo

Numerical simulations of fluid flows is one of the most computationally demanding tasks in applied physics. In order to precisely calculate the dynamics of the system, it would be necessary to determine the momentum, location, and the forces acting on every molecule of the fluid. Considering the enormous number of molecules in a volume of fluid large enough to be significant in most problems, it is not feasible to do this without a supercomputer. Approximation methods have been developded that will allow estimates to be made with a reasonable amount of computations. Examples of this are those involving dimensional analysis and cellular automata.

An alternative is the lattice gas method based on the Ising model [1]. In the Ising model, a lattice site is assigned a value of $+1$ or -1 (spin up or spin down) to represent the local spin orientation. Two neighboring spins interact in such a way that parallel spins have lower energy than antiparallel spins. The energy of the system is then considered to be a function of the spin configuration of that system. The energy of the system is defined by

$$E\{s_i\} = -\Sigma \varepsilon_{ij} s_i s_j - H \Sigma s_i \tag{1}$$

where $\langle ij \rangle$ represents nearest neighboring pair. In a two-dimensional hexagonal lattice, for example, there would be six nearest neighbors for each site. The interaction energy ε_{ij} and external field H are taken to be constants throughout the system.

Ferrofluid Patterns

Experiments done with a ferrofluid and an immiscible, nonmagnetic fluid placed together in a Hele-Shaw cell [2] reveal complex finger patterns forming in the presence of a uniform magnetic field applied perpendicular to the cell. The patterns undergo a phase change as the magnetic field intensity H_0 is changed. Within a certain range of H_0, fingers of the ferrofluid invade the nonmagnetic fluid to form a labyrinth pattern of self-avoiding finers. Under other conditions, the fluids exist in an emulsified state, bulk flow state, or a meniscus state.

In 1986 Rosensweig [3] developed a lattice model similar to the Ising model, which produced patterns much like those formed by a ferrofluid in the presence of a magnetic field. The model uses a hexagonal lattice for the pragmatic reason that the experimental patterns contain nodes, with three branches oriented at 120° with respect to each other at each node. Each site at location (i, j) has a dipole/spin variable $s(i, j)$ such that $s(i, j) = 0$ if the site is occupied (by the magnetic fluid), and $s(i, j) = 1$ if vacant (i.e., occupied by the nonmagnetic fluid). Initially, a flat interface is assumed. A vacant perimeter site is picked at random and the total energy change due to the addition of a magnetic fluid to this site U_t is calculated. The vacant site is actually occupied if $U_t < 0$; otherwise, another vacant perimeter site is picked and the process is repeated. In dimensionless units, $U_t = U_s + U_e + U_a$, where the dipole field energy $U_s = -1$, the dipole-dipole energy $U_e = \alpha_1 \Sigma s(i,j)$, and the interfacial energy $U_a = 2\alpha_2[3 - \Sigma s(i,j)]$. (See Ref. [3] for the origin of these terms.)

Two computer programs have been written to demonstrate these models on the PC. The first, FERROFL.BAS in Appendix A, is written in QuickBASIC 4.5, which produces a dynamic EGA output of the pattern as the lattice is updated. The second, FERROFL.M (not listed here), is written for MATHEMATICA 1.2, which produces a POSTCRIPT image of the equilibrium state that can be rendered on most video consoles and laser printers using MATHEMATICA'S utilities. Figures 1–3 repesent results from our program, in agreement with those in Ref. [3].

Appendix A: Computer Program in QuickBASIC

```
'(***************************** FERROFL.BAS **************************)
RANDOMIZE TIMER
DEFINT C, I-Z: DEFDBL A, E
w = 27: DIM P(w + 3, w + 3)
SCREEN 9: WINDOW (0, 0)-(w + 3, w + 13)

'(*** Defining initial occupied sites ***)
FOR y = 1 TO 2
    FOR x = 1 TO w
        P(x, y) = (x MOD 2) XOR (y MOD 2)
    NEXT
NEXT

'(*** Displays Grid ***)
FOR x = 1 TO w
    FOR y = 1 TO w
        IF P(x, y) THEN
            CIRCLE (x, y), .5, 9: PAINT (x, y), 9
        END IF
    NEXT
NEXT

'(*** Input Values for Constants ***)
VIEW PRINT 1 TO 5
INPUT "ENTER ALPHA 1: ", a1
INPUT "ENTER ALPHA 2: ", a2
LOCATE 4, 1: PRINT "TOTAL COUNT ="

DO
    DO
        DO
            LET x = INT(w * RND + 1)
            LET y = INT((w - 1) * RND + 2)
        LOOP UNTIL ((y MOD 2) XOR (x MOD 2)) AND (P(x, y) = 0)
        count = P(x, y + 2) + P(x, y - 2) + P(x - 1, y - 1)
        count = count + P(x + 1, y + 1) + P(x - 1, y + 1) + P(x + 1, y - 1)
    LOOP UNTIL count
```

14.3.3 Ising-like Model of Ferrofluid Patterns

```
        SELECT CASE count
            CASE 1: LET E = a1 + 4 * a2 - 1
            CASE 2: LET E = 2 * a1 + 2 * a2 - 1
            CASE 3: LET E = 3 * a1 - 1
            CASE 4: LET E = 4 * a1 - 2 * a2 - 1
            CASE 5: LET E = 5 * a1 - 4 * a2 - 1
            CASE 6: LET E = 6 * a1 - 6 * a2 - 1
            CASE 0: LET E = 6 * a2 - 1
        END SELECT
        IF E < 0 THEN
            CIRCLE (x, y), .5, 9: PAINT (x, y), 9
            LET P(x, y) = 1: LET tot = tot + 1
            LET Eising = Eising + count
            LET Energy = Energy + E
        END IF
        LOCATE 4, 15: PRINT tot
LOOP UNTIL INKEY$ = "q"
```

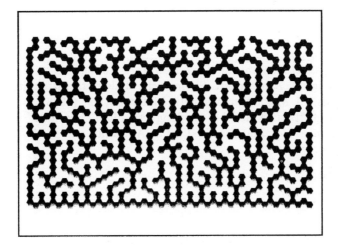

Fig. 1. The Labyrinth phase. $\alpha_1 = 0.5$, $\alpha_2 = 0.1$.

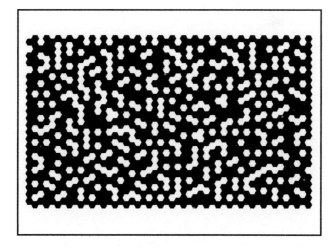

Fig. 2. The emulsification phase.

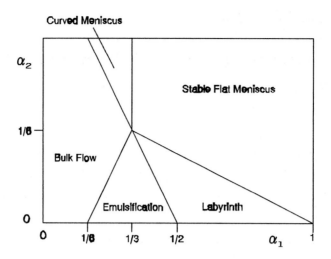

Fig. 3. Phase diagram on the (α_1, α_2) plane.

References

1. K. Huang, *Statistical Mechanics* (Wiley, New York, 1987).
2. R. E. Rosensweig, M. Zahn and R. Shumovich, "Labyrinth Instability in Magnetic and Dielectric Fluids," J. Magnetism Magnetic Mater. **39**, 127 (1983).
3. R. E. Rosensweig, "Lattice Model of the Magnetic Fluid Labyrinth," in *Physics of Complex and Supermolecular Fluids*, edited by S. A. Safran and N. A. Clark (Wiley, New York, 1987).

14.4 Cellular Automata

14.4.1 One-Dimensional Totalistic Cellular Automata
Rocco D. Pochy

One of the simplest of cellular automata is that with one dimension, called a line automaton. It simply consists of a strip of automata that change their states based on a set of rules. At every iteration of the "state clock," the state of the line of automata changes. Even with the simplest of rules, this system is capable of demonstrating the complex behavior of dynamical systems.

A line automaton can be defined by two parameters and a rule of change. The two parameters are: (i) k, the number of states each automaton can be in, and (ii) r, the radius of influence, i.e., the number of neighbors on each side of the automaton that can affect its change.

The rule of change specifies how each automaton changes its state. One set of such rules is called *totalistic*, in which the new state of each automaton is a function of the sum of all the present automata states within r, with that automaton in the center. To be specific, let us assume that $k = 3$ and $r = 2$; the three possible states of each automaton is represented by the set $K \equiv \{0, 1, \ldots, (k-1)\} = \{0, 1, 2\}$. Let us further assume that the present state of the line automata is given by

$$\ldots 2\ 1\ 0\ 0\ 1\ 1\ 0\ 0\ 1\ 2\ 1\ 0\ \mathbf{2}\ 1\ 0\ 0\ 2\ 1\ 0\ 0\ 0\ 1\ 1\ 2 \ldots$$
$$\uparrow\ \Uparrow\ \uparrow$$

For any automaton, the sum of the numbers within its radius of influence s may vary from 0 to $10 = (k-1)(2r+1)$. Therefore, the rule of change can be specified by a lookup table such as this one below.

$$\begin{array}{c}10\ 9\ 8\ 7\ 6\ 5\ 4\ 3\ 2\ 1\ 0 \\ \hline 0\ 0\ 1\ 2\ 0\ 2\ 0\ 1\ 1\ 2\ 0\end{array}$$

Each number in the lower line gives the new state if s is equal to the number above it. Consequently, the new state of that automaton indicated by the double-line arrow above will be 0 according to this rule.

We therefore see that a totalistic rule can be specified by a code consisting of a sequence of $N\ [\equiv (k-1)(2r+1)+1]$ integers, with each integer chosen from the set K. In the program below, one is asked to input k (from 1 to 5), r (from 1 to 319), and the rule code (a sequence of N integers, with $N \leq 128$). If less than N digits are given in the rule code, then the ungiven digits in the rest of the sequence will be assigned 0 automatically by the program. Furthermore, the program automatically start with a line consisting of numbers chosen randomly from the set K. One particular result is shown in Fig. 1.

Fig. 1. History of a line automata. Time increases from top to bottom.

The Program

This program allows the user to enter k, r and a rule code. Initial random state is drawn along the top of the screen. Periodic boundary condition is used. The time evolution of the line automaton is seen as it progresses down the screen. The language used is Microsoft QuickBASIC 4.0.

```
'***** AUTOMATA.BAS *************************

SCREEN 9

CONST line.max = 640
CONST max.rule = 128
CONST start.line = 15
CONST end.line = 349

DIM old.array(line.max)
DIM new.array(line.max)
DIM rule(max.rule)

'** Define Palette Display Colors
PALETTE 0, 0            '** 0 = black
PALETTE 1, 63           '** 1 = white
PALETTE 2, 2            '** 2 = green
PALETTE 3, 4            '** 3 = red
PALETTE 4, 13           '** 4 = yellow
```

14.4.1. One-Dimensional Totalistic Cellular Automata

```
RANDOMIZE
LOCATE 2, 1: INPUT "Enter K value:": k
LOCATE 3, 1: INPUT "Enter R value:": r
LOCATE 4, 1: INPUT "Enter CODE:"; code$


CLS
'** Break CODE string into rule array
length = LEN(code$)
j = 0
FOR i = length TO 1 STEP -1
   x = VAL(MID$(code$, i, 1))
   rule(j) = x
   j = j + 1
NEXT

LOCATE 1, 1: PRINT "CODE:";
FOR i = length TO 0 STEP -1
   PRINT RIGHT$(STR$(rule(i)), 1);
NEXT

LOCATE 1, 20: PRINT USING "K = #   R = #"; k; r

'** Initialize Old Array and Display
FOR i = 0 TO line.max
   old.array(i) = INT(RND * (k - 1) + .5)
   PSET (i, start.line), old.array(i)
NEXT

display = start.line + 1

WHILE (display < end.line AND INKEY$ = "")

   '** Process Line
   FOR i = 0 TO line.max
     sum = 0
     FOR j = i - r TO i + r
       n = j
       IF n < 0 THEN n = line.max + n
       IF n > line.max THEN n = n - line.max
       sum = sum + old.array(n)
     NEXT
     new.array(i) = rule(sum)
   NEXT

   '** Copy New to Old and Display
   FOR i = 0 TO line.max
      old.array(i) = new.array(i)
      PSET (i, display), old.array(i)
   NEXT

   display = display + 1

WEND

WHILE (INKEY$ = ""): WEND
```

14.4.2 Two-Dimensional Cellular Automata: Formation of Clusters

Rocco D. Pochy

On simple example of a two-dimensional cellular automata is the one with two states on each site and the use of the majority rule. Specifically, a Moore neighborhood is used. The new state in each cell is the one possessed by the majority of its eight neighbors; if there is no majority, the state remains unchanged.

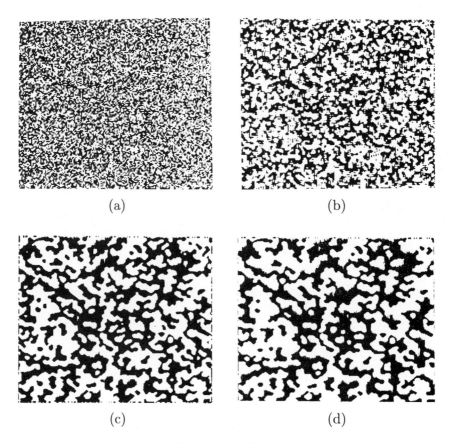

Fig. 1. Formation of clusters in a two-dimensional cellular automaton. Moore neighborhood, and the majority rule is used. The initial random state (a) evolves after 5 (b) and 10 (c) time steps, to the final stable state (d).

In the program below, one begins with a random mix of states (Fig. 1(a)). As time increases one sees more clusters formed (Figs. 1(b) and (c)). The boundaries of the clusters smooth with time, until a stable state is reached (Fig. 1(d)). For different mixture of states used initially, a stable state is always reached.

14.4.2. Two-Dimensional Cellular Automata: Formation of Clusters

The Program

The language used is Microsoft C v5.0. Each cell is either black or white. Each cell is tested with its eight neighbors. If more neighbors are black than white (white than black), the cell becomes black (white). Otherwise, the cell remains what it is. The boundary condition is set so that the states of cells on the four edges of the lattice are frozen throughout.

```c
/* Program: FREEZE.C ---------------------------------------- */
#include <stdio.h>
#include <stdlib.h>
#include <graph.h>

/*------------------------------------------------------------*/
extern void initialize(void);
extern void update(unsigned long);
extern void done(void);

/*------------------------------------------------------------*/

#define     SIZE      170        /* Size of the Matrix        */
#define     BLACK     0
#define     BLUE      1
#define     GRAY      6
#define     WHITE     15
/*------------------------------------------------------------*/

static char     GRID[2][SIZE][SIZE];   /* Matrices of Change  */

/*------------------------------------------------------------*/
main()
{
    unsigned long int i = 1;

    initialize();

    while (kbhit()==0)
    {
        update(i);
        i++;
    }
    done();
}

void initialize(void)
{
    register int i,j,value;
    int          seed;

    printf("ENTER RANDOM SEED (0-32000):");
    scanf("%d",&seed);
    srand(seed);

    /* Set up graphics screen */
    _setvideomode(_MRESNOCOLOR);
```

```c
        _settextposition(1,35);
        printf("ITER:");

        /* Generate a random matrix */
        for (i=0;i<=SIZE;i++)
        {
           for(j=0;j<=SIZE;j++)
           {
            if (rand()>16384)
            {
               value = 0;
               _setcolor(BLACK);
            }
            else
            {
               value = 1;
               _setcolor(WHITE);
            }

            GRID[0][i][j] = value;
            _setpixel(i,j);
           }
        }
   }


   void update(unsigned long int gen)
   {
      float              junk, remain;
      register int       a,b;
      register int       i,j,n;
      register int       left,right,top,bot;
      char               value;

      junk   = gen/2.0;
      remain = junk - (int)(junk);

      if (remain==0.0)
      {
         a = 1;
         b = 0;
      }
      else
      {
         a = 0;
         b = 1;
      }

      for (i=1;i<SIZE;i++)
      {
         top = i + 1;
         bot = i - 1;

         for (j=1;j<SIZE;j++)
         {
            right = j + 1;
            left  = j - 1;

            /* Get Number of Neighbors */
            n = 0;
            n = GRID[a][top][left] + GRID[a][top][j] + GRID[a][top][right] +
                GRID[a][i][left]   +                   GRID[a][i][right]   +
                GRID[a][bot][left] + GRID[a][bot][j] + GRID[a][bot][right];

            if (n>4)             /* If more than 4 "White" neighbors */
            {
```

14.4.2. Two-Dimensional Cellular Automata: Formation of Clusters

```
          _setcolor(WHITE);
          _setpixel(i,j);
          value = 1;
        }
        else
        {
          if (n<4)           /* If more than 4 "Black" neighbors  */
          {
            _setcolor(BLACK);
            _setpixel(i,j);
            value = 0;
          }
          else               /* Equal number of White and Black   */
          {
            value = GRID[a][i][j];
            if (value==0)
              _setcolor(BLACK);
            else
              _setcolor(WHITE);

            _setpixel(i,j);

          }
        }
        GRID[b][i][j] = value;
      }
    }
    _settextposition(2,36);
    printf("%u",gen);
}

void done(void)
{
    _setvideomode(_DEFAULTMODE);
}
```

15 Theoretical

15.1 Curve Length and the Scaling Parameter
Thayer H. Watkins

The fractal dimension of a curve is measured by means of the relationship between curve length (as estimated by applying a measuring rod to the curve) and the length of the measuring rod. This relationship may have to do with characteristics of the spectrum rather than the curve being a fractal.

Introduction

In determining the fractal dimension of a curve such as a coastline, one estimates the relationship between the measured length L and the length ε of the "measuring rod." For a nonfractal curve, L approaches a finite limit as ε goes to zero. For a fractal curve, L increases without bound as ε goes to zero, but after some point the functional dependence of L on ε approaches $c\varepsilon^{1-D}$, where D is called the fractal dimension of the curve. The dimension D is thus given by unity minus the limit of $d[\ln L(\varepsilon)]/d(ln\varepsilon)$ as ε goes to zero.

Although for nonfractal curves $d[\ln L(\varepsilon)]/d(\ln \varepsilon)$ approaches zero in the limit, there is a dependence of curve length on ε such that $L(\varepsilon)$ increases relatively rapidly for decreasing ε, when ε approaches a critical range of values associated with the spectrum of the curve. For a simple sinusoidal curve this critical range is between 1/2 and 1/4 of the wavelength.

The analysis here will be limited to a restricted case that more readily lends itself to mathematical analysis. Firstly, the curves that will be considered are those given by a function $y = f(x)$, over an interval $[0, S]$. (The length of the interval S will be referred to as the span of the curve.) Secondly, instead of fitting a measuring rod of fixed length between points on the curve, the interval $[0, S]$ will be divided up into subintervals of length b (except for the last subinterval) and the distance determined from the corresponding points on the curve (Fig. 1). That is,

$$L(b) = \sum_{j=1}^{n} \{b^2 + [f(jb) - f((j-1)b)]^2\}^{1/2}$$
$$+ \{(S - nb)^2 + [f(S) - f(nb)]^2\}^{1/2} \qquad (1)$$

where n is the integral part of S/b.

The length of the curve, i.e., the limit of $L(b)$ as b goes to zero, when it exists, can be expressed as the integral $\int_0^s \{1 + [f'(x)]^2\}^{1/2} dx$. But in general, this integral cannot be evaluated analytically.

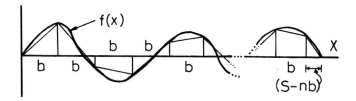

Fig. 1. Sketch of the measuring process of a curve.

The Simplest Case: A sinusoidal Curve

When $f(x) = a\sin(2\pi x/\lambda)$, a sinusoidal curve of wavelength λ and amplitude a, the value of $L(b)$ can easily be computed. This function has the special property that $f(0) = 0$ and it will be assumed for convenience that S is an integral number of λ so $f(S) = 0$. These are not essential assumptions but they conveniently eliminate some messy details that clutter up the analysis.

When b is equal to an odd multiple of $\lambda/4$, the value of $L(b)$, because of symmetry, can be computed from simple geometry. When b is an even multiple of $\lambda/4$, the scaled curve length is just S. It is more reasonable to look at $L(b)/S$, though this ratio does depend on S.

The graph of $L(b)/S$ vs b for $S = 25$, $\lambda = 1$ and $a = 1$ is given in Fig. 2. The functional relationship, even for this simple case, is complicated. Particularly surprising is the sensitivity of the relationship in the vicinity of the scaling parameter b equal to $\lambda/2$ and to $\lambda/4$. Ignoring these anomalous cases, the relationship involves two different regions and an interval of transition between them. For $b > \lambda$ the curve length is approximately equal to the span of the curve, i.e., $L(b)/S = 1.0$. For $b < \lambda/4$, $L(b)/S$ is approximately 4 for this case. The region from λ for $\lambda/4$ is a zone of transition.

In Fig. 2, the lower boundary of $L(b)/S$ is obviously 1.0 and it is attained at the even multiples of $\lambda/4$. This is because at these points the function is equal to zero and hence $f(jb) - f((j-1)b)$ is zero. However, if b differs slightly from such a value, then after enough subintervals, the function is being evaluated at points other than where it is equal to zero and hence $f(jb) - f((j-1)b)$ is not zero. This is why $L(b)/S$ increases quite rapidly from what seems to be small deviation of b from the even multiples of $\lambda/4$. In order for a deviation of b from such points to be small, the deviation times S/b must be small.

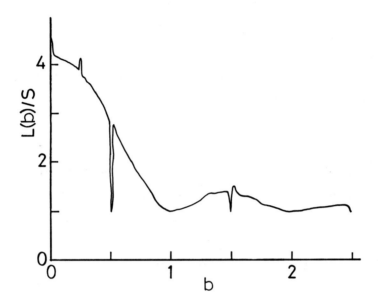

Fig. 2. Curve length $L(b)/S$ vs scaling parameter b.

The values of $L(b)/S$ at the odd multiples of $\lambda/4$ are given by $(1 + a^2/h^2)^{1/2}$. This curve is not, however, an envelope of the $L(b)/S$ function.

If the relationship in Fig. 2 were expressed in terms of the logarithms of the variables, it would look like a step function. The height of the step depends on the amplitude of the sine function. A rough approximation of the increment gives $[1 + (4a/\lambda)^2]^{1/2} - 1$. It is arrived at by computing the difference between $L(b)/S$ at $h = \lambda/4$ and $h = \lambda$. The location of the step depends on the wavelength. The rapid increase in the curve length occurs for a scaling length b between λ and $\lambda/4$; so the location of the step could be taken to be roughly at $\lambda/2$.

Conclusions

If the curve $f(x)$ has components of different wavelengths, the curve of $\ln L(b)$ vs $\ln b$ could look like a stairway of step functions. The slope of this stairway depends on the relationship between the wavelength and amplitude of the components of the curve — the so-called spectrum of the curve. Thus the measurement of fractal dimension using maps of coastlines are probably measuring properties of the spectral density function over a range of wavelengths, with the hypothesis being that the spectral density function has similar properties in the limit as wavelength goes to zero.

15.2 Analysis of the Back-Propagating Neural Network for the XOR Problem

Victor M. Castillo

The downfall of the perceptron is that it is only able to solve problems that are linearly separable. More recent work on artificial neural networks have included the development of the back-propagating network that can solve highly nonlinear problems. The exclusive-OR (XOR) is a simple example of a problem that is not linearly separable. A five node neural network using the generalized delta Function to adjust the weights and a sigmoid transfer function has been used to solve this problem. In this paper, the set of weights that satisfy each of the XOR cases is evaluated for fixed points. The stability of these fixed points is determined and compared to those of the other cases. General conclusions of the system are then made.

Introduction

The first useful artificial neural network was the perceptron. The perceptron was developed by Frank Rosenblatt in 1957. This simple two-layered network was used to recognize characters of the English alphabet, even those with imperfections. The popularity of this computational method grew until 1969 when Minsky and Papert [1] published their book on the perceptron. By careful analysis of the dynamics of this system, they were able to prove that the perceptron was limited to making associations between similar patterns only. They showed that if the outputs were not linearly separable, that the problem could not be solved by the perceptron. Their classic example was the eXclusive-OR (XOR) problem (shown in Table 1).

Table 1. The XOR problem.

Input		Output
X_1	X_2	O
0	0	0
0	1	1
1	0	1
1	1	0

In 1982 Rumelhart and McClelland [2] showed that a multilayered neural network using an error propagation routine could be used to learn about complex relationships. Since then, neural networks have become the topic of much research. The back-propagating model has been used to solve a great variety of problems.

15.2 Analysis of the Back-Propagating Neural Network for the XOR Problem

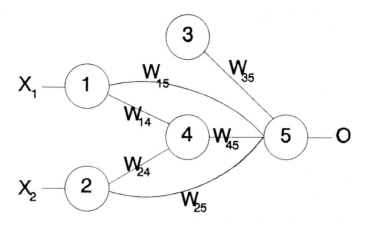

Fig. 1. A back-propagating network for the XOR problem.

Dynamics

The back-propagating neural network uses a two cycle procedure during the learning phase. The first cycle involves transforming the input pattern into an output pattern. The second involves calculating the error and propagating an error signal back throughout the network. The magnitude of an error signal at a particular node is proportional to how much that node influences the output pattern. The connection weights are then modified a small amount in a way that decreases the total error.

Feed Forward

Initially, the network is at some random state (configuration of weights). The first step is to allow the network to transform the input pattern to the output pattern so that the error can be measured. This is accomplished by the distribution of information to the individual nodes, and simple transformations at node level. The nodes at the input level linearly transforms the input pattern to the second layer. A single bias node is also part of this input layer and just outputs a +1 value. The output of these nodes is then weighted by the variable connection weights between the first and the second layer. The nodes of the second layer (single node in this case) do a more complex transformation. A sigmoid transformation is common. The output of this layer is then weighted before continuing to the output layer. The output layer does a final transformation (sigmoid) before evaluating the error value.

In general, the net input to the ith node,

$$p_i = \sum_{j<i} W_{ij} a_j \qquad (1)$$

The output of the ith node,

$$a_i = f(p_i) = (1 + \exp^{-p_i})^{-1} \qquad (2)$$

For the system in Fig. 1,

$$a_1 = X_1, a_2 = X_2, a_3 = 1$$

$$p_4 = W_{14}X_1 + W_{24}X_2$$

$$a_4 = f(p_4) \tag{3}$$

$$p_5 = W_{15}X_1 + W_{25}X_2 + W_{45}a_4 + W_{35}$$

$$a_5 = f(p_5)$$

Feed Back

The next cycle propagates an error signal back through the network so that the weights may be adjusted. The error signal to the output node is just the square of the difference between the actual output O (a_5 in this case), and the desired output D. The magnitude of the error term for a given node is related to the weights of the connections between it and the output. Once the error terms for the nodes are calculated, the connection weights are modified using the generalized delta rule.

Let E be the total net error (this is the square of $D - a_5$). The change to a particular weight must be proportional to how much a small change in that weight has on the global error. The learning rate constant is used to slow the change so that the convergence is smoother. The local error term b_i is the error contribution of the ith node to the total net error.

$$\Delta W_{ij} = -\eta \frac{\partial E}{\partial W_{ij}} = -\eta \frac{\partial E}{\partial p_i} \frac{\partial p_i}{\partial W_{ij}} = -\eta \frac{\partial E}{\partial p_i} a_i = \eta b_i a_i \tag{4}$$

where

$$b_i = -\frac{\partial E}{\partial p_i} = -\frac{\partial E}{\partial a_i} \frac{\partial a_i}{\partial p_i} = q_i \frac{\partial a_i}{\partial p_i} = q_i a_i (1 - a_i) \tag{5}$$

and

$$q_i = -\frac{\partial E}{\partial a_i} = -\sum_{i<k} \frac{\partial E}{\partial p_k} \frac{\partial p_k}{\partial a_i} = \sum_{i<k} b_k \frac{\partial p_k}{\partial a_i} \tag{6}$$

but since $p_k = \sum_{i<k} W_{ik} a_i$ and $\partial p_k / \partial a_i = W_{ik}$, then

$$q_i = \sum_{i<k} W_{ik} b_k \tag{7}$$

Therefore,

$$\Delta W_{ij} = \eta a_i a_j (1 - a_i) \sum_{i<k} W_{ki} b_k \tag{8}$$

15.2 Analysis of the Back-Propagating Neural Network for the XOR Problem

For the system in Fig. 1,

$$E = (D - a_5)^2, \quad \frac{\partial E}{\partial a_5} = -2(D - a_5) \qquad (9)$$

The generalized delta rule leads to the set of weights that satisfy,

$$\begin{aligned}
f_1 &: \Delta W_{14} = 2\eta a_5(1 - a_5)(D - a_5)W_{45}X_1 = 0 \\
f_2 &: \Delta W_{24} = 2\eta a_5(1 - a_5)(D - a_5)W_{45}X_2 = 0 \\
f_3 &: \Delta W_{15} = 2\eta a_5(1 - a_5)(D - a_5)X_1 = 0 \\
f_4 &: \Delta W_{25} = 2\eta a_5(1 - a_5)(D - a_5)X_2 = 0 \\
f_5 &: \Delta W_{35} = 2\eta a_5(1 - a_5)(D - a_5) = 0 \\
f_6 &: \Delta W_{45} = 2\eta a_5(1 - a_5)(D - a_5)a_4 = 0
\end{aligned} \qquad (10)$$

We will now investigate the dynamics for each case, viz.,

$$\begin{aligned}
\{0,0\} &\to \{0\}, \{0,1\} \to 1, \\
\{1,0\} &\to \{1\}, \{1,1\} \to 0
\end{aligned} \qquad (11)$$

The Jacobian Matrix Elements

At this point, we will find the conditions that satisfy the equations in Eq. (10) and therefore give us the fixed points of the system. Then the eigenvalues of the Jacobian matrix are evaluated at these points to determine if they are stable or not.

The matrix elements of the Jacobian are as follows.

$$(J_{ij}) = \begin{pmatrix}
\partial f_1/\partial W_{14} & \partial f_1/\partial W_{24} & \partial f_1/\partial W_{15} & \partial f_1/\partial W_{25} & \partial f_1/\partial W_{35} & \partial f_1/\partial W_{45} \\
\partial f_2/\partial W_{14} & \partial f_2/\partial W_{24} & \partial f_2/\partial W_{15} & \partial f_2/\partial W_{25} & \partial f_2/\partial W_{35} & \partial f_2/\partial W_{45} \\
\partial f_3/\partial W_{14} & \partial f_3/\partial W_{24} & \partial f_3/\partial W_{15} & \partial f_3/\partial W_{25} & \partial f_3/\partial W_{35} & \partial f_3/\partial W_{45} \\
\partial f_4/\partial W_{14} & \partial f_4/\partial W_{24} & \partial f_4/\partial W_{15} & \partial f_4/\partial W_{25} & \partial f_4/\partial W_{35} & \partial f_4/\partial W_{45} \\
\partial f_5/\partial W_{14} & \partial f_5/\partial W_{24} & \partial f_5/\partial W_{15} & \partial f_5/\partial W_{25} & \partial f_5/\partial W_{35} & \partial f_5/\partial W_{45} \\
\partial f_6/\partial W_{14} & \partial f_6/\partial W_{24} & \partial f_6/\partial W_{15} & \partial f_6/\partial W_{25} & \partial f_6/\partial W_{35} & \partial f_6/\partial W_{45}
\end{pmatrix}$$

where $i, j = 1, 2, \ldots, 6$.

Fixed Points and Stability Analysis

For each input/output relationship required by the system, the fixed points of the system will be determined. The value of the maximum eigenvalue of the Jacobian matrix evaluated at that point will then be given. A positive eigenvalue indicates an unstable state, a maximum of zero indicates a neutrally stable point, and a negative value indicates a stable fixed point (an attractor of the system).

■ For $\{0,0\} \to \{0\}$, one has $X_1 = 0$, $X_2 = 0$, $D = 0$. This alone leads to, $\Delta W_{14} = \Delta W_{24} = \Delta W_{15} = \Delta W_{25} = 0$. So we are left with, $\Delta W_{35} = -2\eta a_5^2(1-a_5) = 0$, and $\Delta W_{45} = -2\eta a_5^2 a_4(1-a_5) = 0$, which are satisfied by $a_5 = 0$ or 1. The Jacobian matrix is then reduced to

$$J = 2\eta \begin{pmatrix} 0 & \cdots & & 0 \\ \vdots & \ddots & & \vdots \\ & & e^{-p_5} & a_4 e^{-p_5} \\ 0 & \cdots & a_4 e^{-p_4} & a_4^2 e^{-p_5} \end{pmatrix} \quad (12)$$

So $\mathrm{Max}(\mathrm{Re}(e)) = 2\eta \exp(-p_5)(1 - a_4^2)$. For $a_5 = 1$ we get a positive real eigenvalue. For $a_5 = 0$ we get an eigenvalue of 0. This shows that the weight configurations that lead to the incorrect solution are unstable.

■ For $\{1,0\} \to \{1\}$, one has $X_1 = 1$, $X_2 = 1$, $D = 1$. This leads to $\Delta W_{24} = \Delta W_{25} = 0$, and $\Delta W_{14} = 2\eta a_5(1-a_5)^2 W_{45} = 0$, $\Delta W_{15} = 2\eta a_5(1-a_5)^2 = 0$, $\Delta W_{35} = 2\eta a_5(1-a_5)^2 = 0$, and $\Delta W_{45} = 2\eta a_5(1-a_5)^2 a_4 = 0$. So again we must have $a_5 = 0$ or 1. Consequently, $J = 0$ for all cases since the term $(D - 2a_5 - 2a_5 D + 3a_5^2)$ goes to zero for $a_5 = 1$.

■ For $\{0,1\} \to \{1\}$, on has $X_1 = 0$, $X_2 = 1$ and $D = 1$, which leads to $\Delta W_{14} = \Delta W_{15} = 0$, $\Delta W_{24} = 2\eta a_5(1-a_5)^2 W_{45} = 0$, $\Delta W_{25} = 2\eta a_5(1-a_5)^2 = 0$, $\Delta W_{35} = 2\eta a_5(1-a_5)^2 = 0$, and $\Delta W_{45} = 2\eta a_5(1-a_5)^2 a_4 = 0$. This also is satisfied by $a_5 = 0$ or 1, but as in the previous case both solutions are neutrally stable.

■ For $\{1,1\} \to \{0\}$, no weights drop out. One has $\Delta W_{14} = 2\eta a_5^2(1-a_5)W_{45} = 0$, $\Delta W_{24} = 2\eta a_5^2(1-a_5)W_{45} = 0$, $\Delta W_{15} = 2\eta a_5^2(1-a_5) = 0$, $\Delta W_{25} = 2\eta a_5^2(1-a_5) = 0$, $\Delta W_{35} = 2\eta a_5^2(1-a_5) = 0$, and $\Delta W_{45} = 2\eta a_5^2(1-a_5)a_4 = 0$. This leads to $a_5 = 0$ or 1, which again leads to eigenvalues of zero indicating that both solutions are neutrally stable.

Conclusions

For this neural network model, the fixed points in weight space can be determined analytically. These points represent the set of weights that satisfy the generalized delta rule, and do not necessarily represent a state that yields the desired output. An analysis of the stability of these fixed points by evaluating the eigenvalues of the Jacobian matrix for this system shows that the fixed points that lead to spurious solutions are unstable or metastable. Unfortunately, however, the analysis does not show any stable fixed points at all. Computer simulations of this model (with the program listed in Appendix A) showed that this system converges to yield a correct solution in every case tried.

Appendix A: Computer Program for Back-Propagation

```
'(***********************************************************************
'(*    BACKPROP.BAS (in BASIC)                                          *
'(***********************************************************************

' Graphics Setup
CLS : SCREEN 9 : win = 2000 : WINDOW (0,0)-(win,100)

' Dimension Arrays
DIM x(2), p(4), a(4), b(4), W(4,4), dw(4,4)

' Set Learning Rate and Randomize Weights
eta = 1
RANDOMIZE TIMER
FOR i = 1 TO 4
      FOR j = 1 TO 4
            W(i,j) = RND*8 - 4
      NEXT
NEXT

' Start Learning
READ x(1), x(2), d
      DO
        ' Forward Propagation
        p(3) = W(1,3)*x(1)+W(2,3)*x(2)
        a(3) = (1+EXP(-p(3)))^-1
        p(4) = W(3,4)*a(3)
        E = (d-a(4))                              'error term

        ' Feedback
        b(4) = E
        b(3) = W(3,4)*b(4)*a(3)*(1-a(3))
        dw(3,4) = eta*b(4)*a(3)                   'update in parallel
        dw(2,3) = eta*b(3)*x(2)
        dw(1,3) = eta*b(3)*x(1)

        ' Update Weights
        FOR i = 1 TO 4
              FOR j = 1 TO 4
                    W(i,j) = W(i,j)+dw(i,j)
              NEXT
        NEXT

        ' Graph Results
        LINE (t-1, ABS(Esave)*100)-(t,ABS(E)*100),10
        LINE (t-1, ABS(wsave)*100)-(t,ABS(w(3,4))*100),10+SGN(w(3,4))
        Esave = E : wsave = W(3,4)
        t = t+1
   LOOP UNTIL ABS(E) < .01
SOUND (1000),.1 : PRINT "t = "; t : PRINT; "w = "; W(3,4) : SLEEP

' Input and Output
DATA 0,0,0
```

Reference

1. M. L. Minsky and S. S. Papert, *Perceptrons* (MIT, Cambridge, 1969).
2. D. E. Rumelhart and J. L. McClelland, "Learning Internal Representations by Error Propagation," in *Parallel Distributed Processing*, Vol. 1 (MIT, Cambridge, 1986).

16 Experimental

16.1 Instabilities of Finite Water Columns
Mark C. Fallis, Michael M. Masuda,
Rodney C. LeRoy and Nejat Neisan

Water in vertical laminar flow exiting from a tube is stopped by a horizontal barrier. When the separation between the tube and the barrier h is large, the water-air interface of the water column is smooth. But for small flow rates and h smaller than a threshold value, the interface becomes unstable and bifurcates into a wrinkled column. This wrinkling of the interface is related to surface tension (capillary) effects, and is accompanied by a meniscus-forming instability just above the barrier. This latter instablility is a hysteresis process as h is decreased or increased. Experimental results of these phenomena, measured quantitatively for the first time, are presented. Among other things, the average wavelength of the interfacial wrinkles is found to increase as a power law as h decreases.

This wrinkling instability of a water column can be demonstrated easily with any faucet. First turn on the faucet slightly to allow for a smooth, narrow water column to flow downward, then use one finger to stop the column at some long length, and gradually raise your finger to shorten the water column.

Experimental Procedure

The apparatus setup is shown in Fig. 1. To maintain a constant flow rate, we use a five-liter Nalgene container of distilled water with an overflow valve near the top to keep the water level constant (~ 50 cm deep), and thus to maintain a constant hydrostatic pressure at the base where water can exit through a valve. Water exits the container and passes through a spigot and a flow meter. The flow from the meter exits from a vertical metal tube (0.320 cm inner radius) in air above an aluminum rod (10 cm long and 0.5 cm in diameter).

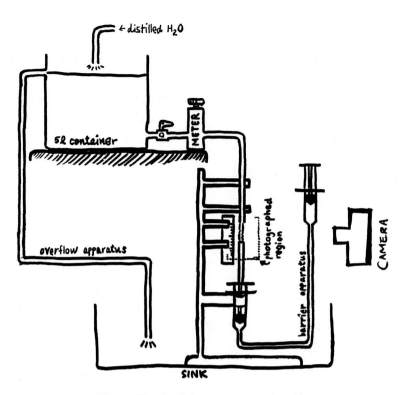

Fig. 1. Sketch of the experimental setup.

The spigot is used as an on/off valve and the meter is used to fine tune the flow rate. The meter (a gauge originally designed for air flow) is calibrated by correlating the markings on it to actual flow rates obtained by measuring the time it takes to fill a graduated cyclinder with 100 ml of water. The meter does not measure a wide range of flow rates; by using different sized ball bearings in it we can increase the range of the meter. To conserve water, water in the system is continously recycled via a water pump (not shown in the figure) which draws water from the catch basin and returns it to the container.

We photograph the column impacting on the barrier using a zoom lens. Because a meniscus forms on the barrier causing the wrinkled column to jitter slightly, a black velvet backdrop is used as the background to eliminate unwanted glare. A vertical ruler is placed near the water column and is included in the pictures. A 40 watt light bulb is used to illuminate the water column and ruler. From the photographs, the average wavelength of the wrinkles, λ, is measured by counting the number of "waves" and dividing by the length of the region where the waves appear. We use an average because the wavelength of the wrinkles is not strictly sinusoidal but varies slightly along the column length.

The separation between the tube opening and the barrier h is adjusted by compressing one of the two syringes which are filled with water and connected by a rubber tube filled with water, so that as one syringe is pushed in the other one, which holds the barrier, is extended upward. The change in h is measured directly off the syringe using a Venier calipers and double checked by making direct measurements on the photographs.

16.1 Instabilities of Finite Water Columns

Results

We have studied several flow rates using the method outlined above. For a fixed flow rate Q, as h is decreased the smooth water column starts to wrinkle with a finite (average) wavelength λ for $h \leq h_c$. The wavelength (amplitude) of the wrinkles increases (decreases) from bottom up along the column. For $h \leq h_m$ ($\ll h_c$), a meniscus appears just above the barrier (see Figs. 2 and 3).

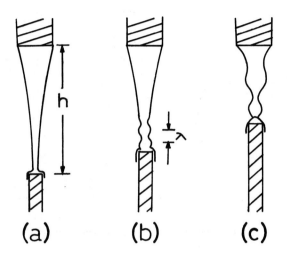

Fig. 2. Sketch of the wrinkling instability as h is decreased. (a) $h > h_c$. (b) $h \ \ h_c$. (c) $h < h_m$ ($\ll h_c$).

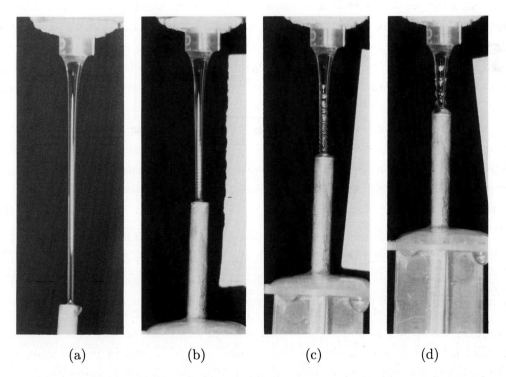

Fig. 3. Typical photographs of the wrinkling instability. $Q = 2.10$ ml/s; $h_c = 63.35$ mm; $h_m = 38$ mm and the meniscus is less than 2 mm high. (a) $h = 68.30$ mm. (b) $h = 41.27$ mm. (c) $h = 30.25$ mm. (d) $h = 18.66$ mm.

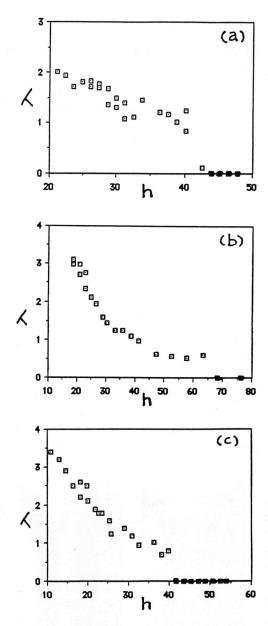

Fig. 4. Variation of λ vs h, as h is decreased. (a) $Q = 1.93$ ml/s; $h_c = 43.94$ mm. (b) $Q = 2.00$ ml/s; $h_c = 65.8$ mm. (c) $Q = 2.97$ ml/s; $h_c = 41.6$ mm.

In Fig. 4, the variation of λ as a function of h is presented for three different flow rates. For $Q = 1.25$ ml/s, the water column breaks up into droplets at a distance below the tube opening smaller than the critical h. A log-log plot of λ vs h is shown in Fig. 5, with the data from different Q included. One obtains $\lambda \sim h^{-\alpha}$ with $\alpha = 1.57$.

In fact, as h is decreased, there is a series of small changes at the bottom of the water column before the meniscus appears. When h is just above h_c and when Q is small enough for the effect to be observed clearly, concentric circular "waves," numbered 1, 2 and 3 in Fig. 6(a), form on the surface of the barrier around the foot

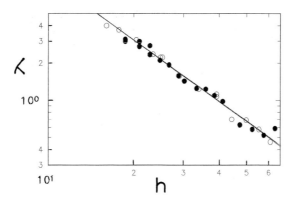

Fig. 5. Log-log plot of λ vs h, as h is decreased. The solid dots represent the data from $Q = 2.10$ ml/s; the open circles are from other Qs.

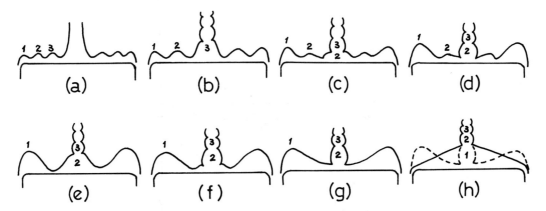

Fig. 6. The formation of circular "waves" above the barrier preceding the appearance of the meniscus, as h is decreased gradually from (a) to (h). In (h), the solid line represents the meniscus; the dotted line represents the situation just before the meniscus formation.

of the column. (The lower part of the water column itself has waves also, but they are too small to be counted.) As h is gradually reduced, the waves on the column become more prominent, but more interesting is that the circular waves on the barrier become more spaced, only allowing two circular waves to form [Figs. 6(b) and 6(c)]. The innermost circular wave, #3, *travels up into the foot of the water column.* As h is further reduced, the innermost circular wave, #2, travels up into the column, leaving one circular wave, #1, on the surface of the barrier [Figs. 6(d) and 6(e)]. Initially it assumes the usual sinusoidal shape of a circular wave, but as h is reduced further to h_m, the sinusoidal shape [Figs. 6(f) and 6(g)] is abandoned for a simpler and more stable configuration — the meniscus configuration [Fig. 6(h)]. The meniscus continues to be a stable configuration until h is zero.

After the meniscus is formed the process is then reversed by gradually increasing h. The meniscus continues to be a stable configuration up to a certain h, say, h_o at which the meniscus is spontaneously replaced by the "two concentric circular waves" configuration seen earlier [Fig. 6(b)]. In other words, the surface above the barrier apparently skips the "one circular wave" configuration of Figs. 6(e)–6(g) as h is increased. The prolonged existence of the meniscus as h is increased shows that the water column behaves differently than if h is decreased — a hysteresis effect. It will not be surprising if hysteresis effect also shows up in the λ vs h graph, but the measurement of λ as h is *increased* has yet to be carried out. Furthermore, the diameter of the barrier surface will also be able to quantitatively affect the experiments. This "simple" experiment has become complicated with the advent of the meniscus observations.

Discussions

Closer observations show that the $\log \lambda$ vs $\log h$ curve for $Q = 2.10$ ml/s in Fig. 5 does seem to have a discontinuous drop at $h \sim 45$ mm, indicating a hysteresis loop. But more data is needed to confirm this.

In analogy to the discussion of the varicose instability of a free-falling (infinite) water column [1], for a finite column here, we calculate the surface area of a column of radius $r(z)$, with

$$r(z) = a_1 + a_2(z) \sin[k(z - h)] \tag{1}$$

and $k = 2\pi/\lambda$. Here the z axis is vertically downward, and

$$a_2(z) = a \exp[(z - h)/z_0] \tag{2}$$

a_1, a and z_0 are constants. z_0 represents the damping length of the wave measured from the bottom of the water column. Equations (1) and (2) mimic the distorted surface as observed [see Fig. 2(b)]. Conservation of volume is used in comparing the two columns; a_2 is assumed to be small and $z_0 \ll h$ is assumed, for simplicity. We find that the surface area of this distorted column is *less* than that of an undistorted column of radius a_1. Since surface energy is equal to the product of surface tension and the surface area, the distorted column of smaller energy is physically preferred. The observed wrinkling instability can thus be understood as a capillary instability. Note that the sudden appearance of a finite wavelength at the critical control parameter h_c is analogous to the case of the undulatory rolls in electroconvective nematic liquid crystals [2], and in many other systems.

We noticed that the use of a curved barrier is able to eliminate the meniscus formation on the barrier. It has also been observed that a sharp object inserted into the water column induces a larger local instability as compared to the flat barrier for the same h. Other parameters of interest may include (i) the viscosity of the flowing liquid — since viscosity usually plays a stabilizing role, we expect the instability to diminish for higher viscosities; (ii) tube diameter, and (iii) barrier size, curvature and resilience.

This work is supported by the Allied-Signal Award of the Society of Physics Students.

References

1. S. Chandrasekhar, *Hydrodynamics and Hydromagnetic Stability* (Dover, New York, 1981), Chap. XII.
2. R. Ribotta, A. Joets and Lin Lei (L. Lam), "Oblique Roll Instability in an Electroconvective Anisotropic Fluid," Phys. Rev. Lett. **56**, 1595 (1986). (Reprinted in Section 12.3.)

16.2 Viscous Fingering in Optical Cement Displaced by Water
James W. Hillendahl

Introduction

Fractals have been studied in numerous experiments. Many such experiments involve viscous fingers because they are easy to make and are of fundamental importance in both basic and applied research. For example, Nittmann *et al.* [1] examined the fractal dimension of the boundaries of viscous fingers in linear Hele-Shaw cells containing a polymer solution, a non-newtonian fluid, displaced by water. The solution and water are miscible. Their findings suggest that for a wide range of experimental conditions the fractal dimension is constant and reproducible.

In our experiments presented here, viscous fingers were produced using water to displace epoxy contained between two closely spaced, parallel glass plates in radial geometry. The experiments demonstrate a simple and direct dependence of the fractal dimension and the number of fractal lobes of the *boundary* of viscous fingers on a single control parameter, i.e., the injection time of a fixed amount of water. The functional dependence of the fractal dimension on the injection time is determined empirically.

Experiments

Twelve runs were made for this experiment. We started with 24 plates of 3 mm thick clear glass, each eight square inches. The plates were thoroughly cleaned, then arranged in pairs to make 12 sets. A 1 mm diameter hole was made in the center of the upper plate of each set. A silicone rubber sealant was applied to each of these holes. After hardening, a small hole was pierced through the sealant, providing an orifice that sealed itself when not in use. The edges of the bottom plate were taped with 0.050 mm (2 mil) Mylar tape to provide a uniform spacer between the plates when they were put together.

With the plate sets separated, a calibrated syringe was used to place 10 ml of epoxy in the center of the lower, undrilled plate. This epoxy, or "optical cement," was Norland Optical Adhesive NOA-65 (from Norland Products, New Brunswick, NJ). This highly viscous cement has a viscosity of 1145 cp and a density of 1.231 g/cm^3. The optical cement was dyed with felt pen ink for contrast prior to application.

The two plates in each set were placed together to form a cell, taking care to avoid making bubbles in the epoxy. The plates were clamped along all edges and allowed to sit for 15 min. The pressure of the epoxy thus equilibrated, making a

16.2 Viscous Fingering in Optical Cement Displaced by Water

uniform layer 0.050 mm thick between the clear glass plates. Small gaps were left in the spacer to allow excess optical cement to escape.

The fingers were formed in each cell by injecting water into the epoxy between the clamped plates through the hole in the silicone sealant in the upper plate. Thus the fingers grew radially outward from the center of the plates. A small, calibrated syringe was used for the injections. The total volume of water injected was kept constant at 1.0 ml for all sets of plates. An electronic timer was used to vary the injection times t from 5 to 60 seconds. (Consequently, the average injection rate of water is proportional to t^{-1}.) No water escaped from the plates during or after the injection due to the good sealing properties of the silicone. The injection process was easily controlled since there was no mixing or chemical interaction between the two fluids, and both are incompressible. The fingers were permanently solidified by curing the optical cement under an ultraviolet lamp. These experimental procedures are summarized in Fig. 1.

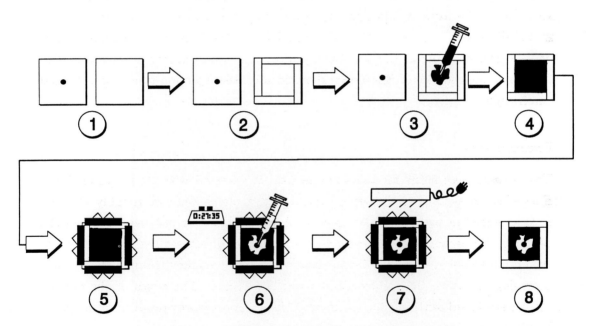

Fig. 1. Sketch of the fabrication method: (1) Make hole in top plate. (2) Add Mylar spacer. (3) Apply epoxy. (4) Assemble and remove bubbles. (5) Add clamps. (6) Inject water. (7) Cure epoxy using ultraviolet lamp. (8) Remove clamps. Finger complete.

The solidified fingers were photographically duplicated by contact printing — laying the fingers directly on the photographic paper and exposing the paper to the transmitted light. Upon development, the photographs showed exact, full-scale images of the cured fingers with very sharp contrast and no distortion. Several of these photographs are shown in Fig. 2. The central portion of each finger is not visible because the sealant used on the injection hole was opaque.

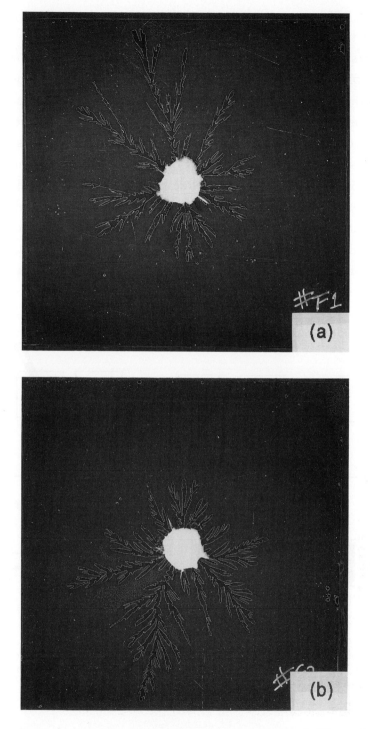

Fig. 2. Photographs of viscous fingers: (a) Sample F1, $t = 5$ s. (b) Sample F2, $t = 9$ s. (c) Sample F5, $t = 25$ s. (d) Sample F7, $t = 35$ s. (e) Sample F12, $t = 60$ s.

16.2 Viscous Fingering in Optical Cement Displaced by Water

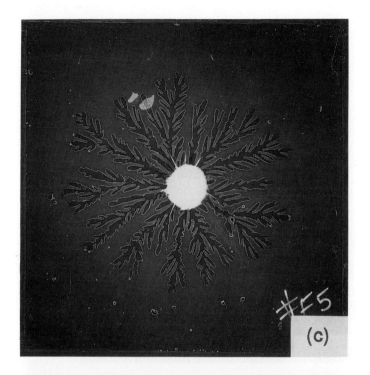

(c)

(d)

Fig. 2. (*Continued*)

Fig. 2. (*Continued*)

Results

The boundaries of the cured fingers form fractals. These fractals were examined visually for trends in their characteristic features. For discussion purposes, each fractal was broken down into "arms" covered with "lobes." This provided qualitative information about the fractals.

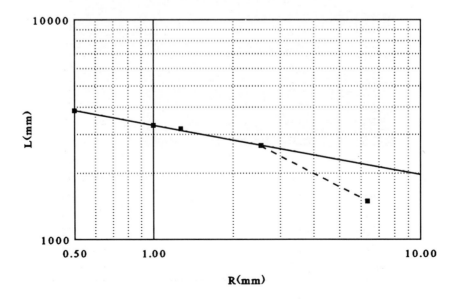

Fig. 3. A typical log-log plot of curve length L vs box size R.

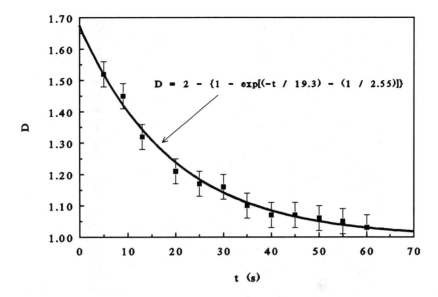

Fig. 4. Fractal dimension D vs injection time t.

The fractal dimension D of the boundary of each cured finger was determined using the box counting method. Tracing paper with grids of progressively smaller sizes were overlayed on the boundary curve. For each grid size R, the number of boxes subtended by the curve was counted. The length of the curve L was calculated by taking the product of the number of boxes and the length of one side of a box. For each fractal, the log-log graph of L vs R is plotted; a typical result is shown in Fig. 3. The plots were linear for small R, and D was calculated as unity minus the slope of this linear part. The D vs t data was then fitted with a curve (Fig. 4).

The data for lobes, arms, and fractal dimension are summarized in Table 1.

Table 1. Experimental data.

Sample Number	Injection Time (s)	Number of Arms	Number of Lobes	Fractal Dimension
F1	5.0	8	many	$1.52 \pm .02$
F2	9.0	7	|	$1.45 \pm .02$
F3	13.0	6	|	$1.32 \pm .02$
F4	20.0	6	|	$1.21 \pm .02$
F5	25.0	7	|	$1.17 \pm .02$
F6	30.0	9	|	$1.16 \pm .02$
F7	35.0	7	|	$1.10 \pm .02$
F8	40.0	7	|	$1.07 \pm .02$
F9	45.0	6	|	$1.07 \pm .02$
F10	50.0	9	|	$1.06 \pm .02$
F11	55.0	7	|	$1.05 \pm .02$
F12	60.0	7	few	$1.03 \pm .02$

A visual examination of the boundaries of the cured fingers provided clear evidence of trends in the characteristic features of the lobes and arms. The arms followed no general trends. The lobes, however, showed a distinct trend in size and number. The number decreases and size increases with increasing injection time. Thus the samples which were made slowly had only large, smooth features, while those made quickly had an abundance of fine, lacy structure. This is consistent with intuitive expectations based on inertia and other basic principles.

The fractal dimension decreases monotonically with injection time, as shown in Fig. 4. This result is consistent with intuitive expectations from visual examination of the boundaries.

An examination of the curve in Fig. 4 suggests the functional form $D = 2 - \{1 - \exp[-f(t)]\}$, where t is in seconds; $f(t)$ is a dimensionless function of t only, and $f(t) \geq 0$. This single-parameter model gives a D which is naturally bounded between 1 and 2. Attempting a fit of the form $f(t) = t/A$, with the constant A having dimensions in seconds, fails. Thus we try the form $f(t) = t/A + 1/B$, with the constant A in seconds and the quantity $1/B$ is small and dimensionless. This approximates the data very well, within the limits of experimental error, and takes the form

$$D = 2 - \{1 - \exp[(-t/19.3) - (1/2.55)]\} \qquad (1)$$

with D and t as before. It defines a characteristic time of 19.3 s which in some way describes the nature of the physical process involved in the creation of the fractals.

Conclusions

The fractal dimension of the boundaries of viscous fingers and the number of lobes of these boundaries depend directly on a single control parameter, the injection time t. The fractal dimension decreases monotonically with increasing t, following the exponential function given in Eq. (1). The simplicity of this empirical equation is encouraging to those attempting theoretical models using first principles.

References

1. J. Nittmann, G. Daccord and H. E. Stanley, "Fractal Growth of Viscous Fingers: Quantitative Characterization of a Fluid Instability Phenomenon," Nature **314**, 141 (1985).

16.3 The Fractal Nature of Shock-Wave Induced Fractures

Richard G. Klingler

The theory of fractal dimensions is applied to quantitatively describe the fracture patterns in plexiglass generated by an almost instantaneous concussive force, viz., a concentrated detonation wave.

Introduction

Recently, the formation of irregular fractal patterns under nonequilibrium conditions has attracted much interest. Such patterns provide important information about materials under stress as well as the processes which cause the irregular patterns. The use of such patterns to characterize a material or process is limited by the fact that there is no simple quantitative method of describing the patterns, until the appearance of Mandelbrot's concepts in fractal geometry.

In this work, one particular example of such irregular patterns which appeared in fractures of plexiglass plates, was generated by shock waves. The fractal dimensions of these fracture patterns were calcualed.

Experimental Procedures and Results

The fractal medium for this study is plexiglass plates in the form of 2 inch diameter, 0.1 inch thick discs for the first three experiments (Tests A, B and C) and a sold

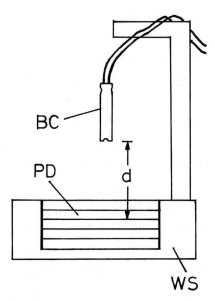

Fig. 1. Sketch of experimental setup. BC: blasting cap; PD: plexiglass discs; WS: wooden support stand. The five plexiglass discs starting from the top are numbered A1 to A5 for Test A, and similarly for Tests B and C.

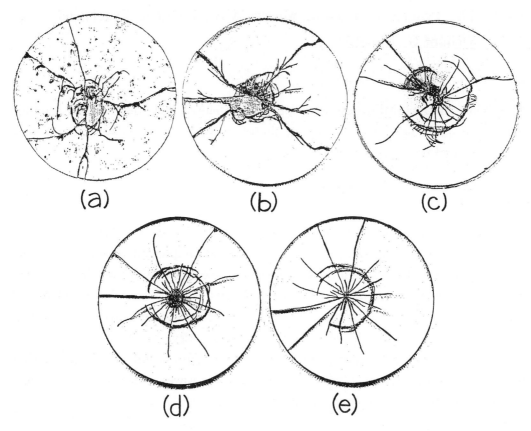

Fig. 2. Photocopies of the fractured plexiglass discs in Test B. (a) B1, $d = 6$ inches; (b) B2, $d = 6.1$ inches; (c) B3, $d = 6.2$ inches; (d) B4, $d = 6.3$ inches; (e) B5, $d = 6.5$ inches.

plexiglass cylinder for the fourth and final experiment. For each of the first three expriments, five discs were stacked and placed in a wooden support. A #8 blasting cap was positioned 4 inches above the top disc for the first test, 6 inches for the second test, and 8 inches for the third test, respectively, before detonation (Fig. 1). Each fractured disc was then photocopied. The results from Test B are shown in Fig. 2. The patterns from Test A are slightly more compact, and those from Test C are slightly less compact.

Each photocopied picture was then recorded on a graph paper [Fig. 3(a)], and N_ε, the minimal number of boxes of box size ε required to cover the pattern, is counted by hand. The fractal dimension D of each pattern is given by the negative of the slope of the log N_ε vs log ε curve [Fig. 3(b)], since $N_\varepsilon \sim \varepsilon^{-D}$ is expected. The results for all three tests are shown in Fig. 4.

For the sake of description, let us treat the five discs in each test as representing the cross sections of a solid plexiglass cylinder (of thickness 0.5 inch). Upon detonation, a three-dimensional fracture pattern is created in the cylinder. Each cross sectional plane has a different distance d from the origin of the external force causing the fracture. It receives the concentrated shock wave in the same location on the disc albeit with a different intensity. (The shock wave intensities were not

16.3 The Fractal Nature of Shock-Wave Induced Fractures

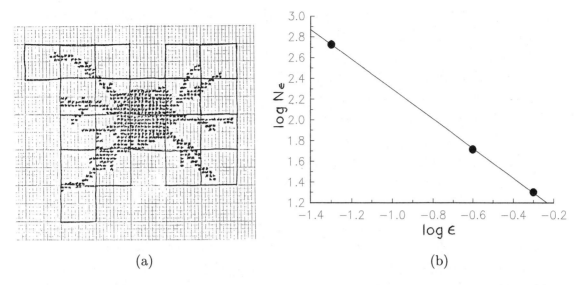

Fig. 3. Results from the photocopied picture of disc B2. The number of boxes N_ε counted from (a) is plotted vs box size ε in (b).

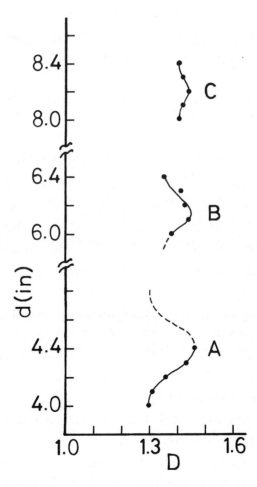

Fig. 4. Variation of fractal dimension D as a function of d, the distance between the disc from the shock wave origin. The dots represent experimental results; the broken lines in Tests A and B are inferred from the results of Test C.

measured in these experiments.) From Fig. 4, for Test B, we see D increases and then decreases as d varies from the top to the bottom of the cylinder. The curve for test A also shows increasing D as d increases, but the cylinder is too short to show the decreasing part. In contrast, Test C provides the full amount of required medium for the whole curve to show up. In this case, the separation of the top disc is at a larger distance from the shock wave origin; this additional distance allows the shock wave to dissipate somewhat before an impact with the plexiglass. The maxium D and range of D (minimum D) within each test decrease (increases) from Test A to Test C. It seems that there is a well defined trend in these fracture processes.

From the results in Figs. 2 and 4, it can be assumed that a concentrated detonation wave will produce a fracture pattern in a *solid* plexiglass with a fractal dimension between 2 and 3. (But there is no way that one can do a box counting with such a fractal embedded in a 3D space.) In fact, such a fractal was actually generated in our fourth test, with the use of a solid plexiglass cylinder. The resulting fractal pattern was compared visually with the patterns formed in the discs and found to be very similar, with the minor exception of the very long radial cracks found in the discs but not in the cylinder. The strong visual similarity between the patterns in the discs and solid cylinder validates the use of the stack of discs to represent the cross sections of a cylinder.

Discussions

The fractal dimension was found to be a useful quantity in characterizing the fracture patterns. Further studies would require thinner plexiglass discs providing more cross sections, and more discs allowing deeper fractures to be observed. Several computer simulation studies have been done on fractures [1,2]. This type of simulation could also provide more insight into the fractal nature of shock-wave induced fractures.

References

1. A. T. Skjeltorp and P. Meakin, "Fractures in Microsphere Monolayers Studied by Experiment and Computer Simulation," Nature **335**, 424 (1988).
2. E. Louis and F. Guinea, "The Fractal Nature of Fracture," Europhys. Lett. **3**, 871 (1987).

Epilogue
The Real World

More often than not, people become physicists because they love physics. But that does not mean that physicists do not love or care about other things in the real world such as the environment, art and literature, and social justices. In fact, most of them do! For example, after the June 4, 1989 Tiananmen massacre, six of the 21 most wanted student leaders of the democracy movement were physicists, according to the October 1996 issue of *APS News*. [For a description of this movement, see, for example, *Massacre in Beijing: China's Struggle for Democracy*, edited by Donald Morrison (Warner Books, 1989).]

As we emphasized in this book, almost all real-world problems are nonlinear problems. Nonlinear science, being applicable to both natural and social systems, offers physicists the chance to do physics even when their topic of interest happens to come from the social arena or the humanities.

In the natural sciences, a system may become nonlinear when the stress applied to the system is no longer small, as in the case of a simple pendulum. Or, the problem may be nonlinear from the outset, for example, in the study of solitons. By facing the nonlinearity head on and taking advantage of it, scientists are able to improve the efficiency of liquids mixing, tame chaotic systems, or greatly enhance the information superhighway through the use of undistorted optical pulses. In other words, new technologies and industries are generated by going nonlinear.

In the social sciences, recognition of the ubiquitous nonlinear behavior of social systems could offer new insights and new directions in solving real-world problems. For example, doubling the budget may not double the success in a project, but halving the number of administrators may triple the output of an institute. And, things may work suddenly beyond a turning point through simple perseverance or concentration of resources, a bifurcation phenomenon, as witnessed perhaps in the decrease of crime rates in New York City. More importantly, the mere act of placing social and natural systems on the same platform in the study of complex systems highlights the conviction that social phenomena can be and *should* be scientifically studied. In particular, theories of social developments, like theories in the physical science, should not be embraced wholeheartedly and put into full practice before they are proved to be valid. For those in power, who are ill-educated in science and cannot wait to do real experiments with their populations — that is, with real men, women and babies — let them try it out in computer simulations, which, hopefully, may soon become feasible. After all, with the advent of nonlinear science, these days more realistic computer simulation models of financial markets, traffic flows and city governance are emerging.

Once again, a reminder: In the real world, nonlinearity is not an option; it is the way of life!

Lui Lam
Department of Physics, San Jose State University
San Jose, California 95192-0106, USA
Email: *luilam@email.sjsu.edu*

Appendix A1

Computer Program for Active Walk

The program here generates the tracks of an arbitrary number of probabilistic active walkers (the PAW model), with the possibility of branching, on a square lattice, using the W_1 landscaping function for each walker (see Section 13.4). The user is prompted to enter the parameters: The background cone height V_0 (written as V0 on screen), which could be positive or negative, if the cone background is desired. The three parameters for the W_1 function: W_0 (W0 on screen), r_2 (r2 on screen), and ρ ($\equiv r1/r2$, rho on screen). The branching factor γ (Gamma on screen), with no branching for $\gamma = 1$ and sure branching for $\gamma = 0$. Number of initial walkers, and their separations.

The tracks of the walkers are shown on screen as they develop with time. The program stops by itself when one of the tracks reaches the lattice boundary, or when all walkers reach local minima in the landscape. An example of the outcome is shown in Fig. 5.4.

```
/*   AW96 (in Microsoft C)   */
/*   Probability active walk (PAW) model, self-avoiding walk,
     with branching possibility, and choice of a cone background.

     Energy look_up table includes negative values
     Potential look_up table U, 200X200 array
     Occupancy look_up table O, 200X200 array */

#include <graph.h>
#include <conio.h>
#include <stdlib.h>
#include <stdio.h>
#include <time.h>
#include <math.h>

#define round(x) (int)(x+.5)
#define metric(x1,y1,x2,y2) sqrt(pow(x1-x2,2)+pow(y1-y2,2))
#define rnd (float)rand()/RAND_MAX
#define PI 3.14159
#define MAXWALKERS 200  /* sets the maximum number of coexisting walkers */
#define LUT_RES 100
#define LUT_RES_INV (float)1.0/LUT_RES
#define SIZE 200
```

```c
    int x[MAXWALKERS], y[MAXWALKERS];
float p[5], du[5];
 float huge U[SIZE][SIZE];
short int huge O[SIZE][SIZE];
   int nw;   /* Current number of walkers   */
   int dx[5] = { 0,  1,  0, -1,  0 };
   int dy[5] = { 1,  0, -1,  0,  0 };
   int lowhigh[5];
   float E[LUT_RES*(SIZE/2)+110];  /* the +110 is for spillover  */
   int LAG=0, zz={0};
float rho, A, R_MIN, BF;
 char dummy;
   int PEAK_ENERGY, VIEW, RADIUS, SEP;

void initialize(void);
void Init_LUTneg(void);
void Init_LUTpos(void);
void changenergy(int xo, int yo);
void get_deltas(int xi, int yi);
void sortlowhigh(void);
float get_probs(int xc, int yc);

        /*
                0 - Start at some location(s)
                1 - Calculate the potential diff., du, between neighbors
                2 - Assign a step probability (to all lower neighbors)
                    proportional to du.
                3 - Check for branching
                4 - Step to randomly chosen neighbor
                5 - repeat 1
        */

/**********************************************************************/
int main(void)
{
int dir, i, stop, j, n = 1;
float total_p, random_number;
int z[MAXWALKERS]={0};
initialize();   /* initializes graphics, initial positios, energy, more */

do
{
        for(i=0; i<nw; i++)    /* i referes to which walker is active */
        {
        p[0] = 0;      /* initial probabilities */
        p[1] = 0;
        p[2] = 0;
        p[3] = 0;
        p[4] = 1;      /* default probability
                          in case no step is possible */

        lowhigh[0] = 0; /* initialize to allow sorting by energy */
        lowhigh[1] = 1; /* to be done by function sortlowhigh */
        lowhigh[2] = 2;
        lowhigh[3] = 3;
        lowhigh[4] = 4; /* lowhigh[4] is never changed
                           Default for no step */

/* get du, delta potential, for neighboring positions */
        get_deltas(x[i], y[i]);

/* sorts the potential difference from low to high energy.*/
        sortlowhigh();

/*  For each neighbor, assign a probability and return sum of probabilities*/
        total_p = get_probs(x[i], y[i]);
```

Appendix A1   Computer Program for Active Walk

```c
        /*  Choose a random direction according to the calculated probabilities  */
            random_number = rnd * total_p;
        for( j = 0 ; j <= 5 ; j++ )
            if( random_number < p[lowhigh[j]] )
            {
                    dir = lowhigh[j];
                    j = 5;
            }
/*  Check for branching */

if((dir==lowhigh[0])&&(dir != 4))  /***if dir was lowest in energy***/
        {
            if(du[lowhigh[1]] < BF*du[lowhigh[0]])
            {
                if(nw < MAXWALKERS)
                {
                    double w=.522;
                    x[nw] = x[i]+dx[lowhigh[1]];
                    y[nw] = y[i]+dy[lowhigh[1]];
                    changenergy(x[nw],y[nw]);
                    _setpixel( x[nw], y[nw]);
                    O[x[nw]][y[nw]] = nw+1;
                    nw =  nw + 1;
                }
            }
        }
/*  Step in chosen dir */
        x[i] = x[i] + dx[dir];
        y[i] = y[i] + dy[dir];

/*  Show the track on the screen   */
        _setpixel( x[i], y[i]);

/*  Update occupancy table */
        O[x[i]][y[i]] = i+1;

/*  Update energy table */
        if (dir != 4)
            changenergy(x[i],y[i]);

/*  check for walkers stopped */
        if (dir == 4)
        z[i]=1;
        stop = 0;
        for ( j = 0; j < nw ; j++ )
        {stop+=z[j];}

        if(x[i]==1||x[i]==SIZE-3||y[i]==SIZE-3||y[i]==1||stop==nw)
            {zz=1;
                break;}

    }       /*end i loop*/
_settextposition(2,1);

/*printf("# of Walker: %d",nw);*/

}while(zz==0);       /*  End do  */
printf("\a");
dummy=getch();
_setvideomode(_DEFAULTMODE);
return 0;

}  /*  End main()   */
/*******************************************************************/
void Init_LUTneg(void)
```

```c
{
    int r;

    float intercept_down = PEAK_ENERGY;
    float slope_down = -(PEAK_ENERGY+1)/R_MIN;
    float slope_up = (float)1.0/(RADIUS-R_MIN);
    float intercept_up = RADIUS/(R_MIN-RADIUS);

    for( r = 0 ; r <= RADIUS * LUT_RES + 100 ; r++ )
    {
        E[r] = slope_down * r * LUT_RES_INV + intercept_down;
        if( r > LUT_RES * R_MIN )
            E[r] = slope_up * r * LUT_RES_INV + intercept_up;
        if( r > LUT_RES * RADIUS )
            E[r] = 0;
    }
}
/*********************************************************************/
void Init_LUTpos(void)
{
int r;

float intercept_down = PEAK_ENERGY;
float slope_down = -(float)(PEAK_ENERGY)/(float)RADIUS;

for( r = 0; r<=RADIUS*LUT_RES + 100; r++)
{
    E[r] = slope_down * r * LUT_RES_INV + intercept_down;
    if(r> LUT_RES * RADIUS)
        E[r] = 0;
}
}
/*********************************************************************/
void changenergy(int xo, int yo)
{
int i,j;
double r;

for( i = xo-RADIUS-10 ; i < xo+RADIUS+10 ; i++ )
    for( j = yo-RADIUS-10; j<yo+RADIUS+10; j++)
        if((i>0)&&(i<SIZE)&&(j>0)&&(j<SIZE))
        {
            r = metric( xo, yo, i, j );
            if( r < RADIUS ) U[i][j] += E[round( r * LUT_RES )];
        }
}
/*********************************************************************/

/*********************************************************************/
void sortlowhigh(void)
{
int swapflag = 1, i, j;

    while(swapflag == 1)
    {
        swapflag = 0;
        for(i=0; i<3; i++)
        {
            if( du[lowhigh[i]] > du[lowhigh[i+1]])
            {
                j = lowhigh[i];
                lowhigh[i] = lowhigh[i+1];
                lowhigh[i+1] = j;
                swapflag =  1;
            }
        }
    }
```

Appendix A1 Computer Program for Active Walk

```c
}
/***********************************************************************/
void initialize(void)           /* initialize everything */
{
int i,j,seed, qq, conenergy=0;
double w=.522;
for(i=0;i<SIZE;i++)
      for(j=0;j<SIZE;j++)
      {
            O[i][j] = 0;          /* initialize occupancy table */
            U[i][j] = 0;          /* initialize potential table */
      }
printf("do you wish to start with a cone(1=yes, 0=no):");
scanf("%d", &qq);
      if ( qq == 1 )
         {printf("cone height V0:");
          scanf("%d", &PEAK_ENERGY);
            conenergy = PEAK_ENERGY;
            RADIUS=100;
            Init_LUTpos();
            changenergy(SIZE/2,SIZE/2);       /* make a hill in the center */
         }
printf("enter W0:");
scanf("%d", &PEAK_ENERGY);
printf("enter r2:");
scanf("%d", &RADIUS);
printf("enter rho(.01-.99):");
scanf("%f", &rho);  /*  rho=.2;*/
printf("enter seed number:");
scanf("%d", &seed);
printf("enter # of walkers:");
scanf("%d", &nw);
SEP=0;
if (nw > 1 )
    {printf("enter separation of walkers:");
       scanf("%d", &SEP);}
printf("enter gamma(between 0 and 1, gamma of 1 means no branching):");
scanf("%f", &BF);

A=R_MIN= (rho*RADIUS);
VIEW= RADIUS;
/*randomize;*/
srand(seed);          /*  seed the random number generator */

      Init_LUTneg();     /*   initialize the look up table  */

x[0] = (SIZE/2)+SEP;    /*    Starting     */
y[0] = (SIZE/2);        /*       point     */


/* by changing nw, more than one walker can start at the same time */
x[1] = (SIZE/2);
y[1] = (SIZE/2)+SEP;
x[2] = (SIZE/2)-SEP;
y[2] = (SIZE/2);
x[3] = (SIZE/2);
y[3] = (SIZE/2)-SEP;
x[4] = (SIZE/2);
y[4] = (SIZE/2);
x[5] = (SIZE/2)+SEP;
y[5] = (SIZE/2)-SEP;
x[6] = (SIZE/2)+SEP;
y[6] = (SIZE/2)+SEP;
x[7] = (SIZE/2)-SEP;
y[7] = (SIZE/2)+SEP;
```

```
        x[8] = (SIZE/2)-SEP;
        y[8] = (SIZE/2)-SEP;

        for(i=0; i<nw; i++)
        {                       /* have the starting walkers affect the potential */
                changenergy(x[i],y[i]);
                /*_setpixel(x[i]*4,y[i]*4);*/   /* show walkers */
        }


        for(i=0; i<SIZE; i++)
        {
                O[0][i] = 1;    /* set boundries as occupied, but don't change */
                O[199][i] = 1;  /* the energy around the border. */
                O[i][0] = 1;    /* this keeps the walkers in bounds */
                O[i][199] = 1;  /*      changenergy(100,i); */
        }

            /*  initialize the graphics */

            _setvideomode(_VRES16COLOR);
            _setvieworg(60,40);
            _setviewport(200,15,(SIZE*2)+200,(SIZE*2)+15);
            _setwindow(0,0.0,0.0,SIZE,SIZE);
            _setcolor(15);
            _rectangle(_GBORDER,0,0,(SIZE),(SIZE));/*+30)0,0,639,440); */

            for (i=0; i<nw; i++)
                    _setpixel(x[i],y[i]);

_settextposition(1,1);

/* print parameters on screen */
printf("square lattice AW96\n");
printf(" r2 = %d\n", RADIUS);
printf(" rho = %.3f\n", rho);
printf(" W0 = %d\n", PEAK_ENERGY);
printf(" Gamma = %1.3f\n", BF);
printf(" seed = %d\n", seed);
printf(" # of\n walkers = %d\n", nw);

if ( nw > 1 )
    {printf(" separation = %d\n", SEP);}

    printf(" cone height V0 = %d\n", conenergy);

}/* end initialize */
/*********************************************************************/
void get_deltas(int xi, int yi)
{
int j, l, m, c;
int ddx[3] = { -1, 0 , 1 };
int ddy[3] = { -1, 0 , 1 };

/* Find potential differences of nearest neighbors   */

for( j = 0 ; j < 4 ; j++ )
{
        du[j] = U[xi+dx[j]][yi+dy[j]] - U[xi][yi];
        c = 0;
        for(l=0; l<3; l++)
                for(m=0; m<3; m++)
                        if(O[xi+dx[j]+ddx[l]][yi+dy[j]+ddy[m]] != 0)
                                c = c + 1;
        if(c>2)
                du[j] = 0;
}
```

```c
}
/***********************************************************************/
float get_probs(int xc, int yc)
{
int j;
float total_u;

total_u = 0;

for( j = 0 ; j < 4 ; j++ )
{
    if( O[xc+dx[lowhigh[j]]][yc+dy[lowhigh[j]]] == 0 )
    {
        if(du[lowhigh[j]] < 0 )
        {
            total_u -= du[lowhigh[j]] ;
            /*  Subtracting makes total_u positive   */
            p[lowhigh[j]] = total_u ;
        }
        else p[lowhigh[j]] = 0;
    }
    else p[lowhigh[j]] = 0;
}

return total_u;

}
```

Appendix A2

Publications from Nonlinear Physics Group of SJSU

To show that undergraduate and graduate students can contribute significantly to research in nonlinear physics, as a tribute to our students who succeeded in doing that, and as an evidence that a meaningful research program can be maintained by an instructor while teaching nonlinear physics, a selected list of publications from our Nonlinear Physics Group at San Jose State University is presented here.

The first three reviews summarize our work on pattern formation in electrodeposits and on active walks. Both topics involve research in experiments, computer simulations and theory. The names of undergraduate authors are bolded; those of graduate students are in bold and underlined.

Reviews

1. L. Lam, "Electrodeposition Pattern Formation: An Overview," in *Defect Structure, Morphology and Properties of Deposits*, edited by H. Merchant (Mineral, Metals & Materials Society, Warrendale, PA, 1995).

2. L. Lam, "Active Walkers Models for Complex Systems," Chaos Solitons Fractals **6**, 267 (1995).

3. L. Lam, "Chapter 15. Active Walks: Pattern Formation, Self-Organization and Complex Systems," in *Introduction to Nonlinear Physics*, edited by L. Lam (Springer, New York, 1997).

Research Papers

4. V. M. **Castillo**, R. D. **Pochy** and L. Lam, "Pattern Changes in Electrodeposit of $CuSO_4$," in *Applications of Statistical and Field Theory Methods to Condensed Matter*, edited by D. Baeriswyl, A. R. Bishop and J. Camelo (Plenum, New York, 1990).

5. L. Lam, R. D. **Pochy** and V. M. **Castillo**, "Pattern Formation in Electrodeposits," in *Nonlinear Structures in Physical Systems*, edited by L. Lam and H. C. Morris (Springer, New York, 1990).

6. M. A. **Guzman**, R. D. **Freimuth**, P. U. **Pendse**, M. C. **Veinott** and L. Lam, "Experiments on Electrodeposit Patterns," in *Nonlinear Structures in Physical Systems*, edited by L. Lam and H. C. Morris (Springer, New York, 1990).

7. L. Lam, "Unsolved Nonlinear Problems in Liquid Crystals," in *Nonlinear and Chaotic Phenomena*, edited by W. Rozmus and J. A. Tuszynski (World Scientific, Teaneck, 1991).

8. M. K. **Pon** and L. Lam, "Stability of Dense Morphologies in Electrodeposit Pattern Formation," in *Nonlinear and Chaotic Phenomena*, edited by W. Rozmus and J. A. Tuszynski (World Scientific, Teaneck, 1991).

9. L. Lam, R. D. **Freimuth** and H. S. Lakkaraju, "Fractal Patterns in Burned Hele-Shaw Cells of Liquid Crystals and Oils," Mol. Cryst. Liq. Cryst. **199**, 249 (1991).

10. R. D. **Pochy**, A. Garcia, R. D. **Freimuth**, V. M. **Castillo** and L. Lam, "Electrodeposit Tree Patterns in Linear Cells: Experiment and Computer Models," Physica D **51**, 539 (1991).

11. L. Lam and J. Prost, "Chapter 1. Introduction," in *Solitons in Liquid Crystals*, edited by L. Lam and J. Prost (Springer, New York, 1992).

12. L. Lam, "Chapter 2. Solitons and Field Induced Solitons in Liquid Crystals," in *Solitons in Liquid Crystals*, edited by L. Lam and J. Prost (Springer, New York, 1992).

13. L. Lam and C. Q. Shu, "Chapter 3. Solitons in Shearing Liquid Crystals," in *Solitons in Liquid Crystals*, edited by L. Lam and J. Prost (Springer, New York, 1992).

14. C. **Larsen** and L. Lam, "Chaos and the Foreign Exchange Market," in *Modeling Complex Phenomena*, edited by L. Lam and V. Naroditsky (Springer, New York, 1992).

15. R. D. **Freimuth** and L. Lam, "Active Walker Models for Filamentary Growth Patterns," in *Modeling Complex Phenomena*, edited by L. Lam and V. Naroditsky (Springer, New York, 1992).

16. L. Lam, R. D. **Freimuth**, M. K. **Pon**, D. R. **Kayser**, J. T. **Fredrick** and R. D. **Pochy**, "Filamentary Patterns and Rough Surfaces," in *Pattern Formation in Complex Dissipative Systems*, edited by S. Kai (World Scientific, River Edge, 1992).

17. D. R. **Kayser**, L. **Aberle**, R. D. **Pochy** and L. Lam, "Active Walker Models for Filamentary Patterns and Rough Surfaces," Physica A **191**, 17 (1992).

18. R. D. **Pochy**, D. R. **Kayser**, L. **Aberle** and L. Lam, "Boltzmann Active Walker and Rough Surfaces," Physica D **66**, 166 (1993).

19. L. Lam and Y. S. **Yung**, "Optical Solitons in Liquid Crystals," in *Modern Topics in Liquid Crystals*, edited by A. Buka (World Scientific, River Edge, 1993).

20. L. Lam and R. D. **Pochy**, "Active Walker Models: Growth and Form in Nonequilibrium Systems," Comput. Phys. **7**, 534 (1993).

21. L. Lam, "Chapter 10. Bowlics," in *Liquid Crystalline and Mesomorphic Liquid Crystals*, edited by V. P. Shibaev and L. Lam (Springer, New York, 1994).

22. Y. S. **Yung** and L. Lam, "Frequency and Temperature Dependence of Refractive Indices of Liquid Crystals," in *Novel Laser Sources and Applications*, edited by J. F. Becker, A. C. Tam, J. B. Gruber and L. Lam (SPIE Optical Engineering Press, Bellington, WA, 1994).

23. L. Lam, "Instrinsic Abnormal Growth," *Overseas Chinese Physics Association Newsletter* **1**(11), 13 (1994).

24. L. Lam, "Active Walks," in *Lectures on Thermodynamics and Statistical Mechanics*, edited by M. Costas, R. Rodriquez and A. L. Benavides (World Scientific, River Edge, 1994).

25. L. Lam, M. C. **Veinott** and R. D. **Pochy**, "Abnormal Spatiotemporal Growth," in *Spatiotemporal Patterns in Nonequilibrium Complex Systems*, edited by P. E. Cladis and P. Palffy-Muhoray (Addison-Wesley, Menlo Park, 1995).

26. V. M. Castillo, M. C. **Veinott** and L. Lam, "Neural Network for Classification of Active Walker Patterns," Chaos Solitons Fractals **6**, 67 (1995).

27. R. P. Pan, C. R. Sheu and L. Lam, "Dielectric Breakdown Patterns in Thin Layers of Oils," Chaos Solitons Fractals **6**, 495 (1995).

28. G. Marshall, S. Tagtachian and L. Lam, "Growth Pattern Formation in Copper Electrodeposition: Experiments and Computational Modelling," Chaos Solitons Fractals **6**, 325 (1995).

29. L. Lam, R. W. **Koepeke** and T. Y. Lin, "Active Walks and Soft Computing," in *Rough Sets and Soft Computing*, edited by T. Y. Lin (Society of Computer Simulation, San Diego, 1995).

30. L. Lam, "Solitons in Liquid Crystals: Recent Developments," Chaos Solitons Fractals **5**, 2463 (1995).

31. L. Lam, M. C. **Veinott**, D. **Ratoff** and L. Lam, "Noise-Induced Abnormal Growth," in *Fluctuations and Order: A New Synthesis*, edited by M. Millonas (Springer, New York, 1996).

Acknowledgments

This book results from three kinds of interrelated activities that kept me busy in the last ten years. The first was the planning and teaching of the two courses on nonlinear physics. These optional courses kept drawing more students compared to other graduate courses in my physics department. I like to thank the 28 students who were brave enough to enroll in the first offering of PHYS 255N, Fall 1988.

Second, my research activities with the participation of undergraduate and graduate students. The students registered in PHYS 180 and PHYS 298 and became members of our Nonlinear Physics Group. There were many happy hours when we worked and discovered things together. For these and especially for those stressful 24 hours before I had to board a plane for a conference somewhere, I sincerely thank my student collaborators, in particular, Victor Castillo, Kin-Chung Chan, Rolf Freimuth, Mark Guzman, Daniela Kayser, Rocco Pochy and Mike Veinott.

Third, in the last four years, I have been giving an expository lecture on nonlinear physics for the general audience, which brought me to various campuses in Mexico, the United States, Hong Kong and Taiwan, including the Taiwan Provincial Senior High School of Hsinchu. These lectures deeply influenced the writing of Part I in this book. I am much grateful to Rosalio Rodriguez for inviting me to Universidad Nacional Autonoma de Mexico in January 1994, which marked the beginning of these adventures; to the colleagues and friends who hosted my visits; and to the students and teachers who made up the audiences and asked stimulating questions. Many fellow scientists generously provided encouragement and helpful discussions. Crucial financial support came from Research Corporation, Tucson, and the Research Experiences for Undergraduates Program of the National Science Foundation.

Part I is an expanded writing of Chapter 1, from *Introduction to Nonlinear Physics*, edited by Lui Lam (Springer, New York, 1997). I thank George Cladis for kindly providing Fig. 3.2. Credit for Fig. 4.4 goes to NASA/JPL. The *beamtree* picture in Fig. 5.1 is the courtesy of Stanford Linear Acceleration Center. Figure 5.4

is adapted from Chapter 15 of *Introduction to Nonlinear Physics*; the electrodeposit pattern appeared originally in Y. Sawada, A. Dougherty and J. P. Gollub, Phys. Rev. Lett. **56**, 1260 (1986); the retinal neuron is a redrawing of Fig. 1.0, credited to R. H. Masland, in *Fractals and Disordered Systems*, edited by A. Bunde and S. Havlin (Springer, New York, 1991); the surface reaction picture is due to Ru-Pin Pan, Chia-Rong Sheu and L. Lam, Chaos Solitons Fractals **6**, 495 (1995). Figure 7.3 is from A. Lomnicki, *Population Ecology of Individuals* (Princeton U.P., Princeton, 1988).

Finally, I thank the staffs at World Scientific for their infinite patience and assistance in association with this book. The long time taken to produce this book is completely my responsibility, not theirs.

Index

A
Abnormal growth, 32, 207, 210
Active walk, 38, 41, 215, 220
    computer program, 321
Active walk model, 27, 38, 41, 207
Aggregate, 13, 39, 51, 166
AIDS, 207
Algae colony, 27
American Physical Society, 2, 319
Amplitude equation, 29, 32
Anderson, P.W., 193
Anisotropy, 166
Ant, 38, 208
Artificial life, 36
Attractor
    ordinary, 17
    strange, 17, 20, 111, 225
AWM. *See* Active walk model

B
Bak, Per, 15, 45, 64, 203
Baker, G.L., 44
Baldwin, William A., 257
Barnsley, Michael, 225
Biased random walk model, 27, 271
Biological system, 36, 40
Bödefeld, Sabine, 215
Box dimension, 11, 237
BPAW model, 29, 39, 41, 210
Branching, 27, 41, 209
Briggs, J., 44
Broken symmetry, 38, 193
Bullough, R.K., 147
Bunde, A., 44

C
Castillo, Victor M., 226, 237, 275, 279, 294
Catastrophe, 203

Cellular automata, 35, 38, 197, 283, 286
    definition, 35, 197
    history, 35
    one-dimensional, 283
    two-dimensional, 286
Chance, 38, 41, 217
Chaos, 8, 17, 37, 92, 104, 111, 114, 139, 211, 225, 257, 260
    application, 20, 21
    controlling, 20, 125
    definition, 17
    history, 17, 92
    quantum, 21, 132
    transition to, 102
Chaos game, 225, 229
    generalized, 84
Chemical reaction, 17
Chemical system, 21, 36, 129
Chemical wave, 29, 187
Cloud, 13, 27
Coevolution, 217
Coin toss, 139
Cold fusion, 40
Colloid, 13
Combs, Allan, 21
Communication, 130
Complex adaptive system, 38
Complex Ginzburg-Landau equation, 32
Complex system, 20, 36, 37, 193, 197
    definition, 37
Complexity, 139, 196, 197
Computer, 9, 11, 20, 35, 37, 203, 225, 226
Conservation law, 35
Conservative system, 21
Contingency, 39, 217
Conway, John, 35
Crime rate, 319
Crutchfield, James P., 92

D

Dendrite, 159, 166
Deterministic system, 17
Devil's staircase, 63
Dielectric breakdown model, 13, 53, 166, 245
Diffusion-limited aggregation model, 9, 13, 52, 51, 166, 240
Dispersion, 23
Dissipative system, 17
DNA, 1, 38, 195
Drake, J.L., 76
Dripping faucet, 17, 100, 104

E

Earthquake, 38, 204
Ecology, 38
Economic system, 9, 38, 140, 205, 218
Electroconvection, 29, 172, 176
Electrodeposit, 27, 32
Electronic circuit, 17
Electronics, 21, 129
Evolution, 1, 7, 38, 217

F

Fallis, Mark C., 226, 301
Fang, L.Z., 58
Farmer, J. Doyne, 92
Feigenbaum, Mitchell, 17
Ferrofluid, 279
Fitness landscape, 217
Fluid flow, 35, 73, 172, 176
Ford, Joseph, 139
Fourier transform, 8
Fractal, 11, 17, 27, 51, 56, 58, 63, 197, 220, 229, 231, 237, 240, 245, 291, 315
    definition, 11
    dimension, 12, 213, 291
    history, 13
    self-affine, 13
    self-similar, 11
    support, 76
Fractal basin boundary, 115
Fracture, 13, 315
Freimuth, Rolf D., 76, 225, 231, 260
Frisch, Uriel, 35

G

Galaxy, 13, 58
Game of life, 35, 36, 204
Glass, L., 44
Glass fiber, 25
Gleick, James, 44, 225
Gollub, J.P., 44
Gomes, M.A.F., 56
Gould, Stephen, 41

Gouyet, J.-F., 43
Grebogi, Celso, 111
Growth process, 12, 207
    dendritic, 163, 166
    diffusive, 27, 51
Gutzwiller, Martin C., 132
Guzman, Mark A., 271

H

Haken, Hermann, 37
Halvin, S., 44
Hamiltonian system, 21, 139
Hasslacher, Brosal, 35
Hayles, N.K., 21
Heartbeat, 17
Heart tissue, 21, 130
Hele-Shaw cell, 8, 27
Hillendahl, James W., 308
History, 38, 41
Homogeneous function, 14
Huang, Yun, 152

I

Immunology, 38
Increasing return, 38, 218
Information, 38, 195
    superhighway, 319
Initial condition, 8, 17
    sensitive dependence, 8, 20
Instability, 29, 172, 301
Integrable system, 8, 24
Interface, 8, 13, 27
Interfacial tension, 166
Intrinsic abnormal growth, 32
Ion transport in glass, 38
Irreproducible growth, 32, 207
Irreproducible experiment, 40

J

Joets, A., 172, 176

K

Kaplan, D., 44
Klingler, Richard G., 315
Korteweg-de Vries equation, 23, 24
Kruskal, Martin, 25

L

Lam, Lui, 27, 44, 76, 176, 207, 215 (see also Lin, Lei)
    address, 320
Lam, P.M., 152
Landscape, 27, 38, 216
Landscaping rule, 39, 207
Langer, J.S., 159
Langton, Christopher, 36

Language, 38, 195, 200
Laplace equation, 8
Laser, 21, 129
Lattice gas, 35
Lattice gas automata, 36
Leaf vein, 27
Lee, T.D., 157
LeRoy, Rodney C., 301
Life, 7, 35, 38, 41
Life science, 21
Life's tape, 41, 217
Lifshitz point, 176
Lightning, 27
Lin, Lei (*a.k.a.* Lam, Lui), 152, 172
Liquid crystal, 25, 29, 32, 152, 172, 176, 195
Literature, 21, 319
Lock-in, 13, 64
Logic function, 76
Lorentz, Edward, 17

M

Mandelbrot, Benoit, 13
Map, 17, 231
    tent, 257
Markworth, Alan, 21
Masuda, Michael M., 301
May, Robert M., 118
Meakin, Paul, 71
Mechanical system, 21
Molecular dynamics, 35
Monte Carlo approach, 35
Morphogram, 39
Morphology, 29, 39, 41, 166, 207
    dense radial, 209
Morphology change, 29, 33
Mountain, 13
Müller, Stefan C., 187
Multifractal, 13, 71, 76

N

Navier-Stokes equations, 35
Neisan, Nejat, 301
Neural network, 38, 294
Newtonian dynamics, 21, 133, 145
Nittmann, Johann, 166
Noise, 17, 32, 39–41, 159, 166, 210, 217, 245
Nonequilibrium system, 14, 38, 207
Nonintegrable system, 24
Nonlinear diffusion equation, 23, 24
Nonlinear dynamics, 111
Nonlinear excitation, 25
Nonlinear forecasting, 118
Nonlinear Physics Group, SJSU, ii, 328
Nonlinear Schrödinger equation, 148
Nonlinear science, 1, 2, 43, 226, 319

Nonlinear wave, 25
Nonlinearity, 6, 23, 43, 147, 319

O

Oil industry, 36
Optical fiber, 23
Ott, Edward, 111, 125

P

Packard, Norman H., 92
Paper ball, 13, 56
Partial differential equation, simulation, 36
Path dependent phenomenon, 41, 215
Pattern, 41
    branching, 27
    cluster, 286
    compact, 27
    electroconvection, 29
    electrodeposit, 27–33, 207
    filamentary, 27, 209
    surface reaction, 38
    type A, 27
    type A1, 27
    type A2, 27
    type B, 27, 29
Pattern formation, 27, 29, 159, 166, 172, 176, 187, 271, 275, 279, 301, 308, 315
    theory, 29, 159, 176
    unified description, 29, 32
    universal mechanism, 27
Pattern selection, 161
Peat, F.D., 44
Pendse, Prasanna U., 225, 229
Perturbation, 24, 32, 139
Phase diagram, 176
Phase space, 17, 21, 94
Pochy, Rocco D., 207, 226, 240, 245, 257, 271, 283, 286
Polymer, 13, 210
Poincaré, Henri, 17, 93
Poincaré section, 112, 135
Pomeau, Yves, 35
Porous medium, 36
Poundstone, W., 45
Power law, 7, 13, 14, 203
Predictability, 20, 118
Prigogine, Ilya, 37
Principle of
    active walk, 38
    organization, 11
    randomness, 11
    regularity, 11
    self-similarity, 11
Psychology, 21, 38, 193

Q
Quantum chaos, 21, 132
Quantum mechanics, 5

R
Randomness, 17, 92, 139
Rayleigh-Benard convection, 29
Real world, 319
Reduction method, 37, 193
Relativity, 5
Remoissenet, M., 44
Revolution, 5
Ribotta, R., 172, 176
River, 1, 27
Robertson, Robin, 21
Rock, 13
Ross, John, 187
Rough surface, 13, 210
Russell, John Scott, 25

S
Salas Brito, A.L., 104
Sander, Leonard M., 51
Sandpile, 15, 203
Scale invariance, 14, 51
Schroeder, M., 44
Schuster, H.G., 44
Self-assembly, 11
Self-organization, 29, 38
Self-organized criticality, 14, 38, 203, 211, 216
Self-reproduction, 35
Self-similarity, 11, 13, 14
Sensitive zone, 39, 40
Shallow water, 23
Shaw, Robert S., 92
Shen, Juelian, 152
Shu, Changqing, 152, 215
Sierpinski fractal, 231
Sierpinski gasket, 11, 76, 197, 229
Simple pendulum, 7, 94
Sine-Gordon equation, 148
Snowflake, 27, 55, 164
Social science, 1, 5, 15, 21, 37, 193, 215, 319
Society of Physics Students, 226, 306
Solidification, 27, 29, 160
Solitary wave, 23, 147
Soliton, 8, 23, 147, 152, 157
    definition, 23, 24
    history, 25, 147
    optical, 25
    reentrant, 211
Solvability condition, 32
Solvability theory, 162
Spano, Mark, 125

Spin glass, 38
Spiral, 209
Spring, 7
Stanley, H. Eugene, 71, 166
Stepping rule, 207
Stock market, 13, 20
Strange attractor. See Attractor, strange
Sugihara, George, 118
Superconductor, high-temperature, 25, 157
Superposition principle, 7, 8
Surface tension, 275
Symmetry breaking, 38

T
Tang, Chao, 15
Thermal convection, 17, 29
Tiananmen massacre, 319
Time series, 37, 118
Tip splitting, 166
Traffic, 1
Transformation growth, 32, 207
Tree, 13, 27
Turbulence, 205

U
Universality, 17, 115, 215
Universe, 58, 215
Urban growth, 216

V
Vargas, C.A., 104
Vicente, L.A., 104
Vidal, Christian, 187
Viscous fingering, 8, 27, 29, 32, 159, 308
von Neumann, John, 35

W
Waldrop, M.M., 45
Warnecken, Hans-Jürgan, 15
Water column, 28, 226, 301
Water drop, 275
Waterwheel, 17, 225, 260
Watkins, Thayer H., 291
Weather, 17, 20
Weisbuch, G., 44
Wiesenfeld, Kurt, 15
Wolfram, Stephen, 197
Worm, 209

Y
Yepez, H.N. Nunez, 104
Yorke, James A., 111
Yung, Yuk S., 257

Z
Zabusky, Norman, 25